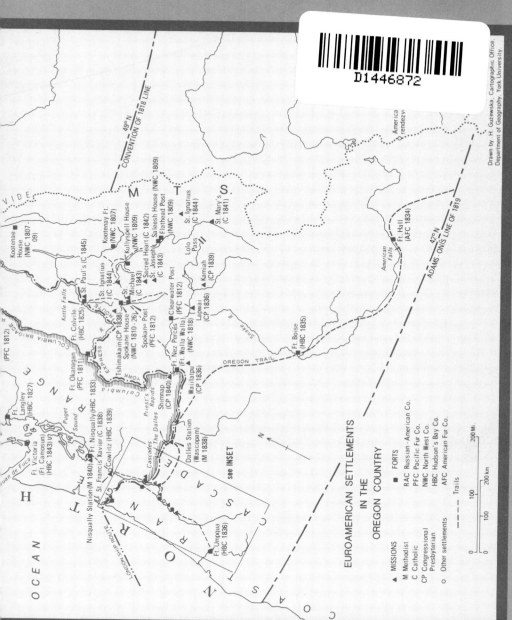

FARMING THE FRONTIER

FARMING *the* FRONTIER

The Agricultural Opening of the Oregon Country 1786-1846

JAMES R. GIBSON

UNIVERSITY OF WASHINGTON PRESS
Seattle and London

Farming the Frontier

The Agricultural Opening of the Oregon Country, 1786-1846

© The University of British Columbia Press 1985
all rights reserved

This book has been published with the help of a grant from the Social Science Federation of Canada, using funds provided by the Social Sciences and Humanities Research Council of Canada.

No part of this publication may be reproduced or transmitted in any form or by any means, electronic or mechanical, including photocopy, recording, or any information storage or retrieval system, without permission in writing from the publisher.

Library of Congress Cataloging-in-Publication Data

Gibson, James R.
    Farming the frontier

    Bibliography: p. 215
    Includes index.

    1. Agriculture — Economic aspects — Northwest, Pacific — History. I. Title
HD1773.A9G53   1985   338.1'09795   85-13350
ISBN 0-295-96297-6

Printed and bound in Canada by John Deyell Company

*for Gail
and all West Sidians*

Contents

Tables	viii
Illustrations	x
Prologue: Opening the Oregon Country	1

PART ONE Post Farming: Hudson's Bay Company Agriculture
                  This Side The Mountains   7

   1. Governor Simpson's Œconomy: The Origins of
      Post Farming   9
   2. Governor Simpson's Reward: The Results of
      Post Farming   29
   3. Physical Extremes: The Problems of
      Post Farming   67

PART TWO Company Farming: Puget's Sound Agricultural
                 Company Operations On The Cowlitz Portage   75

   4. Grain for Alaska and Wool for England: The Origins
      of the Puget's Sound Agricultural Company   77
   5. Success and Failure: The Performance of the
      Puget's Sound Agricultural Company   85
   6. Half Shares and Mean Lands: The Problems of the
      Puget's Sound Agricultural Company   109

PART THREE Homestead Farming: Pioneer Agriculture In
                    The Willamette Valley   125

   7. The Promised Land: The Formation of
      The Willamette Settlement   127
   8. The "Garden of the Columbia": The Success of
      Homestead Farming   139

PART FOUR Mission Farming: Protestant and Catholic Husbandry on the Lower Columbia	149
9. The "Macedonian Cry": The Advent of Missionaries	151
10. The Fruit of the Faithful: The Outcome of Mission Farming	163
11. Divine Testing or Demonic Tempting: The Obstacles to Mission Farming	169
12. From Noble Savage to Sturdy Yeoman: Indian Farming	179
Epilogue: Dividing the Oregon Country	187
Abbreviations	207
Sources for Tables	209
Bibliography	215
Notes	227
Index	259

Tables

1. Hudson's Bay Company Purchases of Provisions from the Red River Colony, 1823-31 — 13
2. Number of Hudson's Bay Company Servants in the Columbia Department, 1821-46 — 20
3. Cultivated Acreage at Fort Vancouver in Various Years between 1829 and 1846 — 36
4. Output of Grains, Peas, and Potatoes at Fort Vancouver in Various Years between 1825 and 1846 — 38
5. Number of Livestock at Fort Vancouver, 1824-29 and 1832-46 — 39
6. Output of Grains and Potatoes at Fort Colvile in Various Years between 1826 and 1841 — 45
7. Number of Livestock at Fort Colvile in Various Years between 1827 and 1846 — 46
8. Output of Grains, Peas, and Potatoes at Fort Langley in Various Years between 1828 and 1835 — 51
9. Number of Livestock at Fort Langley, 1833-46 — 52
10. Number of Livestock at Fort Nez Percés, 1826-29 and 1835-46 — 54
11. Number of Livestock at Thompson's River and Fort Okanagan, 1826-29 and 1834-41 — 55
12. Number of Livestock at New Caledonia's Posts in Various Years between 1828 and 1844 — 57
13. Number of Livestock at Fort Nisqually, 1835-43 — 61
14. Output of Grains, Peas, and Potatoes at Cowlitz Farm, 1840-44 and 1846 — 94
15. Sales of Wheat from Cowlitz Farm by the Puget's Sound Agricultural Company to the Hudson's Bay Company, 1840-44 and 1846 — 95

16. Shipments of Provisions to Sitka by the Hudson's Bay Company, 1840 and 1843-49	95
17. Number of Livestock at Cowlitz Farm, 1839-42	96
18. Number of Livestock at Nisqually Farm, 1840-42	102
19. Number of Puget's Sound Agricultural Company Livestock, 1841-46	103
20. Shipments of Puget's Sound Agricultural Company Wool, Sheepskins, Cattle Hides, and Cattle Horns to London, 1842-47	103
21. Acreages, Sowings, and Harvests of Grains, Peas, and Potatoes at Nisqually Farm, 1841-44	104
22. The Canadian Settlement in the Willamette Valley, 1836	132
23. American Migration to the Oregon Country via the Oregon Trail, 1840-46	134
24. Euroamerican Population of the Willamette Settlement, 1838-45	136
25. Cultivated Acreage and Crop Output at Willamette Station, 1835-39	164
26. Crop Output at Waiilatpu, 1837-41	165
27. Crop Output at Lapwai, 1838-43	166
28. Number of Livestock at Lapwai, 1840-43	167
29. Number of Employees, Cultivated Acreage, and Number of Livestock at the Posts of the Hudson's Bay Company and the Puget's Sound Agricultural Company in the Columbia Department, 1845	188
30. Hudson's Bay Company Fur Returns of the Columbia Department and the Columbia District, 1826-46	201

Illustrations

PLATES

1. Chief Factor John McLoughlin (1784-1857), no date, anonymous (Oregon Historical Society, Portland). 19
2. Plan of Fort Vancouver, 1845, by Mervyn Vavasour (Hudson's Bay Company Archives, Winnipeg, G.1/95). 30
3. View of Fort Vancouver, 1845-46, by Henry Warre (Public Archives of Canada, Ottawa). 33
4. Plan of Fort Colvile, 1845, by Mervyn Vavasour (Hudson's Bay Company Archives, Winnipeg, G.1/193). 42
5. View of Fort Nez Percés, 1846, by Paul Kane (Royal Ontario Museum, Toronto). 53
6. Plan of Camosun Harbour and Fort Victoria, 1845, by Mervyn Vavasour (Hudson's Bay Company Archives, Winnipeg, G.1/198). 62
7. View of Fort Simpson, no date, anonymous (Royal Ontario Museum, Toronto). 65
8. View of New Archangel (Sitka) *circa* 1840, anonymous (Hudson's Bay Company Archives, Winnipeg, F.29/2). 78
9. Plan of the Cowlitz Portage, 1840, anonymous (Hudson's Bay Company Archives, Winnipeg, F.25/1, fo. 13). 89
10. Plan of Cowlitz Farm, 1844-45, anonymous (Hudson's Bay Company Archives, Winnipeg, F.12/2, fo. 57). 91
11. Plan of the Nisqually Plain, 1845, by Mervyn Vavasour (Hudson's Bay Company Archives, Winnipeg, G.1/197). 98
12. View of Nisqually Farm, 1845, by Henry Warre (Royal Ontario Museum, Toronto). 100
13. View of the Willamette Valley, 1845, by Henry Warre (Public Archives of Canada, Ottawa). 128

14. View of Oregon City, 1845, by Henry Warre (Public Archives of Canada, Ottawa). 137
15. View of Waiilatpu Mission, 1846, by Paul Kane (Royal Ontario Museum, Toronto). 155
16. View of Tshimakain Mission, 1843, by Charles Geyer (Washington State University Library, Pullman). 156
17. View of St. Paul Mission, 1846, by Paul Kane (Royal Ontario Museum, Toronto). 158

MAP

1. Euroamerican Settlements in the Oregon Country (endpapers).

Two leading objects of commercial gain have given birth to wide and daring enterprise in the early history of the Americas; the precious metals of the south, and the rich peltries of the north. While the fiery and magnificent Spaniard, inflamed with the mania for gold, has extended his discoveries and conquests over those brilliant countries scorched by the ardent sun of the tropics, the adroit and buoyant Frenchman, and the cool and calculating Briton, have pursued the less splendid but no less lucrative, traffic in furs admist the hyperborean regions of the Canadas, until they have advanced even within the arctic circle. These two pursuits have thus in a manner been the pioneers and precursors of civilization. Without pausing on the borders, they have penetrated at once, in defiance of difficulties and dangers, to the heart of savage countries: laying open the hidden secrets of the wilderness; leading the way to remote regions of beauty and fertility that might have remained unexplored for ages, and beckoning after them the slow and pausing steps of agriculture and civilization.

Washington Irving, *Astoria; or, Anecdotes of an Enterprise beyond the Rocky Mountains*, 1836

Prologue
OPENING THE OREGON COUNTRY

The discovery of a passage by sea, North-East or North-West from the Atlantic to the Pacific Ocean, has for many years excited the attention of governments, and encouraged the enterprising spirit of individuals. The non-existence, however, of any such practical passage being at length determined, the practicability of a passage through the continents of Asia and America becomes an object of consideration. . . . But whatever course may be taken from the Atlantic, the Columbia is the line of communication from the Pacific Ocean, pointed out by nature, as it is the only navigable river in the whole extent of Vancouver's minute survey of that coast: its banks also form the first level country in all the Southern extent of continental coast from Cook's entry, and, consequently, the most Northern situation fit for colonization, and suitable to the residence of a civilized people. By opening this intercourse between the Atlantic and Pacific Oceans, and forming regular establishments through the interior, and at both extremes, as well as along the coasts and islands, the entire command of the fur trade of North America might be obtained, from latitude 48 North to the pole, except that portion of it which the Russians have in the Pacific. To this may be added the fishing in both seas, and the markets of the four quarters of the globe.

ALEXANDER MACKENZIE, 1801

From the middle of the eighteenth century, two powers — Russia and Spain — vied for control of the territory and wealth of the Northwest Coast, that rugged stretch of North American seaboard between Cape Mendocino and Cape St. Elias. Russia had been first on the scene in 1741, when Vitus Bering's second voyage of exploration and discovery revealed the presence of sea otters. Traders quickly began to exploit this lucrative peltry by a series of hunting and trading voyages along the Aleutian chain to the Gulf of Alaska. The first permanent Russian settlement was probably founded at Captain's Harbour on Unalaska Island in 1773, and by 1812 Fort Ross was erected less than one hundred miles north of San Francisco Bay.

The rapid Russian advance soon alarmed Spain, which feared for the security of its exposed colonies on the Mexican isthmus. A naval base was established in 1768 at shallow and pestilential San Blas. It served as the springboard for the subsequent occupation in 1769 of Alta California by a string of Franciscan missions based on Monterey. But this attempt to forestall Russian encroachment led to conflict not with the tsar's subjects but with "King George's men," as the British were known in the Chinook trading jargon of the Northwest Coast. Spanish seizure of a British fort and ships at Nootka Sound in 1789 resulted in the Nootka Convention of 1790, whereby Madrid yielded exclusive sovereignty over the coast north of Nootka, and London gained international recognition of its claim to the Northwest Coast. Spain's retreat culminated thirty years later in the Transcontinental or Adams-Onís Treaty (1819), which divided New Spain and the Louisiana cum Missouri Territory along the forty-second parallel. Thus Spain surrendered all claims north of 42° to the United States. This withdrawal by His Catholic Majesty exposed the main weaknesses of Spanish colonial policy: stubborn adherence to mercantilist exploitation instead of self-sufficient settlement to substantiate territorial claims and official secrecy rather than promotional publicity to broadcast these claims. Spanish tenure was further weakened by Madrid's debilitating involvement in the French Revolution and the Napoleonic wars, especially under the cautious Bourbons.[1]

Great Britain and its former Thirteen Colonies, the upstart United States, cocksure and aggressive after its recent and victorious War of Independence, arrived on the Northwest Coast in the late 1780's, stimulated by the appearance in 1784 of the journal of Captain Cook's third voyage, which publicized the profitable market for dark, silky sea otter fur at Canton. The Americans — "Boston men" in Chinook — were particularly successful. They traded with the coastal Indians, exchanging metals, blankets, guns, and spirits for sea otter skins. The Americans dealt the skins at Canton for china, silks, and teas which they in turn sold in Europe and the United States. British traders were hamstrung by monopolistic conflicts between the South Sea and East India companies, sea otters being procured in the former's preserve and marketed in the latter's. So adept were the American coasters, that even their Russian adversaries (despite a half-century headstart in the maritime fur trade and control of the best hunters of sea otters [the Aleuts] and the most valuable varieties of sea otters [Kurilian-Kamchatkan and Aleutian]) came to rely upon the Yankees for basic necessities. In fact, Russia, like Spain, was overextended in the Americas; personnel were too few and supply lines were too long. These realities were acknowledged by St. Petersburg in 1824 and 1825 when it signed conventions with the United States and Great Britain, respectively. The terms of these accords set a southern limit of 54°40′ N. latitude to Russian dominion on the Northwest Coast.

Meanwhile, the British and the Americans had begun to approach the coast by land as well as by sea, penetrating the intermontane basins of the Columbia and Fraser rivers — the Oregon Country or, from an American perspective, the New or Pacific Northwest. The British arrived first with Alexander Mackenzie of the North West Company, one of Washington Irving's "lords of the lakes and forests." In 1793, by means of the Peace and Fraser river systems, Mackenzie crossed the northern and central interior of British Columbia — what came to be called New Caledonia — and reached the Pacific. His company did not follow with posts until a decade later. By then President Jefferson had countered with the vaunted Lewis and Clark expedition, which was intended to determine the feasibility of a transcontinental commercial communication. It ascended the Missouri, crossed the Pacific slope via the Snake, or Lewis, River, and sighted the Pacific at the Columbia's mouth in 1805. Several other expeditions quickly followed, including those of Simon Fraser, who followed his river to its mouth in 1808, and David Thompson, who reconnoitered the Columbia system in 1807-11. The most important immediate result of these ventures was the disclosure that the Columbia or Oregon River surpassed the Fraser as a route from the coast to the interior; the Fraser Canyon proved more formidable than the Columbia Gorge, the hazardous Columbia Bar notwithstanding. American title to the Columbia had been strengthened by the rediscovery of the river in 1792 by Captain Robert Gray of Boston.[2] It remained for a New York merchant, John Jacob Astor, to try to realize Mackenzie's dream of dominating the North American fur trade by means of a transcontinental-transoceanic strategy centred on the legendary "great river of the west." In 1811-12 Astor's Pacific Fur Company (whose men were mostly Canadian "pedlars of the wilderness") founded several posts along the lower Columbia and one as far north as the junction of the North and South Thompson. From his entrepôt of Astoria at the river's mouth, Astor hoped that through commercial cooperation with the Russian-American Company at its colonial capital of New Archangel (Sitka), he could squeeze the British from the Northwest trade by linking the Columbia gateway, the Pacific sea lane, and the Chinese market. Astor's scheme was thwarted by misfortune and mismanagement as well as by the outbreak of war between Great Britain and the United States in 1812. Seeing the Pacific Fur Company threatened by capture, Astor's agents hastily sold the firm to the North West Company in 1813, and Astoria became Fort George. British dominance was to last three decades, although American "long knives," as the Indians called them, continued to be drawn to the Pacific slope by the prospects of transcontinental dominion and Oriental commerce.

The "Columbian enterprise" soon proved a mixed blessing to the Nor'Westers. Unlike the plains of Rupert's Land, the Oregon Country's mountains and tumultuous rivers slowed transport, while the Indians, with their horses and salmon, had less need or desire to trade furs. Moreover, bison for pemmican and Indian corn were lacking as readymade provisions, and oceanic shipping and Chinese marketing proved troublesome. Nevertheless, the North West Company managed to orient New Caledonia to the Columbia system profitably by means of packhorse/batteau brigades, dominate the Columbia and Snake waterways by founding Fort Nez Percés at the junction of the two rivers, and tap the Snake and Flathead Countries successfully with trapping expeditions.

Meanwhile, the British and American governments were working to settle their territorial disagreements. The Convention of 1818 drew the Great Plains segment of the international boundary along the forty-ninth parallel from Lake of the Woods to the Rocky Mountains. It was no accident that this boundary separated the two great river systems, the Saskatchewan and the Missouri, that were used to reach the "Columbia quarter" by British and American frontiersmen. West of the Rockies, however, the two trunk waterways, the Columbia and the Fraser, flowed north-south, so that an extension of the forty-ninth parallel to the Pacific would inevitably cut the superior Columbia route and thereby prove contentious. Anxious to placate the United States, Great Britain returned Fort George to the Americans (who neglected to reoccupy the post). The 1818 treaty also granted "free and open" access to the Oregon Country to citizens of both nations for ten years. In reality, however, this state of joint occupancy was one-sided, for the rival "Greys" and "Blues" of the North West and Hudson's Bay Companies decided to settle their differences by joining forces. The 1821 merger produced a vigorous "New Concern" that pre-empted exploitation of the Oregon Country, especially under the managerial proficiency of George Simpson (1792?-1860). On "the west side the mountains" the American mountain men were unable to compete with what Canadian fur trader Archibald McDonald called the "great Fur Monopoly of British America North." From the time of his appointment in 1821 as Governor of the Honourable Company's Northern Department (Rupert's Land plus New Caledonia), Simpson laboured energetically and astutely for four decades to promote his company's and his country's interests. For his efforts, the "little emperor of the plains" was appointed Governor-in-Chief of all of the company's territories in North America in 1839 and was knighted in 1841.

By the mid-1820's, Simpson was rationalizing the company's languishing operations in the Columbia Department.[3] A major component of his reorganizational scheme was the expansion of agriculture at the various posts in order to make them as self-supporting and economical as possible. Eventually a separate but affiliated firm, the Puget's Sound Agricultural Company, was created to accommodate the increasing export demand for agricultural products. Retired employees and overland migrants (American and Canadian), as well as Protestant and Catholic missionaries and some Indians, also undertook farming. At its farflung posts, the Hudson's Bay Company succeeded in growing enough grain and vegetables and rearing enough livestock to reduce considerably the former dependence upon local fish and game (sometimes meagre and always monotonous) and costly imports. Frost in the northern interior, drought in the southern interior, and even dankness on the coast were not easily overcome, but some posts, notably Forts Vancouver, Langley, and Colvile, managed to produce sizable surpluses of flour, beef, and dairy products for the less successful establishments. Between Puget Sound and the Columbia River, the Puget's Sound Agricultural Company used wage-earners and share-croppers to grow wheat and rear sheep. Although this corporate venture failed to lure and hold many Canadian or British settlers, who preferred the free and better land of the Willamette Valley, it did succeed in selling flour to Russian Alaska and wool to England and thereby helped to drive American traders from the Northwest Coast and kept at least its middle third for the British crown. The fertile and temperate Willamette Valley witnessed the most successful farming. It proved irresistible to homesteaders, primarily retired Hudson's Bay Company servants until the early 1840's. Their bountiful harvests contributed substantially to the growing export trade of the company, which from its regional base at Fort Vancouver rendered not a little assistance to the valley's pioneers, Canadian and American alike. Protestant and Catholic missionaries were much less proficient in farming and even less proficient in converting their Indian charges to agriculture (let alone Christianity). The missionaries themselves lacked either the time or the skill to farm, and the few Indian converts soon abandoned their crops and herds in the face of increasing Euroamerican pressure on their territory and lifestyle. All of this agricultural activity before the boundary treaty of 1846 forms the focus of this book.

Agricultural development has been neglected by students of the Oregon Country as part of the general disregard of its pre-1846 period by historians (the most notable exception is Frederick Merk).[4] This neglect probably stems from the ethnocentric bent of American historiography, particularly the regional and local variants, and the inaccessibility, until recently, of the archives of the Hudson's Bay Company. It is hoped that the gap in the literature will be partly filled by this book. It is also hoped that readers will be struck by the importance of agriculture in the settlement and development of the Oregon Country from the very beginning of white occupation in the early 1810's, not just from the arrival of the first American wagon trains via the Oregon Trail in the early 1840's. In the intervening period it was primarily Hudson's Bay Company servants from Great Britain and Canada who tested the agricultural resources and proved the agricultural potential of this sprawling and diverse region ("here we have Gods works on a large scale," remarked a veteran "west sidian").[5] Their labourious and largely successful efforts deserve to be better known.

The author is grateful to a number of depositories and individuals for their cooperation in providing source materials, above all to the Hudson's Bay Company, which kindly allowed me to consult and cite their archival microfilm in the Public Archives of Canada (Ottawa) and originals in the Provincial Archives of Manitoba (Winnipeg) and to reproduce several maps and sketches, and to the Oregon Historical Society (Portland) under the directorship of Thomas Vaughan, whose able staff generously rendered both assistance and hospitality. I am also indebted to the Bancroft Library (Berkeley), the Beinecke Rare Book and Manuscript Library of Yale University (New Haven), the Provincial Archives of British Columbia (Victoria), the Houghton Library of Harvard University (Cambridge, MA), the Huntington Library (San Marino, CA), the Library of Congress (Washington, DC), the Metropolitan Toronto Central Library, the Missouri State Historical Society (St. Louis), the Oregon State Archives (Salem), the Public Archives of Canada (Ottawa), the Public Record Office (London), the Scott Library (interlibrary loan service) of York University (Toronto), the National Archives and Records Service of the United States (Washington, DC), the University of Washington Library (Seattle), the Washington State University Library (Pullman), and the Royal Ontario Museum (Toronto).

Part One

POST FARMING
Hudson's Bay Company Agriculture This Side The Mountains

The Company now occupy the Country between the Rocky Mountains and the Pacific, by six permanent establishments on the Coast, sixteen in the interior country, besides several migratory and hunting parties, and they maintain a marine of six armed vessels, one of them a steam vessel, on the Coast.

Their principal establishment and depot for the trade of the Coast and Interior is situated ninety miles from the Pacific on the northern banks of the Columbia River, and called Vancouver in honour of that celebrated navigator: in the neighbourhood they have large pasture and grain farms, affording most abundantly every species of agricultural produce, and maintaining large herds of stock of every description: these have been gradually established, and it is the intention of the Company still further, not only to augment and increase them, to establish an export trade in wool, tallow, hides and other agricultural produce, but to encourage the settlement of their retired servants and other emigrants under their protection. The soil, climate and other circumstances of the Country are as much, if not more, adapted to Agricultural pursuits, than any other spot in America, and with care and protection, the British dominion may not only be preserved in this Country, which it has been so much the wish of Russia and America to occupy, to the exclusion of British subjects, but British interest and British influence may be maintained as paramount in this interesting part of the Coast of the Pacific.

Governor John Pelly of the Hudson's Bay Company to Lord Glenelg of the Colonial Office, 1837

1

GOVERNOR SIMPSON'S ŒCONOMY: THE ORIGINS OF POST FARMING

> *It being necessary that the gentlemen, who are engaged in transacting the business of the [Hudson's Bay] Company west of the mountains, and their laborers, should be better and less precariously supplied with the necessaries of life, than what game furnishes; and the expense of transporting suitable supplies from England being too great; it was thought important to connect the business of farming with that of fur, to an extent equal to their necessary demands.*
>
> REVEREND SAMUEL PARKER, 1836

One of the main problems faced by Canadian fur traders "this side the mountains" was the difficulty of food supply. East of the Rockies in the "Buffaloe Countries" or western plains, *coureurs de bois* and *voyageurs* could easily procure bison meat and Indian corn; moreover, they could receive additional provisions fairly readily via York boats from Hudson Bay and Montreal canoes from the St. Lawrence. West of the Great Divide, however, supplies of pemmican and maize were insufficient. Here the North West and Hudson's Bay Companies' men had to rely upon transcontinental and transoceanic provisionment from Eastern Canada and England as well as upon local agriculture. The North West Company did send supply ships around Cape Horn to the Columbia River, but it depended mainly upon lengthy overland supply from depots on the Great Lakes. It was at a disadvantage in competition with the Hudson's Bay Company, which had a shorter and cheaper lifeline via Hudson Bay to the very edge of the forest belt, the habitat of the beaver, whose pelt was the *raison d'être* of both firms. Indeed, one of the Nor'Westers' principal handicaps was their long lines of supply, which incurred much expense and prolonged outlay of capital before the realization of any return.[2] This problem became more acute as the company expanded beyond the Rockies. In accordance with a company decision to supply the Columbia by sea rather than by land, "no sooner had the North Westers inherited the Oregon . . . than ship after ship doubled Cape Horn in regular succession with bulky cargoes to the fulfillment of every demand."³ But supply remained dear.

Even the Hudson Bay route proved costly with the increase in the number of Bay men caused by expansion into the Athabasca Country and violent resistance on the part of the Nor'Westers. So it was decided to circumvent the uncertainty and expense of navigation of the Hudson Straits by producing provisions in Rupert's Land itself with agricultural colonists. This decision was explained in a letter of 18 March 1815 from Governor Joseph Berens of the Hudson's Bay Company to the Earl of Bathurst, the Colonial Secretary:

> The servants of the HB Co employed in the Fur Trade have hitherto been fed with provisions exported from England —— Of late years this expence has been so enormous that it has become very desirable to try the practicability of raising provisions within the Territory itself. Notwithstanding the unfavourable soil & climate of the Settlements immedly adjacent to Hudsons Bay there is a great deal of fertile land in the interior of the Country where the Climate is very good & well fitted for the cultivation of grain.
>
> It did not appear probable that agriculture would be carried on with sufficient care & attention by servants in the immediate Employment of the Company but by establishing independant Settlers & giving them freehold tenures of land the Company expected to obtain a certain supply of provisions at a moderate price. The Compy also entertained expectations of considble eventual benefit from the improvement of their landed property by means of agricultural settlements. Having a due regard to the implied condons of their Charter they deemed it a Duty incumbent on them (as soon as the practicability of agricultural improvement was demonstrated) to give a liberal degree of encouragement to an experiment which independantly of other advantages promise to have the most beneficial effects in the civilization of the Indians. ——
>
> With these views the Compy were induced in the year 1811 to dispose of a large tract of their land to the Earl of Selkirk in whose hands they trusted that the experiment would be prosecuted with due attention as the grant was made subject to adequate condons of Settlemt.⁴

Thus was founded Lord Selkirk's Red River Settlement (Assiniboia) with the hope that "all the surplus produce, such as flour, beef, pork, and butter, articles the Company require, would by means of the colony be obtained more conveniently, cheaper, and with less risk, than by the annual importation of such articles from England."⁵ In addition, the settlement would provide the company with manpower and validate its 1670 charter, which required it to undertake colonization.⁶ This scheme was forcibly opposed by the rival Nor'Westers, for the Red River Country was "the great depot of the North-West

Company for making pemmican,"⁷ that nutritious and portable staple of the company's wintering partners. Twice each year, in early summer and in autumn, the *Métis* (the frontier offspring of Euroamerican fathers and Indian mothers) conducted a wide-ranging bison hunt on the prairie for themselves and the Nor'Westers.⁸ As a Bay man recalled:

> Fort William on Lake Superior was their headquarters, but Red River was their depot for supplies of all kinds of provisions, especially pemmican, a mixture of grease & buffalo meat, powdered very fine, and put in bags of 100 lbs each. This was a most important commodity; 2 lbs would suffice for a man a day; with nothing else, potatoes nor flour, they never had. It was from this point, Red River, that this pemmican was sent hundred[s] & thousands of miles away, and when Lord Selkirk came with a colony to settle here, then the N.W. Co became infuriated at their intrusion, which they felt was a blow at a vital point, altho' the Charter of the H.B. Co, made it all right for them to be there.⁹

The resulting "Pemmican War" culminated in the massacre of the governor of the Red River Colony and twenty-one settlers at Seven Oaks in 1816. Public outrage over this atrocity undermined the North West Company, which had enjoyed the favour of the Colonial Office, and led to the coalition of 1821. The union was really an absorption of the North West Company by the Hudson's Bay Company, whose successful reorganization in the 1810's, involving more incentive (profit sharing), cheaper transport (York boats), and improved supply (Red River Settlement) engineered by Andrew Colvile, Lord Selkirk's brother-in-law, rendered the Honourable Company pre-eminent.¹⁰

Initially Selkirk's colony did not prosper, owing to physical adversities and human shortcomings as well as to Nor'West opposition. Locusts, birds, wolves, floods, frosts, and droughts decimated farming. Locusts, for example, destroyed the settlement's crops three years in a row (1819-21).¹¹ Many colonists became disheartened and left for the United States. Governor Simpson reported the colony's plight to the Hudson's Bay Company's Governor and Committee in London in 1821:

> The Grasshopper I am extremely sorry to say continues its destructive influence; the Crops have been seriously injured and in many parts wholly destroyed yet with care and economy there will be no danger of Starvation during the winter, indeed at Red River if people have but strength and common industry & exertion they may always live well; the plains afford an inexhaustible stock of animal Food and if even the Buffalo disappointed them which is next to impossible, with a few nets any number of people may be maintained in the Lake near the mouth of the River; I cannot

however recommend your sending any more settlers from Europe far less from Canada until we see if the Grasshopper disappears; many are of opinion that this will be their last season as they no sooner take wing than they migrate in large swarms; if this hope should be realized it would be a most fortunate circumstance and dispel the gloom which seems to hang over this truly unfortunate colony. The want of proper management is another reason why the population should not be immediately encreased; the thing is now getting too extended for Mr. McDonell [the governor], his affairs are in a labyrinth of confusion and he is unequal to the conduct of them in their present state, in short till we are enabled to see our way clearly and many obstacles which now present themselves are removed, I would with all deference take the liberty of recommending that no fresh levy should be forwarded.

Red River at present I am sorry to say assumes more the appearance of a receptacle for free booters and infamous characters of all descriptions than a well regulated Colony, there is no law order or regularity every man is his own Master and the strongest and most desperate is he who succeeds best. Mr. West [the missionary] does all in his power (and his exertions are truly meritorious) to improve the morals of the people but with little success, on the contrary their habits are most vicious: they have not exactly committed Murder or Robbery but the next thing to it and frequently threaten both so that the well disposed feel themselves in continual danger. The Population is now getting very considerable and such a mass of renegades and malcontents of all descriptions are not to be constrained without the assistance of civil & military power.[12]

But in spite of the lawlessness and mismanagement of the Red River population, it was the decline of the bison that was most devastating. In the spring of the following year Simpson reported that "the crops were so unproductive, and the Provision Trade during the Summer so trifling, that nothing of consequence could be collected, and the unprecedented and almost total failure of the Buffalo with the encreased population, has made this Colony the most distressing scene of starvation that can well be conceived."[13] Even the wild rice crops, gathered from the local marshes by the Indians, were too irregular to save the situation.

At this juncture Assiniboia's fortunes improved. The coalition had released many superfluous employees, and the decline of the bison herds had prompted many *Métis* hunters to turn to farming. All were men who were used to the rigours of life in the northwest. And the fall harvest of 1822 was "very productive," so that in 1823 the fur trade for the first time tapped the colony for provisions (Table 1), especially flour, which was milled locally and boated to the depot of

Norway House at the northern end of Lake Winnipeg, whence it was distributed to the posts of the Northern Department.[14] The company's Council of Rupert's Land at York Factory regarded these purchases as "the means of curtailing the requisition from England," while the Governor and Committee in London felt that "the Colony will be of no small importance in furnishing provisions for the Fur Trade and will serve as an asylum for retired servants of the Company who must otherwise be maintained at heavy expense at the different inland posts."[15]

Table 1
HUDSON'S BAY COMPANY PURCHASES OF PROVISIONS
FROM THE RED RIVER COLONY, 1823-31

Year	Flour (lbs.)	Barley (bus.)	Corn (bus.)	Peas (bus.)	Butter (lbs.)
1823	0	0	700	0	0
1824	22,400	127	1,000	100	1,200
1825	22,400	127	1,000	100	1,200
1826	22,400	127	1,000	100	1,200
1827	22,400	127	1,000	100	1,200
1828	22,400	127	1,000	100	0
1829	22,400	127	1,000	100	0
1830	56,000	300	200	0	0
1831[a]	56,000	200	40	40	1,680

For sources see *Sources for Tables.*
[a]Plus 1,160 lbs. of pork and 784 lbs. of beef.

Following the great flood of 1826, the Red River Colony prospered. About 1827, 1,052 settlers cultivated 673 acres of land and herded 816 head of livestock.[16] An American fur trader, Joshua Pilcher, told the U.S. Secretary of War that in the spring of 1830 "the colony consists of three or four thousand inhabitants. . . . The soil was excellent, the farms in good order, crops abundant, houses comfortable, and all the variety of stock which belongs to our settlements."[17] Assiniboia's agricultural success was such by the middle of the 1830's that a company servant could declare that there was "want of a sufficient outlet for produce," for the colony was producing 50,000 bushels (1,500 tons) of wheat annually but the company was buying only fifty-six tons of flour each year.[18] In 1836 the settlement numbered three thousand *Métis* and two thousand whites; a decade later the same population stretched forty miles along both sides of the Red River and cultivated five thousand acres.[19] Overproduction was still the foremost problem. Two British army officers found that "the chief cause of Complaint, among the Settlers appears to be the Want of a Market for their produce. The Hudson's Bay Company are not able to take more than a certain quantity of Grain etc."[20] In a similar vein the Governor of Assiniboia, Duncan Finlayson, wrote to a friend in 1842:

> there is a great deal of grumbling on the score of a better market; yet I believe the greater part are satisfied with their lot. —— The last harvest was so very abundant, that I was under the necessity of limiting each farmer to the sale of 8 cwt of flour only —— and there were some who did not sell, even half of that quantity —— notwithstanding this restriction, we have now in store bordering on 3000 cwt of flour —— & if at this moment we wanted five times that quantity we could readily purchase it. —— I am authorized by the Board to keep a years stock on hand, but I have now more than will be consumed in two years, I therefore, think, it would be advisable for you to discontinue the importation of flour from home, as I must represent this state of things to the Company, & they may naturally say, if there be such abundance of flour in the Country, why import some to York?[21]

Beyond the Rockies, however, provisionment was more difficult. Before the 1821 merger, the North West Company tried to circumvent its costly overland supply lines by instituting a policy of living off the land of the Oregon Country — off what was termed "country produce." In May 1811, Daniel Harmon, superintendent of company operations west of the Rockies, undertook his firm's first efforts at agriculture there by planting some barley and vegetables at Stuart's Lake (Fort St. James). His potatoes yielded forty-onefold in 1816 and his barley thirty-twofold in 1818. These returns gave him "sufficient proof that the soil, in many places in this quarter, is favourable to agriculture," although the Indians proved reluctant farmers, and for corn "the nights in this region are too cool, and the summers are too short, to admit of its ripening."[22] Wherever conditions permitted, gardens were soon meeting some of the food needs of the Nor'Westers' cordilleran posts.

These were not the first attempts at farming in the Oregon Country, however. In 1786 Captain James Strange of the British snow *Captain Cook* had vegetables planted at Friendly Cove in Nootka Sound on Vancouver Island.[23] The Spanish established a vegetable garden at Nootka Sound in 1790 and at Núñez Gaona (Neah Bay) on the northwestern tip of the Olympic Peninsula in 1792.[24] Captain Charles Bishop of the British ship *Ruby* had some vegetables sown on an island at the mouth of the Columbia in the middle of 1795 and again at the beginning of 1796; potatoes, radishes, and beans bore fruit.[25] In 1810 crewmen of the American vessel *Albatross* planted a vegetable garden on the Oregon coast. Inland the Pacific Fur Company initiated farming in the very same month at Astoria as the rival North West Company did at Stuart's Lake. The Astorians' supply ship, the ill-fated *Tonquin*, landed thirty pigs, three goats, and one sheep in the spring of 1811, and in May, vegetables (including potatoes, radishes, turnips, and rape) were planted.[26] The pigs multiplied "tremendously," but only potatoes, turnips, and rape matured. At Spokane House the Astorians had a "thriving" garden by 1814.[27] Generally, however, few of the vegetables "came to perfection" at the posts of the Pacific Fur Company. After the North West Company bought Astoria in 1814, wheat and potatoes were planted and pigs and chickens were kept.[28]

These initial agricultural efforts were redoubled with the increased activity on the Pacific slope generated by the coalition and by the farsighted and far-reaching policies of George Simpson, who became Governor of the Northern Department in 1821 and of both the Northern and Southern Departments in 1826.[29] Economy was his watchword and energy his hallmark. In the words of one of his subordinates, "Governor Simpson had a remarkable administrative ability, to which he bent an industry which did the work of two or three other ordinary men."[30] Under him new posts were built and old ones abandoned, new routes were opened and old ones improved, employees were decreased by more than one-half and discipline was tightened, wages were reduced by as much as one-half and perquisites and gratuities were abolished, fur prices were standardized, liquor traffic was halved, and larger York boats were substituted for North canoes as freighters at a saving of one-third in wages.[31] Simpson thoroughly reorganized the sluggish Columbia District, whose returns were "falling off." Soon after his appointment he had been instructed in 1822 by the Governor and Committee to determine the worth of the district.[32] In the summer of 1824, Simpson informed Andrew Colvile, a prominent director of the company, that the "unsettled state" of the Columbia demanded his immediate attention, especially since there was a "general feeling against it [the district]" among the Bay men.[33] The company was chiefly interested in New Caledonia, intending to consolidate that district in order to check Russian encroachment from the north (especially after the 1821 Russian decree that unilaterally set the fifty-first parallel as the southern frontier of Russian America). However, the company also felt that the lower Columbia basin should be kept as a buffer against American traders for the protection of what one longtime servant, John McLean, called "the great beaver nursery" to the north. In accordance with this view the company would methodically strip the "lower country" of its furs over the next quarter of a century. From 1823 annual trapping parties of thirty to forty men (Snake Expedition and Southern Expedition) left the Columbia base in October and returned the following May or June, "trapping clean" the watershed of the Snake River, the Great Basin, and California in the face of antagonistic American mountain men and hostile Indians. Simpson personally toured the Oregon Country in 1824-25. He reported that "I have last winter prepared a scheme for the remodelling of the business of this side the mountain generally and for the extension of its Trade." Simpson concluded that

London should pay attention to the region because its fur trade could bring handsome profits and would relieve the overwrought trade of "the east side of the mountains." "I think," he reported, "there is a Field for commerce on this side the Mountains which has been much neglected and which if properly cultivated would become an Object of the first importance to the Honble Hudsons Bay Compy." He added that the trade of the cordillera, if "properly managed and sufficiently extended," would return double the profits of the trade of any other part of North America.[34]

In order to fulfil this potential, the governor had to reduce the Columbia District's expenses, which he found to be excessive. During his tour of inspection, he observed that the district had been neglected and mismanaged, with much discord and waste. He wrote:

> Everything appears to me on the Columbia on too extended a scale *except the Trade* and when I say that that is confined to Four permanent Establishments the returns of which do not amount to 20,000 Beaver & Otters altho the country has been occupied upwards of Fourteen Years I feel that a very Severe reflection is cast on those who have had the management of the Business, as on looking at the prodigious expences that have been incurred and the means at their command, I cannot help thinking that no economy has been observed, that little exertion has been used, and that sound judgment has not been exercised but that mismanagement and extravagance has been the order of the day.[35]

Simpson was particularly appalled by the expense of provisionment. He assailed the "gentlemen" of the Columbia for relying too heavily on "European Provisions," their imports requiring up to six boatloads annually, whereas only two should have sufficed. The expense of transport was so great that "they may be said to have been eating Gold," and the boatmen alone were "sufficient to run away with a large share of the Columbia profits."[36] Little wonder that American mountain men had cheaper supplies than company traders. The indent of the 1825 outfit for the Columbia District totalled £2,081, nearly one-third of which comprised provisions, including 500 bushels of Indian corn, 100 bushels of Indian corn meal, 3,550 pounds of tobacco, 2,912 pounds of sugar, 2,000 pounds of rice, 2,000 pounds of tallow, 1,120 pounds of molasses, and 644 gallons of spirits.[37] Simpson was convinced that these imports were unnecessary because with little effort and slight expense most could be produced locally. He noted with disgust in the spring of 1825 at Fort Nez Percés (Walla Walla) that in three years the post had consumed 700 horses, besides imported provisions, whereas the river and a garden would have "abundantly" maintained the men with fish and potatoes. So he gave the "rather indolent" Chief Trader Peter Dease, commandant of the fort, ten bushels of seed potatoes and "a long lecture on the advantages to be derived from attention to the Horticultural Deptm^t of the Post."[38] Simpson was scathing in his criticism of company personnel for their failure to promote agriculture:

> Grain in any quantity might be raised here, but cultivation to any extent has never been attempted, indeed throughout the Columbia no pains have been taken to meet the demands of the trade in that way which was a great oversight or neglect as corn in abundance might have been procured at little or no Expence at the Door of every Establishment but those in charge have preferred the less troublesome and more costly mode of Importing them from England Boston or California and employing extra men to deliver it into their Stores. It has been said that Farming is no branch of the Fur Trade but I consider that every pursuit tending to leighten the Expence of the Trade is a branch thereof and that some of our Factors and Traders on the other side are better adapted for and would be more usefully employed on this side in the peaceable safe and easy occupation of Farming than in Councilling Dealing with Indians or exploring new countries for which many of them are totally unfit but it unfortunately happens that in these savage regions Gentlemen sometimes imbibe the exalted notions of Indian Chiefs who consider that to Slaves or inferiors alone belong the less important yet useful and necessary duties of providing for their Daily wants by their own personal exertions; in short they have no notion of acting but think that their business only is to legislate and direct and are not satisfied unless they have a posse of Clerks Guides Interpreters & supernumeries at their disposal while they look on with a pair of Gloves on their hands.[39]

This opinion was contrary to that of Fort George's Chief Factor James Keith, who wrote to London around 1822 that his men "in a short time ... would have to get all their flour & other supplies around Cape Horn as the Country in the Columbia District was not capable of raising them." Apparently this view stemmed partly from the Columbia's reputation as "a bad, barren Country."[40] Simpson's perception of the region was obviously much rosier. But he had to combat the fur trader's view that hunting-trapping and farming were incompatible. Gabriel Franchère, a Canadian Astorian, expressed this antithesis: "The interest of the Hudson's Bay Company, as an association of fur-traders, is opposed to agricultural improvements, whose operation would be to drive off and extinguish the wild animals that furnish their commerce with its objects."[41] Another fur trader, the Bay clerk William Kittson, was more vehement: "I hate every thing that comes in the way of Beaver," he declared in 1830.[42] On an extensive scale, farming would do what Franchère and Kittson and other "rangers of the woods" feared, but the localized cropping and herding that Simpson had in

18   Farming the Frontier

mind would help rather than hurt the fur trade.

The governor lost no time reorganizing, consolidating, and expanding operations in the Oregon Country, and the company's presence there grew from 13 posts and about 200 servants in 1825 to 22 posts, 6 ships, and some 450 servants in 1836. In other words, its strength doubled within a decade. Simpson recommended that the number of servants be reduced from 151 to 87 and that the amount of freight hauled by the Columbia brigade from Fort George to the interior be reduced from 645 to 200 pieces per year.[43] He further recommended that New Caledonia be outfitted from the coast via the Columbia rather than from the bay via the Saskatchewan because the former route would be easier and cheaper.[44] This change was made in 1826, and in 1827 the New Caledonia and Columbia Districts were amalgamated to form the Columbia Department. Agriculture loomed large in Simpson's reforms as the main component of his diversification of company activities. Such variegation would spread corporate risk in case of stiffened American competition, depleted fur bearers, or reduced fur markets. The department's headquarters were transferred in 1824-25 from Fort George, the so-called "Gibraltar of the West," to Fort Vancouver at the confluence of the Columbia and Willamette. Simpson put Chief Factor Dr. John McLoughlin (1784-1857), a former Nor'Wester, in charge of the new base and its department. A "very bustling active man ... of strict honour and integrity," McLoughlin directed the affairs of the Columbia Department for the next two decades with "great exertions indefatigable labours and unremitting attention."[45] His absolute authority — "the Doctor's big stick" — was firm but fair; some of his subordinates termed him the "Sultan" and the "Premier," but most of them called him the "Big Doctor" or simply the "Doctor." Euroamerican settlers, who benefited from both his kindness and his firmness, sometimes referred to him as the "Emperor of the West."[46] Already over forty when he was posted to the Columbia, McLoughlin had long, white hair that earned him the nickname of "white-headed eagle" from the Indians (Plate 1). Under his able management Fort Vancouver quickly became the company's "grand depôt" and "rendezvous" west of the Rockies and the "grand emporium" of its cordilleran trade. The post was more central to trade and had a more convenient landing and more agricultural land than Fort George. By 1828 Simpson found that at the new headquarters farming was already "perhaps more than anywhere else, the main Spring of the business."[47] In 1825-26 Fort Colvile was established just above Kettle Falls on the middle Columbia on a fertile river terrace. It was intended to provision the interior posts, including those of the upper Snake and middle Fraser. At Fort Colvile, which replaced Spokane House, Simpson felt "as much Grain and potatoes may be raised as would feed all the Natives of the Columbia and a sufficient number of Cattle and Hogs to supply his Majestys Navy with Beef and Pork." Fort Langley was the third major agricultural establishment, founded in 1827 on the lush floodplain of the lower Fraser. It was intended to provision the proposed chain of coastal posts which in combination with several small vessels would, Simpson hoped, soon oust American coastal traders who threatened the company's trade because they were able to tap the furs of the interior through Indian middlemen on the coast.[48] Farming was also encouraged at the remaining posts wherever possible.

Plate 1. *Chief Factor John McLoughlin.*

In this way Simpson hoped more easily and more cheaply to meet the food needs of the company's servants, who were engaged for a period of five years for an annual salary and a daily ration.[49] In 1831 John McLoughlin told the Governor and Committee that:

> Fish Venison or Game when their is any and when their is none — one quart of Indian Corn or pease with two ounces Grease pr. Day — this has been the Regular Rations of the place since first Established and Indeed one quart of Corn and two ounces of Grease pr. Day is the Regular Daily Rations of that Kind of provisions wherever it is used throughout the Indian Country.[50]

This ration represented roughly eleven and one-half bushels of "corn" (wheat) per man per year, on the basis of 1 dry quart = 1/32 bushel. (The number of company consumers in the Columbia Department between 1821 and 1846 is shown in Table 2). Thus, the department's wheat requirements ranged from about twenty-one hundred bushels in 1821-22 to about seven thousand bushels in 1841-42.

Table 2
NUMBER OF HUDSON'S BAY COMPANY SERVANTS
IN THE COLUMBIA DEPARTMENT, 1821-46[1]

Outfit	Number	Outfit	Number
1821-22	184	1834-35	456
1822-23	194	1835-36	453
1823-24	203	1836-37	463
1824-25	214	1837-38	447
1825-26	199	1838-39	465
1826-27	199	1839-40	512
1827-28	210	1840-41	554
1828-29	223	1841-42	616
1829-30	242	1842-43	559
1830-31	363	1843-44	572
1831-32	378	1844-45	611
1832-33	411	1845-46	612
1833-34	429		

For sources see *Sources for Tables*.
[1]These figures include regular employees on wages and exclude freemen and deserters. Figures from 1830-31 include men not assigned to a particular post and men entering or leaving the department; figures from 1843-44 include "Goers and Comers" (transporters) in the Snake Country.

Governor Simpson's motives for expanding agriculture in the Columbia Department were summarized in a letter to the Governor and Committee from McLoughlin, who, like most fur traders, disliked farming:

> if it had not been for the great expense of importing Flour from Europe, the serious injury it received on the voyage, and the absolute necessity of being independent of Indians for provisions, I never would have encouraged our farming in this Country, but it was impossible to carry on the trade without it.[51]

Every year one or two supply ships arrived at the Columbia from the Thames with some five hundred tons of cargo each. From 1823 through 1846, thirty-five "London ships" left Gravesend for Fort Vancouver.[52] They usually left England in the autumn, reached Fort Vancouver in the spring (June), departed in the autumn (November), and returned home in the spring. Three ships were maintained: one going, one coming, and one standing in reserve at Fort Vancouver.[53] From 1841 the company sent two supply ships to the Columbia and two to Hudson Bay every other year, the two destinations alternating. The ships reached Fort Vancouver fully loaded but departed three-quarters empty, owing to the bulkiness of supplies relative to skins.[54]

Occasionally cargo was inferior in quality and damaged en route. In 1825, for example, McLoughlin complained to London:

> I have to acknowledge the receipt of yours of the 24th July 1824 with the several documents p. the *William and Ann* which cast Anchor opposite Fort George on the 11th April last and I am sorry to say part of her Cargo was wet, fortunately the Dry Goods are not injured nor indeed is any part of the Cargo except the Flour and Meal. The Barrels of Both are very bad, we will lose about a seventh of the Flour and Meal. The Gunpowder is damper than any we have hitherto had and there is no appearance of Water having reached the Powder Barrels. The Pork and Beef are not so good as we have hitherto had and the Barrels are very bad, the Bricks are of a very inferior quality.

Similar complaints were made by McLoughlin again in 1826 and 1835.[55]

The London ships were outfitted and dispatched one year in advance in order "to provide against loss of ships with outward freight."[56] Actually, few shipwrecks occurred. Of the thirty-five supply ships, only two were lost (and three remained on the Northwest Coast). Both wrecks — the *William and Ann* in 1829 (all twenty-six hands were drowned) and the *Isabella* in 1830 (all of its crew and most of its cargo were saved) — took place at the mouth of the Columbia, where the notorious Columbia or Peacock Bar jutted from what was fittingly called Cape Disappointment. This bar — the "foaming and tumultuous breakers" on the shoals at the river's mouth — could be crossed only when both tide and wind were favourable, and in winter ships sometimes had to wait up to two months for such an opportunity. (Hence the lateness of the discovery of the Columbia River, which was overlooked by both Captains Cook and Vancouver.) John Townsend, an American ornithologist, described the bar in 1834:

> A wide bar of sand extends from Cape Disappointment to the opposite shore, — called Point Adams, — and with the exception of a space, comprehending about half a mile, the sea at all times breaks furiously, the surges dashing to the height of the mast head of a ship, and with the most terrific roaring. Sometimes the water in the channel is agitated equally with that which covers the whole length of the bar, and it is then a matter of

imminent risk to attempt a passage. Vessels have occasionally been compelled to lie in under the cape for several weeks, in momentary expectation of the subsidence of the dangerous breakers, and they have not unfrequently been required to stand off shore, from without, until the crews have suffered extremely for food and water. This circumstance must ever form a barrier to a permanent settlement here; the sands, which compose the bar, are constantly shifting, and changing the course and depth of the channel, so that none but the small coasting vessels in the service of the company can, with much safety, pass back and forth.[57]

But even the company's small vessels experienced difficulty. In 1844, for instance, the barque *Cowlitz* took eighty days to cross the bar![58] Not a few ships — both British and American — foundered in the breakers. An American churchman declared that "perhaps there have been more lives lost here, in proportion to the number of those who have entered this river, than in entering almost any other harbor in the world." Certainly the Columbia's mouth was the most dangerous part of the long voyage from England.[59]

Even after entering the river, ships faced a difficult voyage upstream to Fort Vancouver. It usually took two weeks but occasionally up to four to overcome the calms, contrary winds, and shifting sand bars on this stretch. In 1841 one-fifth of the duration of the voyage of the barque *Columbia* from Fort Vancouver to New Archangel and back was spent in the river. And in 1845 the barque *Vancouver* took a month to sail from Fort Vancouver downstream to Baker's Bay, where it waited another month and a half to cross the bar.[60] Incoming supply ships were sometimes delayed so much that the Columbia brigade was late leaving Fort Vancouver for the interior. All of these difficulties of overseas supply via the Horn had been instrumental in the Pacific Fur Company's loss of Astoria to the Nor'Westers in 1813, and they were influential in Simpson's decision to boost local agriculture. Although the London ships would still be necessary to bring manufactures and take furs, the Governor and Committee hoped that the Columbia Department would be able to produce enough food to meet most of its own needs, so that the ships would have to bring fewer provisions not only for the department but for themselves as well (enough for a one-way rather than a return trip).

In addition, post farming would alleviate the final leg of the supply line, the Columbia brigade system of transport from the departmental headquarters and entrepôt (initially Fort George and subsequently Fort Vancouver) to the interior forts. Supplies were batteaued upriver as far as Fort Okanagan, transferred to pack horses and carried via Thompson's River to Fort Alexandria, and distributed from there by boat on the Fraser system. Furs were returned via the same route. Both inward and outward transport were threatened by river conditions (rapids in particular, notably the Dalles, "a long & intricate chain rushing with great force thru a number of narrows & crooked channels, bounded by huge masses of perpendicular rock")[61] and unfriendly Indians. But the upstream haul was especially onerous, so that any lightening of inbound freight would be particularly welcome. That could be accomplished by producing more food at the interior posts and transporting less from the mouth of the Columbia. As Chief Factor Alexander Kennedy noted around 1825, "the provisions sent normally from Fort George to the interior including that for the voyage forms more than half of the boats cargoes which is a considerable drawback on the profits and independant of the cost it requires a greater number of men to transport it."[62]

In addition to reducing the amount of river transportation, Governor Simpson reasoned that post farming would obviate the lengthy and costly necessity of overland supply from York Factory. Asserting that "the great advantage of outfitting New Caledonia from this side [of the Rockies] instead of from York Factory is that the business can be done with greater facility and less expense," Simpson estimated that by ending transcontinental provisionment, the number of employees in New Caledonia could be reduced by 50 per cent and the profits on furs raised by 25 per cent.[63] This view was in line with the company's policy of minimizing transport costs by promoting self-sufficiency, including agriculture, especially at remote posts. The policy was particularly relevant to the Oregon Country because it was the remotest locale of company operations.

Simpson hoped, too, that post farming would circumvent dependence upon uncontrollable foreign sources such as Indians and American vessels for provisions. In moving Fort George to Fort Vancouver, his "main object" was "that of rendering ourselves independent of Foreign aid in regard to the means of subsistence."[64] Company posts had been relying heavily upon fish, game, and berries, obtained mostly from Indians, again in order to minimize imports. For example, as late as 1829, two-thirds of Fort Prince of Wales's (Fort Churchill's) provisions consisted of "country produce," mainly fowl, venison, and fish.[65] This dependence was particularly marked in the distant Columbia Department. However, the company did not feel secure relying upon outsiders for many of its basic necessities. In 1822 Simpson advised his superiors "to keep the Posts independent of the Natives for they will exact high prices when they can."[66] Moreover, McLoughlin believed that the abundance of fish and roots in his department harmed the fur trade because they made the Indians independent of the "great company" for provisions, which would otherwise be traded for furs.[67]

Generally salmon were plentiful, as noted by Lieutenant Charles Wilkes, commander of the U.S. Exploring Expedition (1838-42):

> It will be almost impossible to give an idea of the extensive fisheries in the rivers and on the coast: they all abound in salmon of the finest flavor, which run twice a year from May until October, and appear inexhaustible; the whole [Indian] population live upon them.[68]

As the Catholic missionary Father Pierre-Jean De Smet noted, just "as the buffalo of the north, and deer from north to east of the mountains, furnish daily food for the inhabitants of those regions, so do these fish supply the wants of the western tribes."[69] The principal Indian fisheries were located at the Dalles and Kettle Falls on the Columbia and in the Fraser Canyon and Iron Canyon on the Fraser. Salmon were more numerous in the Fraser but inferior in quality and size to those of the Columbia.[70] In both rivers, however, the salmon runs were not infrequently late or light. The bourgeois of Stuart's Lake reported in 1815 that salmon generally failed "every second year and completely so every fourth year."[71] John McLean, clerk at the same post in 1834, noted:

> The salmon (the New Caledonian staff of life) ascend Frazer's River and its tributaries, from the Pacific in immense shoals, proceeding towards the sources of the streams until stopped by shallow water. Having deposited their spawn, their dead bodies are seen floating down the current in thousands; few of them ever return to the sea; and in consequence of the old fish perishing in this manner, they fail in this quarter every fourth year.[72]

New Caledonia's salmon run was "abundant" in 1825, 1829, 1833, 1837, 1841, and 1845 but scanty in the intervening years. The failures of 1827 and 1828 were both described as "unprecedented"; hundreds of Indians starved and fur returns dropped.[73] The company's servants also suffered, for west of the Continental Divide dried salmon, or *bardeau* [shingle], replaced pemmican. Salmon was more nutritious than pemmican but less preservable, less portable, and less liked. (Clerk Frank Ermatinger expressed the general feeling when he lamented "the misery of Damned Dried Salmon.")[74] The company's ration was three pounds of pemmican, ten pounds of salmon, four dried salmon, fifteen pounds of whitefish, or twenty pounds of carp per man per day! This ration totalled 25,000 dried salmon annually for the ten men (plus women and children) of Stuart's Lake in New Caledonia in the early 1810's.[76] At Fort Okanagan in 1826, thirty-eight whites and several Indians consumed 18,411 dried salmon, 739 fresh salmon, and 19 fresh trout, as well as 1,268 pounds of fresh venison, 400 pounds of horse meat, 25 pounds of fresh bear meat, 4 pounds of dried beaver meat, 97 partridges, 74 ducks, 64 badgers, 29 rabbits, 16 dogs, 13 geese, 4 swans, 6 beaver tails, 1,171 quarts of berries, 192 quarts of roots, 20 quarts of nuts, 570 eggs, 40 bushels of potatoes, and 25 pounds of suet.[77] At Thompson's River, where Chief Factor Archibald McDonald asserted in 1827 that "dried salmon is the Staff of life," another thirty-eight whites and some Indians ate more than 12,000 dried and 700 fresh salmon yearly.[78] In the same year it was reported that Fort Nez Percés, one of the medium-sized posts, required 3,000 salmon for fifteen servants.[79] And in the last half of the 1830's, the Bay men of New

Caledonia required 30,000 salmon annually (two-thirds came from the Babine Indians).[80] At Fort Colvile, where fresh salmon averaged sixteen pounds each (enough to feed one man for two days), Clerk John Work reported in 1830 that the supply of "country produce" was "precarious" and "uncertain."[81]

The supply of fish and game, however, was too irregular. Nowhere was this predicament more pronounced than in New Caledonia. Simpson toured the district in 1828 and reported that the food supply

> altho' bad, as to quality, would not afford grounds for complaint, if the quantity was sufficient to satisfy the cravings of Nature; but the people are frequently reduced to half allowance, to quarter allowance, and sometimes to no allowance at all, so that the situation of our New Caledonia Friends in regard to the good things of this Life, is any thing but enviable.[82]

He found McLeod's Lake to be "in regard to the means of living . . . the most wretched place in the Indian Country"; its three men were "starving, having had nothing to eat for several Weeks but Berries, and whose countenances were so pale and emaciated that it was with difficulty I recognised them."[83] The frequent shortage of salmon caused Simpson to complain to London in 1834 that whenever the catch was small in New Caledonia, the men at the posts spent the winter seeking provisions ("which is too frequently the case in this inhospitable region") to the detriment of trade.[84] "No salmon, no furs" was a saying "the west side the mountains."[85]

The governor realized that agriculture would provide not only a steadier but also a more healthful diet than fishing and hunting,[86] and healthier employees were better workers. The servants' diet was particularly wanting in New Caledonia. Here salmon (the principal "stand by"), sturgeon, whitefish, trout, carp, ling, deer, moose, caribou, horses, bears, ducks, berry cakes, and, in emergencies, "giddees" (Indian dogs) were eaten. Salmon and sturgeon were sarcastically known in New Caledonia as "Albany beef."[87] A steady diet of salmon had a diarrheal effect at first, as was plaintively noted in 1831 by Thomas Dears, a clerk at Thompson's River, in a letter to a colleague:

> a wholesome meal of food. which is not what we enjoy in New Caledonia. and if your destination was to have been this way think it one of the most fortunate occurences of your life you did not return. You Know I am generally a slender person what would you say if you saw my emaciated Body now I am every morning when dressing in danger of Slipping through my *Breeks* and falling into my Boots. many a night I go to bed hungry and craving something better than this horrid dried Salmon we are obliged to live upon. it is quite *Medicinal* this very morning one of my men in attending on the calls of nature evacuated to the distance of six feet. this

is a reall fact, and is almost incredibale and often are we trouble this way. excuse me. these hardships are enough to drive me out of the Country.[88]

This debilitating diet could have fatal consequences, as in the summer of 1828, when three men from New Caledonia were drowned in the lower Columbia at Priest's Rapids. Work reported that "this accident was principally owing to the weakness of the starved, dispirited N.C. Men."[89] Scurvy was another consequence. A clerk at Connolly's (Bear) Lake lamented in 1830 in a letter to a friend:

> One might suppose this quarter, at least, was exempt from its power; nevertheless I was very near falling a sacrifice to it shortly after I had the pleasure of writing you last year. I suffered two tedious months —— at length however the Spring came round "with healing in his wings" —— and set me on my legs again —— for that time —— but my health is not yet perfectly re-established and I feel will not so long as I continue a Salmon eater —— How long oh L–d![90]

The paltry rations in New Caledonia contributed to its unenviable reputation among company employees as the Siberia of the Canadian fur trade. Simpson acknowledged in 1828 that "there is not a District in the country, where the Servants have such harassing duties or where they undergo so many privations; to compensate for which, they are allowed a small addition to the Wages of other Districts." A posting to New Caledonia was tantamount to a prison sentence.[91]

In addition to improving the quality of provisions at company posts, Governor Simpson hoped that his agricultural programme would foster Pacific trade. Indeed, his policy of diversification was predicated on the satisfaction of foreign as well as domestic markets. Simpson knew that American Nor'westmen had spawned a prosperous trans-Pacific commerce from the Northwest Coast using sea otter pelts as specie to obtain Chinese commodities. He foresaw the day when enough foodstuffs would be produced in the Oregon Country to provide a surplus for export. "Provisions," Simpson declared, "for which there has been such a hue and cry from Fort George since its first Establishment is not only required but by attention I am of opinion that we might actually make it an article of Trade from this coast say Beef Pork Fish Corn Butter etc etc."[92] Russian America in particular was regarded by the governor as a prime potential market. If the Bay men could supplant the American coasters as New Archangel's suppliers, the Yankee "birds of passage" would be driven from the coast, leaving its trade to the Honourable Company.[93] In fact, this plan was accomplished in 1839, when the Russian-American and Hudson's Bay Companies signed a ten-year agreement whereby the latter supplied the former in exchange for trading rights in the Alaska Panhandle.

Finally, in addition to his trading motive, Simpson expected that agricultural development would attract loyal colonists who would help to secure the Columbia Department for the British crown. He felt certain that the northern bank of the Columbia would remain British territory, and agricultural settlement would strengthen this claim. It was for these "imperial reasons" that Fort Vancouver was purposely founded on the right side of the river. Moreover, Simpson's policy of exhaustive trapping south of the Columbia and ruthless competition on the coast would, he hoped, keep the Americans at bay. Unopposed occupation, backed by British imperial might, he reckoned, would prove unassailable.[94]

For all of these reasons, the Governor and Committee in 1826 approved Simpson's arrangements for farming at the cordilleran posts, expressing confidence that "much benefit will be derived from raising here all the provisions that can be required for the whole of our trade West of the Mountains."[95] Such "benefit" was indeed forthcoming, although the results of Simpson's agricultural programme varied from post to post. Fort Vancouver became a major agricultural centre, and Forts Colvile and Langley developed substantial and productive farms; crops were tilled and herds were tended at the other posts on a smaller scale.

2

GOVERNOR SIMPSON'S REWARD: THE RESULTS OF POST FARMING

When Dr McLoughlin took charge that Department was supplied with provisions principally from Eng. round Cape Horn. Seeing the enormous expense attending this importation to a Country which seemed to him capable of supplying all the Agricultural produce required he set to work to plough and clear up the land, imported Sand. Islanders from Woahoo, as Agricultural laborers, to assist the French Canadians then in the employ of the Compy. and in a short time raised enough of flour & other produce, not only to supply Fort Vancouver his own Establishment, but the outlying Posts throughout the Country as well, thus saving a large sum of money to the Compy. and establishing a useful precedent.

RODERICK FINLAYSON, 1878

FORT VANCOUVER: "THE FARM AT THE SEA"

In fact, Finlayson's recollection was faulty. It was Governor Simpson, not Chief Factor McLoughlin, who foresaw the advantages of making the company's Columbian posts self-supporting; McLoughlin simply put the "great man's" plan into operation. The first step in Simpson's programme of agricultural development was the removal of the Columbia Department's "main depot" from Fort George upriver to Fort Vancouver, which became the centre of post farming in the Oregon Country (Plates 2 and 3). In 1824 the Governor and Committee had directed the chief factors of the department to relocate their headquarters to the northern side of the Columbia at "the most convenient Situation" because the United States was "to have possession of Fort George whenever they please" by the terms of the Treaty of Ghent (1818).[1] The main reason for the shift, however, was the superior agricultural potential of the new post.[2] Its site, called Jolie Prairie by the Canadians and Katchutequa ("the plain") by the Indians, was on the right bank of the river just above its junction with the Willamette (Multnomah) and more than one hundred miles from the sea.[3] It was glowingly described by Simpson in 1825:

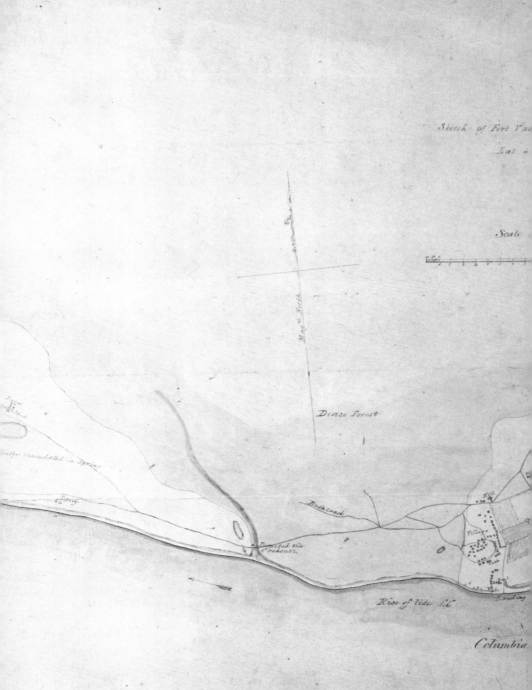

Plate 2. Plan of Fort Vancouver, 1845.

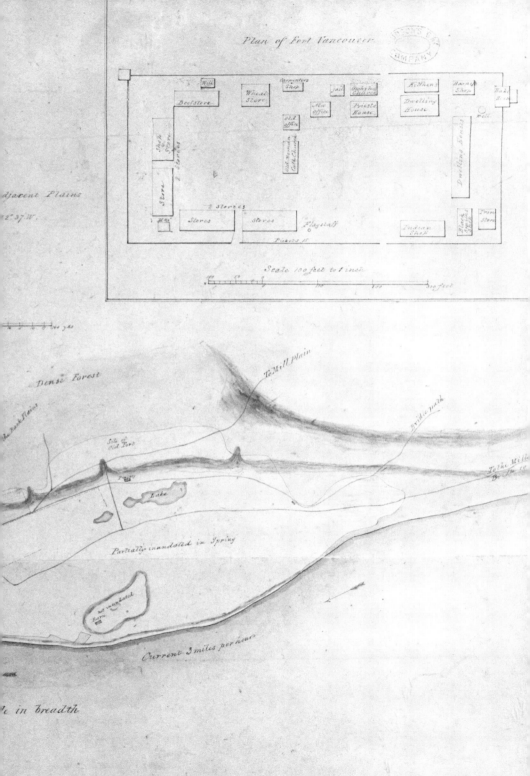

> The Establishment is beautifully situated on the top of a bank about 1¼ Miles from the Water side commanding an extensive view of the River the surrounding Country and the fine plain below which is watered by two very pretty small Lakes and studed as if artificially by clumps of Fine Timber. The Fort is well picketted covering a space about 3/4th of an acre and the buildings already completed are a Dwelling House, two good Stores and Indian Hall and temporary quarters for the people. It will in Two Years hence be the finest place in North America, indeed I have rarely seen a Gentleman's Seat in England possessing so many natural advantages and where ornament and use are so agreeably combined. This point if situated within One Hundred Miles of London would be more valuable to the proprietor than the Columbian Trade.[4]

Fort Vancouver was unquestionably better suited to cultivation and stock-breeding than Fort George. In Chief Factor Alexander Kennedy's words:

> The removal of the Fort from Point George to about 70 miles up the River, must be attended with beneficial effects as the new Post named Fort Vancouver is well situated for Trade & we will be enabled here to raise grain, and a Stock of Cattle etc. that will produce Beef Pork & Butter enough in a little time to supply any demand that will be upon it which never could be the case of Fort George where our Cattle for want of pasture could not have been increased in number, and there was little space of cleared ground for Cultivation.[5]

The site of the new post was the only promising locale for farming on the northern side of the Columbia between the Cascade Mountains and the Pacific.[6]

Simpson reported that "a Farm to any extent may be made there, the pasture is good and innumerable herds of Swine can fatten so as to be fit for the Knife merely on nutricious Roots that are found here in any quantity and the climate so fine that Indian Corn and other Grain cannot fail of thriving."[7] The governor expected the new establishment to produce enough wheat, beef, and pork for the entire department without special personnel; in addition, each post would raise its own potatoes and catch its own fish. "No man," he stated flatly, "is fit for this Country who cannot content himself with such fare."[8]

McLoughlin quickly began efforts to fulfil Simpson's hopes. In the spring of 1825, two bushels of peas and some potatoes were planted; in the autumn one bushel each of wheat, barley, oats, and corn and one quart of timothy arrived

Plate 3. *View of Fort Vancouver, 1845-46.*

from York Factory and were sown "in proper time."[9] Livestock were brought from Fort George and imported from California; pigs and chickens were also brought overland from York Factory by Simpson's party in 1825.[10] Before long farming was flourishing at "the farm at the sea." In 1832 Chief Trader Work wrote to a retired Fort Vancouver clerk, Edward Ermatinger, that "the Doctor's perseverance has made a great change you would scarcely know the place every bit of cleared ground and a great deal more which has been cleared is under cultivation and the quantity of grain produced is immense."[11] In the same year Townsend described the farm:

> Mr. N. and myself walked over the farm with the doctor, to inspect the various improvements which he has made. He has already several hundred acres fenced in, and under cultivation, and like our own western prairie land, it produces abundant crops, particularly of grain, without requiring any manure. Wheat thrives astonishingly; I never saw better in any country, and the various culinary vegetables, potatoes, carrots, parsnips, etc., are in great profusion, and of the first quality. Indian corn does not flourish so well as at Walla-walla, the soil not . . . being so well adapted to it; melons are well flavored, but small; the greatest curiosity, however, is the apples, which grow on small trees, the branches of which would be broken without the support of props. So profuse is the quantity of fruit that the limbs are

covered with it, and it is actually *packed* together precisely in the same manner that onions are attached to ropes when they are exposed for sale in our markets.

On the farm is a grist mill, a threshing mill, and a saw mill, the two first, by horse, and the last, by water power; besides many minor improvements in agricultural and other matters, which cannot but astonish the stranger from a civilized land, and which reflect great credit upon the liberal and enlightened chief factor.

In the propagation of domestic cattle, the doctor has been particularly successful. Ten years ago a few head of neat cattle were brought to the fort by some fur traders from California; these have now increased to near seven hundred. They are a large framed, long horned breed, inferior in their milch qualities to those of the United States, but the beef is excellent, and in consequence of the mildness of the climate, it is never necessary to provide them with fodder during the winter, an abundant supply of excellent pasture being always found.

On the farm, in the vicinity of the fort, are thirty or forty log huts, which are occupied by the Canadians, and others attached to the establishment. These huts are placed in rows, with broad lanes or streets between them, and the whole looks like a very neat and beautiful village.[12]

By 1836, when the settlement numbered from 700 to 800 persons (Euroamerican and Indian men, women and children),[13] it struck Narcissa Whitman, a Protestant missionary, as the "New York of the Pacific Ocean." There were from 550 to 600 whites at the post in 1839. Two years later there were at least fifty log houses outside the fort's walls, and the entire settlement occupied 3,000 acres (nearly five square miles).[14] Visitors were invariably impressed by the establishment's beautiful site, large scale, neat arrangement, thriving industry, and generous hospitality. "King George," as it was called by the Chinook Indians, was at its height in 1845, when it was viewed dispassionately by two British army officers, Lieutenants Henry Warre and Mervyn Vavasour:

> Fort Vancouver is situated on the North Bank of the River at the end of a plain which is partially inundated by the Spring freshets. It is similar in construction to the Posts already described having an enclosure of Cedar Pickets 15 feet high.
>
> It is 220 yards in length by 100 yards in breadth.
>
> At the Northwest Angle is a Square two Storied Blockhouse containing 8 3 lb. Iron Guns . . . .
>
> The Fort was formerly situated on a rising ground in rear of the present site but was removed on account of the inconvenient distance from the River, for the conveyance of Stores Provisions etc. The present site is ill

adapted for defence being commanded by the ground on the rear.

There is a small village occupied exclusively by the Servants of the Company on the West Side extending to the River.

About 5 miles above the fort on a small stream falling into the Columbia is an excellent grist mill and about One mile higher up is a saw mill.

On the plains in the rear of these Mills the Company have about 1200 acres of Land under Cultivation.

There are also large flocks of Sheep & herds of Cattle attached to this Establishment.[15]

Farming was perhaps the main activity at the post; in the fall of 1838 twenty-three of the seventy-four servants — excluding eight invalids — were employed in farm work.[16] But it was not the only industry. In response to Simpson's call for diversification, McLoughlin also developed fishing and sawmilling, especially the latter.[17] There were two sawmills: one on Wappatoo Island, the Island Mill, and one on the northern bank of the Columbia five or six miles upstream from the fort. The latter, which was powered by a spring-fed stream, ran twelve saws day and night; it cut thirty-five hundred feet of one-inch boards in twenty-four hours in 1835.[18] By the autumn of 1837, two hundred thousand feet of lumber had been stockpiled for export.[19] In 1841 the mill worked nine saws, employed from twenty-five to thirty men, mostly Kanakas (Hawaiians), and twelve yoke of oxen, and produced twenty-five hundred feet of lumber daily.[20] The Sandwich (Hawaiian) Islands were the principal market for Fort Vancouver's surplus lumber, fish, and flour, which were exchanged for coffee, sugar, rice, molasses, and salt. During the early 1840's, the post shipped two or three cargoes annually, which netted $10,000.[21] Russian America was the Oregon Country's stiffest competitor for the Hawaiian market; Sitkan timber was better for spars and planks than Columbia timber because it was lighter and equally strong.[22]

This diversification at the fort was advantageous because it kept the company's servants busy throughout the year and also spread the company's risk. McLoughlin pointed out that farming was beneficial in two respects: "the freight it saves us [and] ... as, if the people had not been doing that, they would have been unemployed."[23] McLoughlin's deputy, Chief Trader James Douglas, added:

> in order to receive the full benefit of agricultural experiments, they should, if possible, be combined with other pursuits, to supply occupation, when the men cannot be profitably employed at the farm. Upon such principles the business of this place is conducted: the duties connected with the Depot, the Indian Trade, Farm & Saw Mill never leave us a moment's leisure, at any season, and form together, a system that works to more advantage, that either would produce if disunited from the others.[24]

However, of all of Fort Vancouver's activities, agriculture occupied not only more men but also more space than any other. In 1841, when the entire settlement covered three thousand acres, twelve hundred were cultivated (and an unknown number pastured), which represented a tenfold increase in arable land since 1829 (Table 3). By the end of the period of joint occupancy in 1846, the post's farmland fronted the river for twenty-five miles and reached inland for ten.[25]

Table 3

CULTIVATED ACREAGE AT FORT VANCOUVER IN VARIOUS YEARS BETWEEN 1829 AND 1846

Year	Acreage	Year	Acreage
1829	120	1841	1,200
1832	200-700	1842	1,500
1837	900-1,000	1845	1,000-1,200
1838	861	1846	1,420

For sources see *Sources for Tables*.

This tilled and grazed farmland occupied several "plains" or "prairies" above, below, and behind the fort. These were flattish expanses of parkland with high, thick grass, scattered copses of trees, and occasional small lakes.[26] The original site of Fort Vancouver itself was Jolie Prairie, which covered about three hundred acres.[27] The second site comprised three prairie levels (river terraces). The first or low level, which stood ten feet above the river at low water, included the Fort Plain on the eastern side of the palisade and the West or Lower or Cox's Plain, which began two or three miles west of the fort and extended downstream for five miles.[28] The former, which included nearly five hundred cultivated acres, was the centrepiece of the fort's agriculture, with many stables, granaries, and workshops; the latter had little tilled or fenced land, serving primarily as pasture and containing half a dozen buildings. Below the West Plain was the Vancouver or Camas Plain, whose "rich & luxuriant grass" stretched fifteen miles along and one behind the river. The second level, which lay two hundred feet above low water, comprised the First North or First Back Plain; it began three miles from the fort. The third or high or upland level, which rose three hundred feet above low water, contained the Second North or Second Back Plain, beginning six miles from the fort. The first and second levels had the best soil but were subject to spring flooding. The soil of the flood-free third level was light and gravelly. Half a dozen miles upstream from the fort and a mile inland from the river lay the Mill Plain, which encompassed about one thousand enclosed acres and a dozen buildings. Nearby were a sawmill and a gristmill.

Four more miles upstream was Prairie de Thé. In addition, several islands in the river were farmed, most notably Sauvie's (Wappatoo) Island at the mouth of the Willamette, where the company had four dairies and hundreds of pigs, which foraged at large on wappatoos and acorns.[29]

A wide array of crops and animals were raised on Fort Vancouver's scattered farmlands. The field, garden, and orchard crops included grains, grasses, legumes, roots, melons, squashes, berries, and various fruits (mainly tomatoes and apples). The livestock consisted of cattle, horses, pigs, sheep, goats, chickens, turkeys, and pigeons, which ran both at large and in check. Wheat, oats, barley, peas, and potatoes, plus cattle, pigs, and sheep, predominated. McLoughlin oversaw the farm work, most of which was done by Indians, Kanakas, and Canadians. They were engaged for five years at an annual salary and a daily ration. Few skilled farmers were brought from the British Isles. Mr. and Mrs. William Capendal were among the exceptions; they were hired in 1835 as a field supervisor and a dairy manager, respectively. (They returned to England in 1836 because Mrs. Capendal found circumstances on the Columbia "different to what she expected.") [30] Another example was William Bruce, a gardener who accompanied McLoughlin upon his return from furlough in England in 1839.[31]

Most grains were planted in the autumn and harvested in mid-summer, with harvesting lasting from three to four weeks. Oats, peas, and potatoes were sown in the spring and reaped in late summer.[32] Grains were scythed, cradled, flailed, and winnowed; by 1842 there was a threshing machine at the fort.[33] Fields were ploughed throughout the winter; frequently from eight to ten ploughs and as many harrows were used.[34] Implements were made locally to supplement those imported from England. Crop rotation, fallowing, and manuring were practised to conserve soil fertility.[35] Livestock were brought from California, Canada, England, the Sandwich Islands, and the United States. Cattle were herded in daytime and penned at night, and they browsed year round, so little hay or shelter were required. The natural meadows of the bottomlands of the Columbia and its islands produced two crops every year, one in early spring and another in mid-summer. Cultivation was largely confined to the higher riverine terraces above the seasonal floods. The cattle furnished beef, milk, hide, and tallow, as well as draught; sheep provided mutton and wool. Dairies on the West Plain and Sauvie's Island produced whole milk, butter, and cheese, while gristmills on the Mill Plain (water-driven), at Willamette Falls (water-driven), and at the fort itself (horse-driven) made flour.[36] All in all, this was farming on a solid foundation.

Fort Vancouver soon justified Simpson's faith in its agricultural potential. As an American visitor observed in 1841, farming was done on "a stupendous scale," with a "great abundance" of corn, oats, and potatoes being grown and "vast quantities" of butter and cheese being made, while "every description" of garden vegetables "flourished" and apple, pear, and peach trees fared "well."[37] Harvests — at least of the major crops — were generally bountiful (Table 4).

Table 4

OUTPUT OF GRAINS, PEAS, AND POTATOES AT FORT VANCOUVER IN VARIOUS YEARS BETWEEN 1825 AND 1846

Year	Wheat	Oats	Barley	Corn	Peas	Potatoes
1825	? bus.	? bus.	? bus.	? bus.	9½ bus.	900 bls.
1826[a]	12	6	27	1½	114	600
1827	155	50	389	200	434½	4,000 bus.
1828	1,000-1,300	100-200	500-1,000	200-400	300-400	4,000
1829	1,500	?	200	250	587	4,000-5,000
1830	1,260	[150]	183	600	583	9,000
1831	1,800	[32]	1,200	400	600	6,000
1832[b]	[3,000-3,500]	[1,500-2,000]	[2,000-3,000]	[800]	[2,000-3,000]	[6,000-15,000]
1833[c]	[3,000]	[1,000]	[1,500]	[500]	[3,000]	?
1835	[4,000]-5,000	[1,000]	1,000-[1,200]	?	[1,500]-2,000	1,300
1836	4,000-5,000	1,500-2,000	1,500-2,000	? ?	4,000	0
1837	[5,000]	[1,500]	[1,000]	?	[5,000]	?
1841	4,000	3,500			?	?
1844	4,000	3,000	1,200	?	2,500	?
1845	4,000	5,000	?	?	4,000	?
1846	5,000	1,500-3,000	200-300	?	2,000	6,000

For sources see *Sources for Tables*.
[1]Figures in brackets represent expected amounts.
[a]The 1826 crop was less than expected because of defective wheat, oat, and corn seed; consequently, all of the wheat, oats, corn, and peas and half of the barley were kept for seed in 1827 (HBCA, D.4/120: 54).
[b]Plus 50 bushels of buckwheat (Rich, *Letters of John McLoughlin*, vol. 1, p. 105).
[c]Plus 1,000 bushels of buckwheat (Ibid., vol. 1, p. 113).

Yields were also sizable. In the middle 1830's, the average yield per acre on the best land with manuring and rotation was twenty bushels of wheat, forty of barley, fifty of oats, and thirty of peas, and on the worst land half as much.[38] About the same time, McLoughlin told company physician William Tolmie that wheat yielded fifteenfold and barley from fortyfold to fiftyfold the seed. He also said that corn was the most and oats and peas the least demanding of soil, with wheat, barley, and hay being moderately demanding.[39] And in 1841 it was reported that wheat averaged from twenty to thirty and barley twenty bushels per acre.[40] Lieutenant Wilkes noted that buckwheat returned a "good crop some seasons" and peas, beans, and potatoes yielded "generally abundantly."[41] He added that "fine" strawberries and "very fine" melons, raspberries, and currants were grown, as well as hops "in abundance" and even figs. Apples, pears, and grapes had been tried but "do not yet yield well."[42] Eight years earlier another American officer, Captain Nathaniel Wyeth, had asserted that "carrots are here finer and larger than I have ever before seen"; one that he saw was "3 inches through and of fine flavor."[43] The orchard likewise succeeded. A Methodist missionary, Jason Lee, recorded in 1834 that "the orchard is young but the quantity of fruit is so great that many of the branches would break if they were not prevented by props." "I never before saw trees of the same size so heavily laden with fruit," he added.[44]

Table 5

NUMBER OF LIVESTOCK AT FORT VANCOUVER, 1824-29 AND 1832-46[1]

Year	Cattle	Horses	Pigs	Sheep	Goats	Poultry
1824	17	?	?	?	?	?
1825[a]	27-40	120	?	?	?	?
1826	59-61	52	29	0	0	40
1827	87	162	71	0	0	0
1828	122-150	174	101	0	13	62
1829	172-200	22-50	182-300	0	14-28	0
1832	400-450	107	315	49	116	0
1833	403	98	310	76	94	20
1834	416	81	360	93	75	25
1835	427	120	303	153	90	27
1836[b]	622[c]	198	392	217	59	9
1837	685[d]	155	328	305	25	?
1838	450	151	353	518-578	?	?
1839	848	188	579	?	62	?
1840	1,085	289	575	949	?	?
1841	703	394[e]	307	2,500[f]	20	?
1842[g]	1,181	386	20	?	?	?
1843	593	487	?	?	?	?
1844[h]	1,241	718	831	?	?	?
1845[i]	509	97	798	?	15	?
1846	1,915	517	800	3,000	?	?

40   Farming the Frontier

For sources see *Sources for Tables.*
[1]These figures represent spring (31 March) populations and inlude livestock inventories of Indian Hall (the Indian trading store).
[a]Altogether 120 head of cattle, horses, and pigs (Scouler, "Dr. John Scouler's Journal," p. [126]).
[b]Plus 4 mules. In the fall there were 750 cattle, 200 to 300 horses, and 400 pigs (Spalding, "Letter," n.p.).
[c]Including 50 to 60 cows in the fall (Drury, *First White Women,* vol. 1, p. 103).
[d]Including 40 oxen in the winter (Slacum, "Memorial of William A. Slacum," p. 7) and nearly 100 cows in the summer (Jessett, *Reports and Letters of Herbert Beaver,* p. 80).
[e]Including 300 brood mares (Wilkes, "Oregon Territory," p. 46).
[f]Most of these sheep actually belonged to the Puget's Sound Agricultural Company, which had 1,435 sheep at Fort Vancouver in the autumn of 1841 (HBCA, D.4/110: 22v.).
[g]Plus 4 mules.
[h]Plus 14 mules.
[i]Excluding the cattle and horses at the fort itself. In the fall there were 1,377 cattle, 702 horses, 1,581 pigs, and 1,991 sheep ("Papers Relative to the Expedition of Lieuts. Warre and Vavasour," p.86).

Livestock also multiplied well (Table 5). Little wonder that Canadian painter Paul Kane remarked in 1846 that at Fort Vancouver "they have immense herds of domestic horned cattle, which run wild in unknown numbers; and sheep and horses are equally numerous."[45]

Cattle breeding was especially successful. The animals originated from one cow and one bull at Fort George in 1818; in 1825 seven cows and one bull were brought to Fort Vancouver from Fort George.[46] From this stock sprang Fort Vancouver's large herd, although occasionally McLoughlin augmented it with purchases of animals from Willamette settlers.[47] By 1838 the cattle were too numerous for the limited pasture around the fort so they were split into three herds: one at the fort itself; one on Wappatoo Island at the mouth of the Willamette; and one at McKay's Farm on the south side of the Columbia eighteen miles downriver. During the early 1840's about half of Fort Vancouver's cattle were kept on Wappatoo Island. In the fall of 1841 the island's dairy herd numbered 450 head and the fort's 250.[48] The dairy pens and sheep folds were relocated every year in order to disperse the manure and relieve the pasture. Pasture had to be economized because the livestock grazed throughout the year. Lieutenant Wilkes observed that owing to the mild winters cattle did not require shelter, and "little or no" hay was made; consequently, he added, two-thirds of a farmer's time was saved.[49]

More than enough beef, pork, butter, and wool were produced to meet the post's needs. During the first decade of its existence no cattle were butchered (except one or two bull calves each year for rennet) in order to increase the herd and to provide one hundred oxen for farming and sawmilling. Instead, fresh and salted venison and wildfowl were used as meat.[50] McLoughlin refused to slaughter any cattle until their number reached six hundred head; this figure was met in 1836, when forty oxen and barren cows were salted for the company's vessels. The ships also took Fort Vancouver's salted pork output: twenty barrels in 1829 and forty in 1830. As for butter and wool, in 1838 fifty-six kegs of surplus butter and six hundred pounds of wool were produced.[51]

Not only did Fort Vancouver's lands produce beyond the post's own needs, but they also did so quite soon after their opening. Simpson felt in 1825 that the fort would be self-sufficient in food within two or three years. McLoughlin agreed: "Hitherto it [the means of subsistence] has been principally on Imported provisions — and Salmon procured from the natives — but in future we Expect to dispense with Imported provisions and hope after fall 1828 to have all the provisions we want from our own farm." The chief factor was right. By 1828 his headquarters was raising enough food to meet its needs and forego the importation of grain.[52] That year Simpson reported from Fort Vancouver that "our crops this season were very abundant and we have now a two years stock of Grain on hand, so that we shall not require either Flour or Grain from England in future."[53] The Governor and Committee in London commended McLoughlin for

> the success which has attended your exertions in Agricultural pursuits and raising Stock, and trust you will continue to prosecute these objects, indeed your whole management is marked by a degree of energy zeal and activity highly creditable to yourself, important to the interests of the Service and meeting our warmest Commendations.[54]

By the fall of 1830 Fort Vancouver had enough wheat and flour on hand for two years and enough corn and peas for one. The shipment of flour from England — sometimes forty bins yearly — was discontinued, and by 1835 the departmental capital was able to provision the London ship on its homeward voyage.[55] By at least 1836 the fort was meeting the requirements of Simpson, who in 1829 had told McLoughlin that farming would reach a satisfactory level only after eight thousand bushels of grain were grown, six hundred cattle were kept, and ten thousand pounds of pork were cured annually. Fifty pounds of flour could now be supplied annually to each of the servants of New Caledonia to supplement dried salmon.[56] McLoughlin reported in 1843 that the daily ration per man in his department was three pounds of salted salmon and one and a half pounds of biscuit (or one bushel of potatoes per week); "dry Salmon," he added, "used to be hitherto the principal food of the country." Two years earlier Wilkes had noted that "all their wants here are now supplied with the exception of groceries [such as tea, sugar, coffee, and spices]."[57] This was a significant achievement, as Simpson informed London in 1834:

> The Farm . . . has enabled us to dispense with imported provisions, for the maintenance of our shipping and establishments, whereas, without this

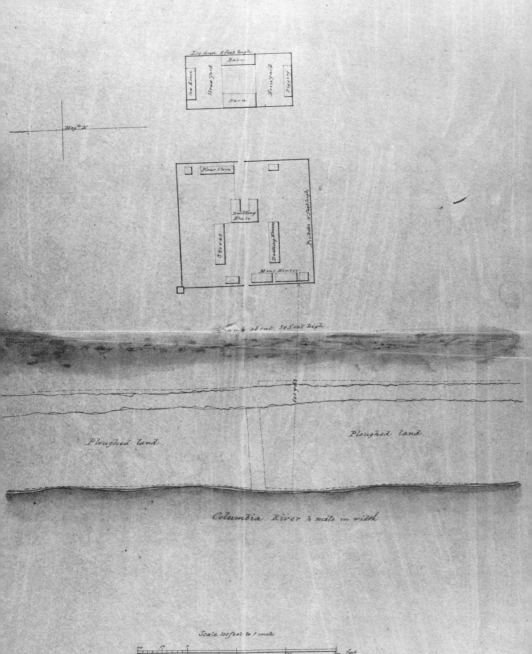

Plate 4. Plan of Fort Colvile, 1845.

farm, it would have been necessary to import such provisions, at an expense that the trade could not afford.[58]

The saving was considerable. McLoughlin stated in 1837 that since its commencement, agriculture had provisioned the fort, the coastal posts, and company shipping to a value of £2,000 annually. James Douglas reiterated the importance and success of farming in a letter to Simpson in 1838 while McLoughlin was on leave:

> The prosperity of the general business, is so intimately connected with the agricultural operations, and depends, so much upon the possession of an ample and regular supply of Provisions, that it long since became a desederatum with us to secure independently of the rising crop, a full years provisions, in advance, and it is now attained, as our barns contain a sufficient quantity of the more useful kinds of grain to meet the home and outward demand, at a reasonable calculation, for the next eighteen months.[59]

Wheat became as common as peltry, and during the early 1840's both were almost the sole currencies at Fort Vancouver.

FORT COLVILE

Farming was also a major activity at Forts Colvile and Langley, although it was conducted on a smaller scale than at Fort Vancouver. Fort Colvile or Colvile House (Plate 4) was founded in 1825 to replace Spokane House, which was abandoned the following year.[60] Simpson explained this decision to McLoughlin:

> A most serious expense and inconvenience is annually incurred Spring and Fall in transporting the Spokane Outfit and Returns between the forks [of the Columbia and Okanagan Rivers] and Establishment which would be avoided by removing the Post to the Kettle Falls, this is a more desirable situation in regard to Farming, Fish, provisions generally as also in respect to Trade and many other points of view, we have therefore determined that the Establishment shall be removed after the arrival of the Brigade.[61]

The new post was located just above Chaudière or Kettle Falls, the highest on the Columbia, on the second alluvial terrace, the first being partially subject to flooding. Simpson reported in 1825 that "we selected a beautiful point on the South side about 3/4ths of a Mile above the Portage where there is abundance of fine Timber and the situation elegible in every point of view. An excellent Farm can be made at this place."[62] The fort was described in 1838 by an American missionary:

Colville, the fort, is about one quarter of a mile from the Columbia River, on a rise of ground which has a barren appearance, and is really so. The top of the ground all around the fort is covered with coarse gravel stones. The area of the fort is twice as large as any I have seen on this side of the Mts. It is built with sticks of timber set up, supported by braces. In and about the fort are quite a number of dwelling houses, three or four large stables & storehouses for grain. He [Chief Trader Archibald McDonald] is well provided with farming tools, carts and sleds, a sleigh and a gig.[63]

Fort Colvile had two main functions: to collect and protect the trade of the upper Columbia, Kootenay, and Flathead regions and to provision the inland posts from Fort Hall to Connolly's Lake. Also, as Archibald McDonald, one of the establishment's commandants, noted, "like Kameloops it is necessary for the Communication," that is, the Columbia brigade system.[64] Certainly, the alimentary role motivated Simpson when he established the post. In the summer of 1825, he informed McLoughlin that:

> Fort Colvile is well adapted for a Farming Establishment and from what I have already seen of Mr. Dease's neatness of arrangement in that way I entertain confident hopes that under his management it will become a very important auxiliary to us in the way of living. Indian corn, Pease, Wheat and Barley I am satisfied would thrive there, Potatoes in any quantity may be raised and the country is so well adapted for the rearing of Hogs that I expect he will very soon be able to furnish any quantity of Pork we may require.[65]

John Work also believed that agriculture at Fort Colvile would be beneficial because it would decrease expenses and increase returns. He wrote:

> Scarcely the two thirds of the annual expenditure of the [Colvile] District goes for furs the rest is consumed in the trade of what is denominated Country produce, provisions and other expenses. The raising grain at Colvile may diminish the expenses for provisions to a trifling amount. A sufficiency of Country produce to supply other places can seldom be procured. Were the demand for these articles to be done away with or lessened, it would be the means of diminishing the expenses and increasing the Returns of Furs. When other means of procuring supplies would discontinue the Natives would be obliged to exert themselves more hunting beaver to procure their wants [than fishing, hunting, or gathering for the posts].[66]

Crops were cultivated at White Mud Farm at the fort itself and at White Meadow Farm a dozen miles to the south. Livestock were pastured on Big Prairie nearby. There were 20 acres "cultivated & enclosed" in 1827, from 60 to 70 in 1829, 80 in 1833, nearly 200 in 1836, from 130 to 135 in 1841, 118 in 1845, and from 350 to 370 in 1846 (340 at White Mud Farm and 30 at White Meadow Farm, plus five sections of pasture). White Mud Farm's alluvial site, with its "rich black loam," was deemed a "highly favored spot" by a longtime company servant. In 1838 the farms employed twenty men, probably two-thirds of the post's engagés. Both autumn and spring wheat were planted. McDonald reported in 1842 that "fall ploughing is a thing . . . that must absolutely be attended to, as no other grain now give much satisfaction here." By then there was a gristmill on Mill Creek from three to four miles from the fort and a dairy at White Meadow Farm.[67]

Figures on Fort Colvile's crop production are fragmentary but sufficient to indicate that output was substantial (Table 6). In 1845 Lieutenants Warre and Vavasour reported that one thousand bushels of wheat were harvested annually at Fort Colvile. The average yield was fifteen bushels per acre.[68]

Table 6

OUTPUT OF GRAINS AND POTATOES AT FORT COLVILE IN VARIOUS YEARS BETWEEN 1826 AND 1841

Year	Grains	Potatoes
1826	15 bus.[a]	410 bus.
1827	200	2,000
1832	2,898[b]	1,600
1836	5,200-5,550[c]	?
	3,500	3,500
1838	1,500[d]	7,000-8,000
1841	800	?

For sources see *Sources for Tables*.
[a]All barley (HBCA, D.4/120: 63).
[b]Including 1,404 bushels of wheat, 870 of corn, 375 of barley, 241 of oats, and 8 of buckwheat, plus 104 of peas (HBCA, B.223/d/49: 49v.).
[c]Including 3,000 bushels of wheat, 1,000 of corn, and 1,200 of other grains (Anonymous, "Documents," (1908), p. 255).
[d]All wheat (Drury, *Nine Years with the Spokane Indians*, p. 75).

Figures on livestock propagation at the post are more complete (Table 7). The cattle, which fared "remarkably well," sprang from two cows and one bull that were brought from Fort Vancouver in 1825. And in the spring of 1826 Simpson took "precious *first calves*" to the upper Columbia from Fort Vancouver with the overland express; at the same time the cattle and horses of abandoned Spokane House were moved to its successor post.[69]

Table 7

NUMBER OF LIVESTOCK AT FORT COLVILE IN VARIOUS YEARS BETWEEN 1827 AND 1846[1]

Year	Cattle	Horses	Mules	Pigs	Poultry
1827	3	50	?	11	13
1828	6	60	?	19	24
1829	7	58	?	44[a]	12
1833	17	56	?	106	0
1834[b]	25	104	?	82	2
1835	31	41	?	81	2
1836	50-60	106	?	88-155	2
1837	53-80	96	?	100-104	0
1838	67	128	7	124	0
1839	84	132	5	150	?
1840	105	151	6	140	?
1841	135-196	166	6	100	48
1842	50	151	7	90	120
1844	81	325	8	93	60
1845	96-100	300-400	?	73	37
1846	?	270	?	?	?

For sources see *Sources for Tables.*
[1]These figures represent spring populations.
[a]Ninety in the fall (John Work, "Work Correspondence," PABC, AB40: 13).
[b]Plus 9 goats.

Fort Colvile successfully fulfilled its agricultural role. The first year's results were only fair, as Chief Factor Dease reported in 1826:

> With regard to our Agricultural pursuits our seeds of different kinds were sown in good time in fact as early as the season would admit of, every thing came up well with the exception of Indian Corn and Wheat, the latter was damaged on the way in, and the former had not come to maturity of course neither fit for seed, our Barley which was stored before I came off yielded 14 for one, 24 bushels of Potatoes were planted and were thriving well, but unfortunately a kind of ground mice got among them and had destroyed more than the half before I came off for this place other vegetables such as cabbage, turnips, etc., I can not complain of, the cattle brought from Vancouver were safe when I came off and thriving well the pasture about being excellent.[70]

Similarly, John Work reported in 1826 that the post's crops "do not appear to realize the expectations that were entertained for them," potatoes faring "pretty well," barley "middling," wheat not "at all," corn and peas "indifferent," and vegetables "so so"; he likewise complained that "moles" had decimated the potatoes. The first potatoes — from five to six kegs — had been planted in the spring of 1825 and had yielded thirteen kegs that fall (at the same time five kegs had produced twenty-eight at Spokane House); all were used for seed the next year, Governor Simpson having ordered "the produce to be reserved for seed, not eat, as next spring I expect that from 30 to 40 Bushels will be planted."[71]

However, crops improved after the first year. Dease predicted that the post would be independent of the Indians for provisions by 1829 and also have a "moderate" surplus for other posts. McLoughlin was more optimistic, forecasting independence by the autumn of 1828.[72] That year Simpson reported that farming was so successful at the post that "it supplies all the Grain required for the interior, rendering it unnecessary to furnish any Provisions at the coast except for consumption on the Voyage; whereas formerly, there were about 30 men employed in the transport of Provisions alone, for the use of the interior."[73] In the same year Work wrote to a friend that "we are also blessed with abundance of potatoes & plenty of barley"; a year later he wrote to the same friend that in 1828 "the farm was very productive."[74] By 1830 Work could report that "the farm which was commenced three years ago is now become so productive as to render the place nearly independant of any other means of subsistance whatever." This success enabled him to tell the neighbouring Indians to bring no more "country produce" to the fort to sell after the spring except fat.[75] It also prompted McLoughlin to write to Chief Trader Francis Heron of Fort Colvile that "I am happy to find you have had so abundant a harvest as to make the Interior entirely independent of this place for Provisions and which was the object in view in establishing a Farm at Colvile."[76] Actually, Fort Colvile was not yet provisioning all of the interior posts, although it could do so, Work felt, with a little more effort:

> At a very trifling additional expense and without interfering with the trade, I have little doubt, it may not only render the place independant of the Indians for provisions, but furnish a sufficiency of grain and pork for the other establishments in the Columbia above Vancouver, and for New Caledonia.[77]

The "additional expense" was outlayed, and by 1834 Simpson was able to inform the Governor and Committee that "Fort Colvile, besides being the most valuable inland establishment in point of trade, is of great importance to the other posts, being the granary or provision Depot of the interior, as it possesses the advantages of soil and climate highly favorable to cultivation."[78] In the late 1830's, Fort Colvile supplied Fort Nez Percés with one hundred hundred-pound sacks of flour annually.[79] "New Caledonia also depends on this place for her flour, pork, corn and meal etc. etc.," noted a member of the United States

48   Farming the Frontier

Exploring Expedition in 1841.[80] In that year from eighty to ninety sacks of grain were packhorsed from Fort Colvile to Thompson's River for New Caledonia by the inward brigade, and in 1842 fifty horses packed probably one hundred sacks of flour over the same route.[81] Fort Colvile's grain was in great demand. In 1836 McDonald wrote to a friend that "you have no idea of the quantity of grain now Consumed here & dragged to the other posts."[82] Six years later McDonald reported to Fort Vancouver that "grain is called for from I may say every point of the compass .... there is not an individual above Vancouver I am in communication with but the cry is for provisions." He estimated in 1841 that Fort Colvile needed two thousand bushels of wheat alone to meet its obligations.[83] The post's farms met an average of 90 per cent of this requirement (that is, 75 per cent in 1832, 150 per cent in 1836 and 1837, 75 per cent in 1838, 40 per cent in 1841, and 50 per cent in 1845).

FORT LANGLEY

Like Fort Colvile, Fort Langley had two purposes: to tap the trade of the lower Fraser Valley and the eastern coast of Vancouver Island and to provision the coastal posts. Simpson went even further, instructing McLoughlin that

> it should be kept on a respectable footing and every pains taken to have a good Farm attached to it, so that in the event of a failure of crops, to any evil happening to our Cattle or Pigs at Fort Vancouver, or our being compelled to quit the Columbia, we may not be entirely destitute of the means of subsistence.[84]

Fishing was also a major activity at Fort Langley; indeed, the post probably derived as much benefit from fishing and farming as from trading. The Fraser teemed with salmon. Chief Factor James McMillan wrote to a friend in 1828 that "we could trade at the door of our fort I suppose a million of dried Salmon if we chose enough to feed all the people of Ruperts Lands."[85] "There is no place on the coast where Salmon is so abundant and got so cheap as at Fort Langley; and if we find a sale for Salmon, it would alone more than pay the expence of keeping up that place," reported McLoughlin to Simpson in 1835. Fishing even came to be regarded as more important than farming. In a letter of 1838 to the fort's commandant, Chief Factor James Yale, Douglas stressed the priority of fishing over farming, declaring that "the Salmon trade must not be sacrificed, as it will always yield, a more valuable return at less trouble risk & expense than the farm."[86] The salmon were salted and barreled (with up to 750 fish per barrel) for shipment as far as the Sandwich Islands in exchange for sugar, coffee, and molasses. It cost the company $4.00 a barrel to produce the salmon, which brought from $10.00 to $11.00 a barrel at Honolulu in 1845. During the 1830's from two to three hundred barrels were packed annually, and in 1845 eight hundred and in 1846 1,530 barrels were cured.[87]

Simpson had both farming and fishing in mind when he had the lower Fraser reconnoitered in late 1824 by an expedition of forty-two men under James McMillan, whom the governor esteemed as one of his best men.[88] Simpson paraphrased McMillan's report:

> The Banks as far as Mr McMillan ascended were clothed with a great variety of prodigious fine large Timber. The Soil appeared to be rich and fertile; good situations for the site of an Establishment in every reach and many beautiful clear spots adapted for Agricultural purposes, but the entrance of the small River falling in from the South down which he went in gaining the Main Stream he particularly recommends as there is an extensive meadow where any number of Cattle & Pigs may be kept and where the Plough can be immediately used at a little distance; and by barring up this small Stream or forming a Weir, a sufficient quantity of Salmon & Sturgeon might be taken at the proper Seasons for the maintenance of the Establishment without rendering it necessary to have recourse to the Natives for the means of subsistance.[89]

The post was founded with twenty-three men in 1827 "with the double object of securing a share of the Coasting Trade which had previously been monopolized by the Americans, and of possessing a Settlement on the coast which would answer the purpose of a Depot."[90] In 1828 land was cleared, and three fields just outside the fort were each planted with thirty bushels of potatoes. They produced "the finest I ever Saw in the Country," according to McMillan.[91] In the spring of 1829, wheat, barley, and peas were also sown. In the same year a "greate prairie" astride a nearby stream was found to be more suitable for cultivation than the land at the fort. Bourgeois Archibald McDonald described this new acreage, which was to form the heart of the post's agriculture:

> In the afternoon, for the first time Since a Came here I went out for a few hours with a Canoe & Eight men, accompanied by Mr. Annance — — our object was to view the nature of the Country here abouts and the practicality of Converting Some of it into Cultivation; which with great advantage Could be done to any extent were it not for the length of the distance from the Fort — — after leaving this, we, about a mile up on the Same Side, entered a Small winding Creek, that in 50 minutes more brought us to a beautiful prairie of at least 3 miles in Circumference & of uncommon rich Soil all round — — The upper or eastward extremity of it extends to within a Very few Acres of the main river, & perhaps another mile above the mouth of the little Creek . . . — — Entering the little river — — on right hand

there is a half dry *Marais* [marsh] of near two miles round, that for the Space of a hundred yards along the Creek, presents Soil of Superior quality requiring however a little Clearing: & even the Marais itself Could be brought to Something with Some labour —— The query is, whether it be worth while going to such labour, or Could our people with Safety be employed at Such distances? —— A road Could be Cut not only to the first opening, but to the Second prairie also —— While on this Subject, we may as well advert to the ground near the Fort —— For the ordinary purposes of the Establishment, ample ground can be wrought under the very eye of our Bastions, but includes all that can be done, unless we look to the tedious result of clearing Strong woods not on the best of Soil, & to the draining of fens and marshes that would be equally labouring & unsatisfactory —— of about 15 acres now open here, 5 of them is low meadow —— 5 fine mellow ground fit for the Plough, and the rest full of ponderous Stumps & roots —— Not many hundred yards off, is another Small marais formerly a Beaver pond, that may afford occasional pasturage for a Couple of horses; & without going to the upper Creek, feed with difficulty Can be procured for more than two or three Cows —— Pigs we Can find room for.[92]

This is the first description of the Salmon River and Langley Prairie, which was called the "Big Prairie" or "Big Plain" by the Bay men.

Farming became more important after fur trading failed to meet expectations. In early 1831, when there were sixteen men at the post, McDonald reported that "Frasers River does not come up to original expectations [for fur trading]." In 1830 cattle, pigs, and chickens arrived from Fort Vancouver aboard the brig *Eagle*.[93] A separate farm was established in 1832 or 1833, when twenty acres were put under cultivation. By 1835 from seventy to seventy-five acres were tilled, thirty at the fort and forty to forty-five seven miles away on Big Prairie.[94] In the spring of 1839 — one year after it had been razed by fire — the post was rebuilt two miles upriver, partly in order to be closer to the bulk of the farmland. In 1841 twenty men (eleven Hawaiians, eight Canadians, and one Iroquois) worked the farm, which lay some three miles from the fort. Two hundred and forty acres were cultivated in 1845.[95] Governor Simpson described the settlement in 1841:

> Fort Langley is situated in Latitude 49° 6′, Longitude 122° 47′, and is intended to collect the trade of the numerous tribes inhabiting the mainland coast and east coast of Vancouvers Island .... The complement of people at this place is an officer and 17 men; the returns in furs amounting to about £2,500 and in salted salmon for market say about 400 barrels to about £800, the profits on the post being about £1,600 per annum. The establishment was destroyed by fire about 18 months ago but has since been rebuilt on a larger scale. There is an excellent farm in the immediate neighbourhood,

the produce of which with fish and venison maintains the establishment, and assists in provisioning some of the others on the coast. This has for a length of time been a very well regulated post, but as the country has been closely wrought for many years, the returns in furs are gradually falling off; but the increasing marketable produce of the fisheries makes up for that deficiency.[96]

Crop production at Fort Langley was modest, although available figures are fragmentary (Table 8). Livestock multiplied slowly, but this was not unexpected, since the cattle were primarily dairy rather than beef animals (Table 9). Pigs, which numbered up to twenty in 1830, propagated faster than cattle or horses.[97] Work reported in 1835 that "the cattle are in fine order, but the pigs are rather lean."[98] These cattle — the precursors of today's large dairy herds in the lower Fraser Valley — supplied considerable butter for export. To increase butter output twenty-nine milch cows were sent to Fort Langley from Fort Vancouver in the fall of 1839.[99] In 1840 the post milked thirty cows and raised 500 bushels of winter wheat (fivefold yield), 250 of spring wheat (tenfold yield), 250 of barley (one-thirdfold yield), 500 of oats (twelve and one-halffold yield), 600 of peas (eightfold yield), and 1,176 pounds of butter (as well as 300 barrels of fish); the yields of winter wheat, barley, and peas were deemed "poor," and only the yield of spring wheat was rated "good." But this output, including salmon and venison, was enough to provision Fort Langley itself and some of the coastal posts.[100] The butter was exported to Russian Alaska and the salmon to the Sandwich Islands.

Table 8

OUTPUT OF GRAINS, PEAS, AND POTATOES AT FORT LANGLEY IN VARIOUS YEARS BETWEEN 1828 AND 1835

Year	Grains	Peas	Potatoes
1828	0 bus.	? bus.	2,000-2,010 bus.
1829	35	20	1,500
1830	100	?	?
1832	76	110	2,300
1835	250	300	?

For sources see *Sources for Tables*.

Table 9
NUMBER OF LIVESTOCK AT FORT LANGLEY, 1833-46[1]

Year	Cattle	Horses	Pigs	Poultry
1833[a]	11	?	55	?
1834[a]	17	5	60	?
1835	13	4	60	?
1836	20	9	80	?
1837	25	8	100	?
1838	32	7	150	?
1839	41	9	360	?
1840	53	10	?	?
1841	93	14	200	14
1842	98	13	200	7
1843	122	13	192	40
1844	157	13	200	40
1845[b]	195	15	180	?
1846	240	18	250	40

For sources see *Sources for Tables.*
[1] These figures represent spring populations.
[a] Plus 5 goats.
[b] Autumn populations.

FORT NEZ PERCÉS AND THOMPSON'S RIVER

At the remaining posts in the Columbia Department agriculture was a minor activity. However, two posts, Fort Nez Percés (Plate 5) and Thompson's River, specialized in horsebreeding. As Nathaniel Wyeth observed in 1833, Fort Nez Percés, or Walla Walla, was maintained by five men "mostly for trading horses and the safety of the communication."[101] The "communication" was, of course, the vital Columbia waterway, while the horses were needed for the trapping ventures of the Snake Party and the Southern Expedition, as well as for general travel purposes.

Townsend described the post in 1834:

> The fort is built of drift logs, and surrounded by a stoccade of the same, with two bastions, and a gallery around the inside. It stands about a hundred yards from the river, on the south bank, in a bleak and unprotected situation, surrounded on every side by a great, sandy plain, which supports little vegetation, except the wormwood and thorn-bushes. On the banks of the little river, however, there are narrow strips of rich soil, and here Mr. Pambrun [the commandant] raises the few garden vegetables necessary for

the support of his family. Potatoes, turnips, carrots, etc., thrive well, and Indian corn produces eighty bushels to the acre.[102]

Plate 5. View of Fort Nez Percés, 1846.

The cultivated land was located two or three miles east of the fort on the banks of the Walla Walla River. Here fifty acres (including two in vegetables) were tilled in 1841 and twelve acres in 1845. In addition, in 1846 another thirty acres were cultivated twenty miles from the fort.[103] Horses were bred at the post and bought from the Cayuse Indians (hence, apparently, the Western synonym for a horse). In the late 1820's some 250 horses were purchased annually from the Indians.[104] There were usually from 50 to 100 horses at the settlement (Table 10).

Thompson's River or Fort Kamloops was founded in 1812 by the Pacific Fur Company as Shewaps [Shuswap's] Fort at the confluence of the North and South Thompson Rivers in order to secure the trade of the Thompson basin. It also became a key station for the Columbia brigade. Simpson described the settlement in 1828:

> The Post of Kamloops, or Thompsons River, is a very unprofitable Establishment, and the principal cause of its being kept up as the people could be employed to more advantage elsewhere, is the danger to which the New Caledonia outfits and returns would be exposed, from the Natives of Thompsons River in passing to and from Vancouver, if we were to withdraw from their country. The outfits of this Post are brought from Vancouver to

Table 10

NUMBER OF LIVESTOCK AT FORT NEZ PERCÉS, 1826-29 AND 1835-46[1]

Year	Cattle	Horses	Pigs	Goats
1826[a]	0	24	0	?
1827	?	16	?	?
1828	?	60	?	?
1829	?	50	?	?
1835	12	75	6	?
1836	15	60	5	8
1837	22	80	7	17
1838	25	90	7	15
1839	33	93	7	15
1840	38	61	13	13
1841	46	66	12	?
1842	12	37	10	?
1843	13	50	?	?
1844	18	90	?	?
1845[b]	23	68	12	?
1846[c]	36	115	38	?

For sources see *Sources for Tables.*
[1]These figures represent spring populations.
[a]Plus 10 poultry.
[b]Plus 19 poultry.
[c]Plus 14 poultry.

Okanagan by Boat, and from Okanagan to the Establishment which is situated on the Banks of Thompsons River a distance of about 300 miles, by Horses. The Natives are upon the whole well disposed towards the Whites, but being numerous, it is considered advisable to keep a larger complement of people than the Trade can well afford to guard against accident. Its present strength is two Clerks and Eleven Men, and the profits were in 1825 £1000, in 1826 £1110 and in 1827 £1300; beyond which, we do not think they will rise this year.[105]

Thompson's River was generally regarded as a "troublesome and most arduous, as well as perilous charge," which was maintained because of its importance to the "Communication." It served as the depot for the posts of New Caledonia to the north. In McDonald's words, "with very few exceptions every year since my last, they [the posts of New Caledonia] were in the habit of getting more or less of their staff of life from Thompson's River."[106] As many as 350 horses were kept for the brigades to haul "goods in" and "returns out" (Table 11). Around 1845, in anticipation of the boundary settlement, the company moved cattle and horses from below to above the forty-ninth parallel, including 200 brood mares to Thompson's River. The animals were pastured on bunchgrass and enclosed in "stockades," or corrals, near the fort that held 300 to 400 head.[107]

Table 11

NUMBER OF LIVESTOCK AT THOMPSON'S RIVER AND FORT OKANAGAN, 1826-29 AND 1834-41[1]

Year	Cattle	Horses	Pigs	Goats
1826[a]	0	72	0	?
1827	?	61	?	?
1828[b]	?	60	1	?
1829	?	104	4	?
1834	5	170	18	12
1835	7	192	18	12
1836	11	215	20	21
1837	14	252	20	21
1838	19	292	17	23
1839	23	320	17	22
1840	31	358	19	25
1841	39 (24)	266 (265)	32 (21)	20 (15)

For sources see *Sources for Tables*.
[1]Fort Okanagan's share in 1841 is indicated in parentheses. The figures represent spring populations.
[a]Plus 10 poultry.
[b]Plus 4 poultry.

Cultivation was apparently confined to potatoes. In the spring of 1842, for example, twenty-three kegs (probably sixty-nine bushels) were planted; they were expected to yield four hundred kegs, which "they have formerly done," provided that the weather was not too dry. In 1845 the post's fifteen servants cultivated only six acres.[108]

NEW CALEDONIA

Governor Simpson's agricultural policy was begun later at the posts of New Caledonia, probably because their physical surroundings were more inimical to farming. It was not until Peter Dease was put in charge of the district in 1831 that cattle were introduced from Fort Vancouver and cultivation was commenced.[109]

Fort Alexandria had been established in 1821 at the lowest point on the Fraser reached by the post's namesake, Alexander Mackenzie, before he turned westward to the Pacific. Here the Columbia brigade changed from horses to boats, so numerous horses were kept — from two to three hundred in the early 1840's, as well as thirty cattle and some pigs and chickens. Cultivation was limited to four acres in 1837 and forty-six in 1845, when ten employees manned the post.[10] Vegetables (potatoes, carrots, turnips, radishes, cabbages, onions) and grains (wheat, barley, oats) were grown. An effusive company employee wrote in 1836 that:

> Fort Alexandria . . . is agreeably situated on the banks of Frazer's River, on the outskirts of the great prairies. The surrounding country is beautifully diversified by hill and dale, grove and plain; the soil is rich, yielding abundant successive crops of grain and vegetable, unmanured . . . . The charming locality, the friendly disposition of the Indians, and better fare, rendered this post one of the most agreeable situations in the Indian country.[111]

The output of potatoes fluctuated considerably — 56 kegs (nine gallons or three bushels each) in 1827, 136 in 1837 (plus 177 of turnips), 643 in 1842, 420 in 1844 (plus 200 of turnips), 632 in 1845 (plus 150 of turnips), and 90 in 1846 (plus 100 of turnips). There was a bumper harvest in 1843: 5,200 large sheaves of wheat, 2,000 of barley, 600 of oats, and 660 bushels (probably 220 kegs) of potatoes. By 1842, Alexandria, as it was sometimes simply called, had a gristmill, which ground the 400 to 500 bushels of wheat that were grown there yearly from 1843 through 1847. Also, many salmon were caught — 52,000 in 1845 alone.[112] The post sent salmon to Thompson's River and even to Fort Okanagan.

Fort George had been founded in 1807 at the junction of the Nechako and Fraser Rivers. It was abandoned in 1824 and reoccupied in 1829. Farming was not begun here until 1836, as reported by the post's commandant, Clerk Archibald McKinlay: "A Farm on a considerable large scale was commenced here by Mr. McLean in 1836 with a view of supplying the District with Flour, but I am sorry to say that all the crops of wheat have hitherto been destroyed by Frost Potatoes, Turnips and other common vegetables however thrive well."[113] Four acres were tilled in 1837 and thirty in 1845, when there were ten men at the post. In 1838 fifteen cattle, including five cows, were kept.[114]

At the remaining posts of New Caledonia, Fraser's Lake (Fort Fraser), Stuart's Lake (Fort St. James), McLeod's Lake (Fort McLeod), Babine Lake (Fort Kilmaurs), Connolly's Lake (Bear Lake), and Chilcotin's Lake (Fort Chilcotin), farming was commonly restricted to small gardens and some cattle. In 1845 these six posts had fifty-seven acres under cultivation and there were ninety-four cattle, thirty-nine horses, and fourteen pigs at Stuart's Lake, the depot of New Caledonia, where a "few" cattle had been introduced in 1830. Here in 1831 four kegs of seed potatoes yielded twelve kegs of tubers on nearly one acre, a "very poor return."[115] It seems that animals fared better than crops, although altogether New Caledonia's posts did not have many livestock (Table 12).

Table 12

NUMBER OF LIVESTOCK AT NEW CALEDONIA'S POSTS IN VARIOUS YEARS BETWEEN 1828 AND 1844[1]

Year	Cattle	Horses	Pigs	Poultry
1828	?	91	?	?
1829	?	110	?	?
1832	?	157	?	?
1833	4	146	?	?
1834	15	184	?	?
1835	25	212	?	?
1836	29	221	?	?
1837[a]	43	205	?	?
1838	44	228	?	?
1840[b]	74	192	4	?
1841[c]	54	198	7	6
1842	80	176	12	10
1844	99	?	13	101

For sources see *Sources for Tables*.
[1]These figures represent spring populations.
[a]Plus 2 goats.
[b]Plus 11 goats.
[c]Plus 6 goats.

Already by the middle 1830's these modest agricultural efforts had alleviated New Caledonia's chronic shortage of food. In 1837 McKinlay wrote, "As regards food the district has much improved of late years, Cattle are now getting numerous at all the Posts except one (Connollys Lake) so that we have milk and butter in abundance, that with plenty vegetables make the salmon very palatable."[116] And another fur trader, Alexander Anderson, recalled that when he had been a clerk in charge of Fraser's Lake in the last half of the 1830's:

> there were good gardens in the vicinity of the fort in which potatoes, turnips and other vegetables with barley and even wheat came to perfection. Wheat however is not cultivated to any extent as it was a precarious crop owing to occasional summer frosts. The pasture round is extremely luxuri-

ous and the few cows which we had yielded copious supplies of milk & butter.[117]

In 1844 Chief Factor Peter Ogden, outgoing bourgeois of New Caledonia, reported that he had done everything possible to encourage farming, with the result that ten men at Fort George and an even larger proportion of Fort Alexandria's servants were supported solely on grain, thereby permitting a slight reduction (of twenty-five bags) in the demand for flour from Fort Colvile in 1843.[118] Salmon remained a necessity, however.

FORT OKANAGAN, FORT BOISE, AND FORT HALL

The remaining interior posts in the Columbia basin, Forts Okanagan, Boise, and Hall, were somewhat more successful agriculturally than those in New Caledonia, whose cold was more insurmountable than the aridity of the Columbia Plateau. Fort Okanagan was built in 1811 near the junction of the Okanagan and Columbia Rivers at the head of batteau navigation on the latter. It was an outpost of Thompson's River and served more for transport than for trade.[119] In Simpson's words:

> This place is maintained almost entirely for the accommodation of New Caledonia and Thompsons River, being the point at which the route from those places strikes upon the Main River, and where it is necessary to keep two or three men throughout the year for the purpose of Watching Boats, Horses, Provisions etc. left here from time to time for the use of those places.[120]

There were seven acres under cultivation in 1845, when only two men ran the post, and four acres in 1846. The few potatoes and cattle that were raised were superior in quality. The governor stated in 1824 that Fort Okanagan "produces the finest potatoes I have seen in the Country"; fifty-five kegs of "fine" potatoes were grown in 1826. A member of Wilkes's expedition asserted in 1841 that "I never beheld finer Cattle in my life than I did there."[121]

Fort Boise (Big Wood Fort or Snake Fort) and Fort Hall (built by Nathaniel Wyeth and sold to the Hudson's Bay Company in 1837), both adobe posts, were founded in 1835 and 1834 on the middle and upper Snake to siphon the trade of the Snake Country, where American mountain men also trapped and traded. Here the Hudson's Bay Company successfully forestalled American competition. In 1845 at Fort Boise there were eight men, who tilled two acres and tended twenty-seven cattle and seventeen horses. At Fort Hall cultivation was initiated in 1836 with a garden and a cornfield. In 1843 it was reported that wheat and turnips were grown "with success" and that there was a "fine band" of horses and a "large herd" of cattle which fared "well." In 1845 the fort's twenty men cultivated five acres and herded 95 cattle and 171 horses. Their efforts were worthwhile, for already by 1841 the post was self-sufficient in grain and vegetables.[122] These efforts and those at Fort Boise enabled the two posts not only to feed their own personnel but also to succour American migrants on the Oregon Trail in the first half of the 1840's. In this respect, however, both of them were overshadowed by Fort Nez Percés and, especially, Fort Vancouver.

FORT UMPQUA

Fort Umpqua was an intermediate post, being located neither on the damp coast nor in the dry interior but halfway between the two, on the edge of the mountain knot at the southern end of the Willamette Valley. In 1820 or 1821 McKay's Fort was founded on the right bank of the upper Umpqua River to prosecute the trade of the surrounding region and to occupy the Willamette freemen, free traders who had left the employ of various fur trading companies and had settled in the valley. In 1832 it was succeeded by Fort Umpqua, which was established by Chief Trader John McLeod farther downriver on the left bank opposite present-day Elkton, Oregon. This "Old Establishment" was replaced in 1836 by a new fort, perhaps at the junction of the Umpqua River and Calapooya Creek. It served as a station for the Southern Party on its trapping and trading forays into California from Fort Vancouver via the Siskiyou Trail. In 1845, eight men cultivated fifty acres and tended sixty-four cattle, forty-six horses, and forty-five sheep. Eighty acres were ploughed in 1846. On this scale the post must have more than met its own food needs, given the alluvial soil and the mild climate.[123]

FORT GEORGE

Agriculture was also undertaken at the company's settlements on the coast, in the face of excessive wetness rather than cold or aridity. Farming began at Fort George as soon as the post was founded as Astoria by the Pacific Fur Company in 1811. The Astorians cultivated vegetables, especially potatoes. In the "excellent kitchen garden" twelve potatoes were planted in 1811, yielding ninety, which in 1812 produced five bushels, two of which returned fifty bushels in 1813. The potatoes weighed as much as three and a half pounds each.[124] By 1817 two hundred acres had been cleared, including twenty in potatoes, and twelve head of cattle and some pigs and goats were kept.[125] In 1818 Fort George comprised seven to nine buildings (including five to six dwellings) and numbered forty-five to fifty-four men (twenty to twenty-five whites and twenty-five to twenty-nine Hawaiians). Although there were eighty acres of potatoes in 1825, an American visitor noted that "very little attention is paid to Agriculture . . . the grand object is procuring skins: —— their principal food during the summer is Salmon in Winter Wild fowl: at present they have many hogs . . . also Cattle."[126] With the transfer of the Columbia Department's headquarters to Fort Vancouver late that year, agriculture declined at Fort George, which was "kept up" as a lookout for ships entering the Columbia, a post for the local Indian trade, and a depot for the salmon fishery. Townsend observed in 1834 that:

> it scarcely deserves the name of a fort, being composed of but one principal house of hewn boards, and a number of small Indian huts surrounding it, presenting the appearance, from a distance, of an ordinary small farm house with its appropriate outbuildings. There is but one white man residing here, the superintendent of the fort; but there is probably no necessity for more, as the business done is not very considerable.[127]

Seven years later another visitor found that "it is of little importance to the Company except as a convenient stopping place near the mouth of the river and as a depot for stores." "A small cleared space near the fort is the only land in the vicinity under cultivation," he added.[128] This "space" amounted to only four acres in 1845 and two in 1846.[129] The post probably met its own food needs in vegetables, meat, and milk but not in grain.

FORT NISQUALLY

Fort Nisqually or Nisqually House was established in 1833 to tap the trade of Puget Sound.[130] It was situated at the end of a large plain (Grande Prairie) along a stream about a mile from the shore. It was described by Abbé Jean Bolduc, a Catholic missionary, in 1843:

> To have an exact idea of this fort, picture to yourself an enclosure of fir logs, on an average, eighteen feet high, enclosing a space one hundred fifty feet on each side and having a small unarmed bastion at the four corners. Inside is a house for the superintendent, a store for trading in furs and several small buildings for the lodging of servitors and voyageurs.[131]

Farming began at the post in the year of its establishment with the planting of corn, peas, potatoes, carrots, radishes, turnips, onions, and cabbages and the pasturing of cattle and horses on the surrounding meadows.[132] From 1835 the post's fur trade declined, so more attention could be given to agriculture. In

1837, 174 bushels of grain (a fivefold yield), 100 of peas (a fivefold yield), and 385 of potatoes (a tenfold yield) were harvested. Twenty servants cultivated 100 acres in 1845.[133] Enough livestock were also raised to satisfy the fort's own needs (Table 13).

Table 13

NUMBER OF LIVESTOCK AT FORT NISQUALLY, 1835-43[1]

Year	Cattle	Horses	Pigs	Poultry
1835	11	6	?	?
1836	21	11	6	?
1837[a]	27	10	24	?
1838[b]	35	16	27	10
1839[c]	37	15	37	20
1840	50	18	35	10
1841	?	21	21	11
1842	?	53	23	13
1843	?	61	36	16

For sources see *Sources for Tables*.
[1]These figures represent spring populations.
[a]Plus 9 goats.
[b]Plus 6 goats.
[c]Plus 2 goats and 728 sheep.

Most of Fort Nisqually's agricultural property was transferred to the Puget's Sound Agricultural Company during the winter of 1840-41.[134]

FORT VICTORIA

As early as 1830 the Hudson's Bay Company considered moving the "Principal Depot" to the vicinity of Puget Sound, owing to "the unhealthy [malarial] state of Fort Vancouver and its distance from the sea."[135] The latter consideration, which originally affected only the London ships, became more important as company shipping to new coastal posts and to new markets in Alaska and Hawaii increased. In 1835 the Governor and Committee encouraged McLoughlin to relocate his headquarters "somewhere inside the Straits of De Fuca."[136] By then the company was perhaps less certain that the Columbia would become the international boundary. In 1836 Simpson instructed McLoughlin to examine the coasts and islands of the strait for a site for a new "main depot." It had to be suitable for trading, hunting, fishing, farming, and shipping.[137] Douglas reported that in the following year:

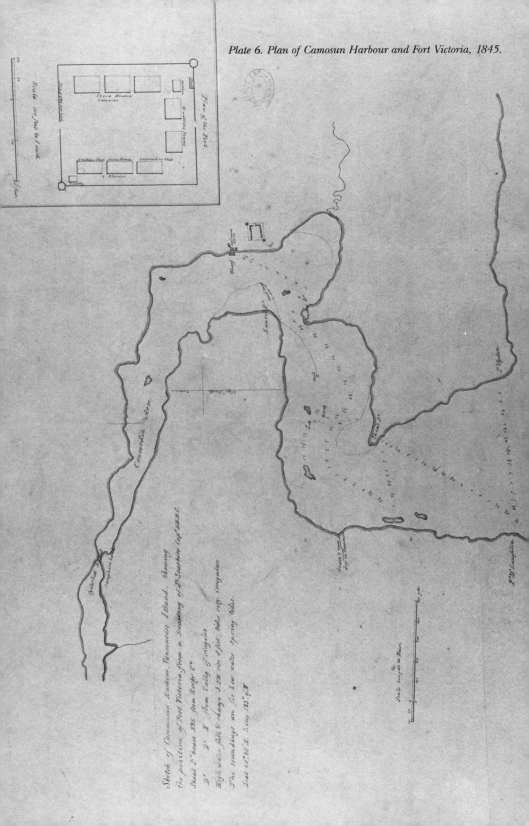

Plate 6. Plan of Camosun Harbour and Fort Victoria, 1845.

Captain McNeill... examined the north [south] end of Vancouvers Island, and ... found an excellent harbour, of easy access with good anchorage, surrounded by a plain of several miles in extent, of an excellent Soil, and intersected with a number of Rivulets; from his account it appears to be an excellent place for tillage and Graising.

Douglas added that "the site ... is, as a whole, decidedly unequalled, by any other known portion of the Coast north of Columbia River."[138] No action was taken to move the departmental capital, however, until the early 1840's, when more American settlers began to homestead the Willamette Valley on Fort Vancouver's doorstep and when expansionist American congressmen became more clamorous on the Oregon question. This American threat was cited by Simpson in a report to London:

> notwithstanding a certain degree of popularity which the Company's officers enjoy, arising from the hospitalities & assistance that have been rendered to almost every American citizen who has come to the country, the Hon. Company, as a body, is looked upon with much jealous rancour and hostility, leading to serious apprehensions on the minds of the Council [of the Northern Department] that, the depot at Fort Vancouver & the other posts within reach of these people, are not safe from plunder. These apprehensions have determined us on giving directions that ... the great bulk of the property in depot at Vancouver be removed to F. Victoria, which is intended to be made the principal depot of the country.[139]

So in 1843 Fort Victoria was built at Port Camosun on the southern tip of Vancouver Island (Plate 6). Douglas declared that the new post "was designed to serve as a general Depot for our Pacific trade, and to become a rendezvous for the shipping." The Governor and Committee hoped that "the dangers of the bar and the loss of time in the navigation of the Columbia River ... will thus be avoided."[140]

While Fort Victoria had been established to facilitate the shipping and the security of the department's headquarters, farming was not overlooked. During its construction, Simpson had been instructed by London to develop farming, fishing, and sawmilling, besides the fur trade. Beaver House added: "we are desirous that tillage and pasture farms should be established as soon as possible, and on such a scale as to meet the demands of the establishment."[141] Warre and Vavasour even asserted in 1845 that "the Situation has been chosen solely for its Agricultural Advantages and is ill adapted either as a place of refuge for shipping or as a position of defence."[142] By then the fort was staffed by twenty men and had 120 acres under cultivation. There was "a Farm of several hundred acres attached where they raise Wheat and Potatoes in abundance."[143] In 1846 the farm produced 800 bushels of wheat, 400 of oats, 300 of peas, and 2,100 of potatoes; also, sixty-three calves were dropped. In addition, a six-acre orchard was planted with apple, pear, and peach trees. The cattle, horses, and pigs, respectively, increased from 23, 7, and 1 in 1844 to 31, 12, and 3 in 1845 and 128, 40, and 6 in 1846, a fivefold to sixfold increase in two years.[144] Output met not only the needs of the post itself but also part of the requirements of the Russian-American Company as contracted by the Hudson's Bay Company in 1839.

FORT SIMPSON

The coastal posts north of the Fraser were established in the early 1830's to thwart both American and Russian rivals in the fur trade. In conjunction with coasting vessels, the new posts succeeded in ousting the Americans and blocking the Russians, although the former were already withdrawing anyway in search of richer waters and the latter were too overextended to offer unyielding opposition. The most important of the coastal posts was Fort Simpson (Plate 7), which tapped the trade of some fourteen thousand natives.[145] It was erected in 1831-32 at the mouth of the Nass River but was moved in 1834 to the nearby Tsimean Peninsula to facilitate shipping. Governor Simpson regarded the fort as "a valuable and important establishment" and as "the depot of the coast." He also reported that for the post's twenty men "the means of living are abundant, consisting principally of fish, venison, and potatoes."[146] The potatoes, however, sometimes failed; in 1835, for example, half of those that were planted did not grow.[147] Moreover, most of the fish and game and even many of the potatoes were supplied by the Indians, who had adopted this cultigen probably as early as the 1810's. In Chief Factor Duncan Finlayson's words:

> The resources in the way of living which Fort Simpson affords, are Deer, Halibut, and Salmon, which, however may be considered as precarious while our dependence is placed on the natives for providing them, as they entertain such hostile feelings towards one another that frequent and fatal disturbances arise which will prevent their fishing or hunting more than is barely sufficient for their daily subsistence. I am therefore afraid that this post can never be maintained with safety without having a six month's stock of provisions on hand. But in peaceable times, with the assistance the Garden is likely to afford, on which Mr. Work has spared neither labor nor attention, and which will this season produce from 1 to 200 bushels of Potatoes, the resources already mentioned are quite ample for the maintenance of the fort.[148]

Plate 7. *View of Fort Simpson.*

This situation did not change, with first Fort Vancouver and then Fort Victoria providing grain and dairy products and the post growing its own vegetables. In the spring of 1837, 33 bushels of potatoes were planted on one and one-third acres, which in the autumn yielded 651 bushels (a 19½-fold yield or 487 bushels per acre); the harvest would have been 50 bushels more if crows had not rustled some of the seeds and plants and Indians had not stolen some of the tubers.[149] The vegetable patches were fertilized with rotten seaweed. John Work told a friend in 1841 that "with much labour I have got the fort in a good state of defence, and a large garden enclosed which yields us plenty of potatoes, turnips, cabbages and other vegetables, which with good venison, fish, wild fowl ... render us comfortable so far as living is concerned."[150] In 1845 Fort Simpson's twenty men cultivated eight acres.[151]

FORT McLOUGHLIN

Fort McLoughlin was founded in 1833 on an island twenty miles up Milbanke Sound. Here farming was limited to potato gardening, but even the potatoes did not yield well because of the poor soil. In 1836, for instance, one hundred bushels were raised but from five to six hundred were needed. In 1837, however, eight hundred bushels were harvested.[152] Simpson reported in 1841 that the post's thirteen men were "principally maintained on country provisions, say fish in great abundance and variety, venison and potatoes."[153]

FORT STIKINE AND FORT TAKU

Fort Stikine or Highfield, which was located on Duke of York's Island near the mouth of the Stikine River, was the northernmost of the coastal posts. It replaced nearby St. Dionysius Redoubt, which had been erected in 1824 by the Russian-American Company and abandoned in 1840 as a result of the *Dryad* affair (see Chapter 4). The post did not impress Simpson in 1841:

> The establishment, of which the site had not been well selected, was situated on a peninsula barely large enough for the necessary buildings, while the tide, by overflowing the isthmus at high water, rendered any artificial extension of the premises almost impracticable; and the slime, that was periodically deposited by the receding sea, was aided by the putridity and filth of the native villages in the neighbourhood, in oppressing the atmosphere with a most nauseous perfume. The harbour, moreover, was so narrow, that a vessel of a hundred tons, instead of swinging at anchor, was under the necessity of mooring stem and stern; and the supply of fresh water was brought by a wooden aqueduct, which the savages might at any time destroy, from a stream about two hundred yards distant.[154]

Agriculture failed here. The governor found that the post's twenty men were "maintained by fish and venison which are procured in great abundance from the natives at a very cheap rate."[155]

Fort Taku or Durham was founded in 1841 on Wrangel Island in Stephen's Passage about 125 miles north of Fort Stikine. The furs of both posts came from the interior via Indian middlemen on the coast. Simpson noted in 1841 that Fort Taku's twenty-four men were "principally maintained on venison got here as at the other establishments on the coast at so cheap a rate from the natives, that we absolutely make a profit in our consumption of provisions, the skin of the animal selling for much more than is paid for the whole carcass."[156] In the same year Simpson recommended that Forts Taku, Stikine, and McLoughlin be abandoned for a saving of £4,000 annually. His recommendation was implemented in 1843.

3

PHYSICAL EXTREMES: THE PROBLEMS OF POST FARMING

It was mostly too cold for gardens.

CHIEF TRADER JOHN TOD
RE NEW CALEDONIA, n.d.

Although farming was generally successful at the Hudson's Bay Company's posts in the Columbia Department, some of the efforts, particularly on the Northwest Coast, in New Caledonia, and in the Snake Country suffered setbacks, both temporary and permanent. Most of the obstacles were physical rather than cultural, stemming primarily from the rough terrain, dense forest, and "weighty rain" of the coast, the cold of the northern interior, and the aridity of the southern interior.[1]

At some posts the soil was unproductive. Even at Fort Vancouver, the department's most successful agricultural operation, four hundred of the five to six hundred acres that were not subject to spring flooding consisted of "poor miserable Dry S[h]ingly soil."[2] It required manuring but cattle could be penned at night on only half of these four hundred acres because there was not enough grass on the other half; consequently, the fields had to be left fallow for four years after each crop.[3] And at the original site of Fort Langley, another successful agricultural post, the soil was "very indifferent," although at nearby Langley Prairie it was "excellent" in the opinion of John Work.[4] So farming was concentrated on the prairie. Fort Nisqally's first commandant, Archibald McDonald, found in 1833 that the plains there were "very dry and Steril."[5] A year later Francis Heron reported that the post's land was "but pure *sand & stones*, on which no crop even approaches to maturity."[6] This problem prompted the company in 1836 to consider replacing both Forts Langley and Nisqually with a new post on Whidbey Island, which was reputed to have "a fine Harbor and excellent Soil." Upon closer inspection, however, it was found that the soil was "very poor," being "only sand but a little vegetable mould on the top."[7] Moreover, the island was, in McLoughlin's words, "not conveniently situated for

Trade."[8] So Forts Langley and Nisqually were retained. At Fort McLoughlin the soil was likewise "very poor."[9] In 1835 Work observed that: "the soil is very wet and composed of black peat moss on a bed of rocks, and appears not well adapted for yielding good crops without much labour and a mixture of clayey gravel or manure which is difficult to be obtained. From the wetness of the soil draining will also be required." At Fort Simpson the soil was also soggy and peaty.[10]

All along the coast, except where large rivers like the Columbia and Fraser had deposited sizable and fertile floodplains and deltas, the soil was low not only in quality but in quantity as well, owing to the looming mountains. Peter Ogden noted in 1831 that Fort Simpson, for example, had "no Ground about it to make a Garden," and the relocation of the post in 1834 did not solve this problem.[11] Fort George at the mouth of the Columbia had no more than fifteen to twenty tillable acres in 1824.[12] By 1841 farming had nearly ceased at this initial entrepôt because the amount of arable land was so limited and the cost of clearing so high.[13] Even Fort Vancouver's productive farm had "limited pastures." In 1838 James Douglas reported that:

> The country in this vicinity is not adapted for herding, on a large scale; the only tolerable pasturage being found upon low marshy lands, bordering the River, which are intersected with, innumerable pools of water, and belts of closely matted forest; where the cattle, like the native deer, take refuge, whenever pursued; and from whence, they are not easily dislodged by their keepers.[14]

The shortage of prime farmland was partly attributable to the great difficulty and expense of clearing the temperate rain forest of the coast before the advent of blasting powder, bulldozers, and chain saws. The gigantic Western red cedar and Douglas fir (Oregon pine) and the thick undergrowth (including such thorny entanglements as blackberry and devil's club) were formidable obstacles, particularly if they had to be removed in the presence of belligerent Indians, as at Fort Langley. James McMillan complained in 1827 that "the great size of the Timber and the thick growth of the underwood have been sadly against us in clearing the ground, the jungle on the banks of the River is almost impenetrable and the trees within are many of them three fathoms in circumference, and upwards of two hundred feet high." The site of the fort itself had been cleared with difficulty in the same year because of "the timber being strong, and the ground completely covered with thick underwood, interwoven with Brambles & Briars."[15] It took one man a whole day to fell a single tree.[16] The situation was even worse at Forts McLoughlin and Simpson, which lacked prairies. At the founding of Fort McLoughlin in the summer of 1833, "when the men first went on shore, it was like entering an impenetrable forest."[17] Work reported from Fort McLoughlin in early 1835 that "we got only about two miles below the fort where Mr. Manson [the commandant] has the men busily employed clearing ground for a garden. And a tedious laborious job it is, the ground is full of stumps and roots many of them so large that it is very difficult and great labour to get them out." Stumping was no easier at Fort Simpson, where Work found that:

> when the trees are cleared off with their branches, the stumps that remain are mostly of a large size and so close together that there is no means of making the ground of any use but by entirely removing them, which from their great weight is very difficult. Indeed the ground is mixed with a complete mass of roots intertwined, and in order to get rid of them the ground has to be all turned over from 1 to 2½ or 3 feet in depth.[18]

Here twenty men took three weeks to stump five-eighths of an acre.[19] Even at Fort Vancouver the thickly wooded uplands were a "dreadful labour" to clear, decried Douglas, requiring men and money "beyond our means."[20]

Once trees had been cleared and prairies had been ploughed or grazed, another problem periodically arose — literally — in the form of annual flooding, especially along the voluminous Columbia and Fraser Rivers. Then neither dams nor dykes controlled their spring freshets, which were fed by vast watersheds with heavy winter precipitation. At Fort Vancouver the Columbia rose up to twenty feet in late spring, and from one-quarter to one-half of the arable land (including three-fifths of the best arable land) was subject to flooding.[21] Of the three plain or prairie levels adjoining the river, the first or lower level was only ten feet above the river at low water. The first and second or middle plains had the best soil, but the former was usually flooded, in spite of a levee, which was breached by seepage. Both plains were "admirably" suited to grazing but not to cropping, since crops were obviously fixed, although flooded wheat might be fit for seed if it were not "overtopt."[22] On the third or upper plain, however, crops fared poorly in the light, gravelly soil. So cropping of the lower plain continued, with the result that it was unusual to harvest every acre that had been sown. The 1830 crop was damaged "very much" by the "extraordinary height" of the river.[23] The flood of 1838 was described by Douglas:

> The periodical river flood, caused us considerable loss: it commenced at an uncommonly early season, rose to the level of the highest bank, and gradually subsided; a second flush came down early in May, and soon rose above its former level. For some days we battled successfully with the flood

> by throwing up repeated embankments, and though the water by its mightly lateral pressure, was forced up in the centre of the plain which is considerably lower than the bank of the River, we still trusted that a change of weather might in the mean time create a diversion in our favour and cause a sudden depression in the River. Our hopes were disappointed, as the irresistable flood baffled all our efforts, and on the 21st May broke over both the natural and factitious banks, and rushing into our fields, laid waste 80 acres, of our most promising crop. This blow came upon us with stunning effect, at the close of seed time, and the opening of the travelling season, when there was no prospect of repairing the loss.[24]

Fortunately, the high water receded within fifteen days, so that by the end of June another crop of barley, buckwheat, peas, and potatoes was sowed and subsequently harvested. Two springs later, however, the swollen Columbia destroyed half of the post's grain acreage.[25]

Fort Vancouver's warmer summers and multiple terraces afforded more protection against flooding than was available at such posts as Fort Langley. At the latter fort, for example, the "summer flush" did "considerable damage" to crops in 1829 and "great damage" in 1830.[26] But the cool, damp climate was more problematical. Planting and harvesting were often hampered by rain, and the growing season was seldom more than tepid. In 1845, for instance, nearly all of the cropland was flooded in June, and in August two-thirds of the recovered and replanted crop was damaged or destroyed by protracted rainfall, "a misfortune of frequent occurrence at that Post," according to Douglas and Work.[27] McDonald reported that

> our Farm is well established with Buildings, and fences, and quite extensive enough ... but the climate is unfavourable and the soil ill adapted to its disadvantages, being low and wet, making the suitable time for ploughing, sowing and harvest exceedingly backward [late].[28]

He added that "our Farm, which though it certainly at times presents a grand appearance yields hardly sufficient to remunerate the labour bestowed upon it; this however is owing chiefly to the unfavourableness of the climate." McDonald's judgment was corroborated by Ogden and Douglas in 1847, when they told Simpson that "the climate of Fort Langley is not well adapted for Agricultural purposes, the wet season setting in early, and there being always more or less difficulty in securing the Crops, in Consequence of variable weather in harvest." Partly for this reason farming at Fort Langley was greatly reduced in 1847 on Douglas's orders.[29] Even winters could be too wet, as on the lower Columbia in 1844-45, when "many" cattle died during the rainiest winter in memory.[30]

Inland just the opposite problem sometimes occurred. Here in the rain shadow of the coastal mountains there was much less precipitation, as well as much more evaporation during the warmer growing season, particularly towards the south. As McDonald reported from Thompson's River to the Council of the Northern Department in 1827, "the intense heat and constant drough[t] of the Summer are much against cultivation in the interior of the Columbia."[31] Fort Colvile's crops largely failed in 1839 and 1840 because of drought, so that barely a year's supply of grain was available in 1840 and 1841.[32] The 1839 drought, which was accompanied by "locusts," or grasshoppers, also struck Forts Boise and Hall. At Fort Nez Percés farming was virtually halted by drought in 1829.[33] Not even Fort Vancouver was immune to aridity. Occasionally, the Hawaiian high pressure system prevailed long enough in summer to cause drought, especially if it also drew hot, dry air from the interior through the Columbia gap. In 1827 "excessive heat" and "severe drought," with the temperature reaching 105°F. in July, reduced the wheat and barley harvest to twelvefold the seed, and the underfed horses could not draw ploughs properly. Again in 1831 there was "Want of Rain in the Summer," and corn, peas, and potatoes failed.[34] In 1835 the potato crop was lowered to one-third of normal by drought, which also ruined Fort Langley's crops.[35] On 14 August 1839, it was reported from Fort Vancouver that "the dryness of the season in this quarter is without example"; by 10 October it still had not rained.[36] Perhaps the post's longest dry spell occurred in 1844, when no rain fell between mid-July and mid-October. The aridity sparked a grass and forest fire that nearly destroyed the establishment on 27 September; three thousand bushels of stored grain and peas were burned.[37]

The summer dryness of the southern interior was more than matched as an obstacle to farming by the summer coolness of New Caledonia. Its high latitude and elevation made for a growing season that was both short and cool with late (June) and early (August) frosts, particularly at those posts which were subject to nocturnal air drainage, such as Fraser's Lake, or were exposed to northeasterly winds, such as Stuart's Lake. Generally only hardy vegetables and livestock survived, as Chief Trader John McLean learned in 1837:

> Four acres of land [at Fort George] were put in a condition to receive seed, and about the same quantity at Fort Alexandria. Seed was ordered from the Columbia, and handmills to grind our grain. Pancakes and hot rolls were thenceforward to be the order of the day; Babine salmon and dog's flesh were to be sent — "to Coventry!" The spring, however, brought with it but poor prospects for pancakes; the season was late beyond all precedent; the fields were not sown until the 5th of May; they, nevertheless, promised well for some time, but cold weather ensued, and continued so long that the

crops could not recover before the autumn frosts set in, and thus our hopes were blasted.[38]

And Archibald McKinlay reported in 1838 from Fort George that

> A Farm on a considerable large scale was commenced here by Mr. McLean in 1836 with a view of supplying the [New Caledonia] District with Flour, but I am sorry to say that all the crops of wheat have hitherto been destroyed by Frost Potatoes, Turnips and other common vegetables however thrive well.[39]

The summer weather was also highly variable, as McLean found in the last half of the 1830's: "I have experienced at Stuart's Lake, in the month of July, every possible change of weather within twelve hours; frosts in the morning, scorching heat at noon; then rain, hail, snow."[40] Farming was not even able to meet local needs. Peter Ogden told Wilkes at Fort Vancouver in 1841 that New Caledonia, where Ogden spent the years 1835-44, was "unsusceptible of cultivation" because of summer frosts, so that the company servants at the six district posts lived "the greater part of the year" on dried salmon and obtained all of their "stores flour etc." from Forts Colvile and Vancouver.[41] The Columbia brigade thus continued to transport provisions from the lower Columbia to the upper Fraser.

Cold winters were generally less troublesome than cool summers in the Oregon Country, but occasionally they did take a heavy toll of livestock, which often ran at large year round. The winter of 1827-28 at Fort Colvile, for example, was "unusually severe" and about twenty piglets died. The same winter was "uncommonly severe" at Fort Langley.[42] The winters of 1836-37 and 1846-47 were perhaps the coldest on the lower Columbia. During the latter, "the coldest season ever experienced" at Fort Vancouver, 267 of 270 horses perished at Fort Colvile and the Nez Percé Indians lost half of their horses.[43] This disastrous reduction hampered brigade transport in particular.

Yet more obstacles to successful farming were diseases, pests, and weeds. For instance, nearly all of Fort Vancouver's sheep died of an intestinal ailment during the winter of 1842-43; "several" pigs were killed by wolves at Fort Vancouver during the winter of 1826-27; most of Fort Alexandria's crops were foraged by unfenced cattle in 1837; a "frightful pest" of caterpillar grubs destroyed Fort Langley's vegetable crop in 1830; grasshoppers attacked crops at Fort Colvile in 1839; a poisonous root killed pigs at Fort Vancouver in 1825 and 1827; and "fern and other weeds" choked Fort Langley's cropland in 1835.[44] These problems were relatively minor, however, despite the absence of antibiotics, insecticides, and herbicides.

More widespread and frequent were some man-made difficulties, although they were less immutable than the physical obstacles. The shortage of agricultural labour loomed largest. Among Hudson's Bay Company employees, the Oregon Country, particularly New Caledonia, had a repelling reputation for hardship and remoteness. The climate was extreme, the fishy diet was monotonous, the posts were isolated from "civilization," and the terrain was rugged for travel. Few servants volunteered to serve in the Columbia Department, and not a few were demoted or banished there for misconduct "the east side the mountains." So the company's Columbia posts were often short of men, and particularly of able men, and operations suffered accordingly. Moreover, with the restricted diet and the raw climate the men were vulnerable to illness, both physiological and psychological, and accidents. McMillan told McLoughlin in September 1827 that fully one-quarter of his men at Fort Langley were incapacitated by sickness (mainly venereal disease) and mishaps, and at Fort Simpson in the mid-1830's, one-quarter to one-third of the men were on the wintertime sick list.[45] At Fort Colvile in 1840 and 1841 farming was hampered by a shortage of hands, especially healthy and diligent workers, for such labourious tasks as railing and scything. Bourgeois McDonald complained in August 1842, the height of harvest time, that "all our grain is now overripe, & scarcely a man ... in a fit condition to put his hand to any efficient labour yet."[46] Furthermore, the agricultural labour was part-time and even secondary, for the men had other demanding duties such as trading, hauling, fishing, felling, and hunting, some of which were more pressing. Farming did not receive their undivided attention, which was essential in places like New Caledonia, where the physical conditions of crop growing and stock rearing were so marginal as to leave little room for delay or error. The demands of other duties could postpone tilling, seeding, watering, weeding, thinning, and so forth to the point of little or no return.

And when a return was realized, as was usually the case, it might be burned in storage. Almost all of the buildings at the company's posts were wooden, and partially dried grain or hay was subject to spontaneous combustion. A company servant recalled that "barns at that time were either nin est [non-existent] or rude affairs and loss was often sustained by the wheat heating in the close moist climate."[47] Fire levelled Fort Langley in the spring of 1840 and Fort Nez Percés in the autumn of 1841. In late summer of 1844 Fort Vancouver was nearly engulfed by flames.

In spite of these diverse obstacles, which were mostly temporary or sporadic, agriculture largely succeeded at the Hudson's Bay Company's posts in the Oregon Country. It was too wet along the coast and too cold in the northern interior, except for vegetable gardening, but these failures were more than offset by surplus yields at the other posts, especially at Forts Vancouver, Colvile, and Langley, where there was less rain, more heat, better soil, and more and abler hands. The surpluses supported these posts and the others and even an export trade. At the end of the 1830's, this export trade in agricultural produce was dramatically expanded as the result of collusion between the Hudson's Bay and Russian-American Companies, a relationship which led to the formation of the Puget's Sound Agricultural Company.

Part Two

COMPANY FARMING
Puget's Sound Agricultural Company Operations on the Cowlitz Portage

The fur trade is the principal branch of business at present in the country situated between the Rocky Mountains and the Pacific Ocean. On the banks of the Columbia river, however, where the soil and climate are favourable to cultivation, we are directing our attention to agriculture on a large scale, and there is every prospect that we shall soon be able to establish important branches of export trade from thence in the articles of wool, tallow, hides, tobacco, and grain of various kinds.

Governor George Simpson to Governor John Pelly, 1837

4

GRAIN FOR ALASKA AND WOOL FOR ENGLAND: THE ORIGINS OF THE PUGET'S SOUND AGRICULTURAL COMPANY

> *We consider it both desirable and necessary to put our farming establishments on a more regular and systematic footing than they have heretofore been as they now promise to form an important branch of export business.*
> GOVERNOR AND COMMITTEE TO CHIEF TRADER JAMES DOUGLAS,
> 31 OCTOBER 1838

SIMPSON'S PLAN TO SUPPLY RUSSIAN AMERICA

So successful were the agricultural operations of the Hudson's Bay Company at its posts "this side the mountains" that by the late 1830's there was sufficient output not only for the Columbia Department itself but for export markets as well. Herbert Beaver, the aptly named chaplain and missionary at Fort Vancouver, noted in 1838 that considerable surplus grain was stored there, and at the Willamette Settlement, for want of a market.[1] But much earlier Governor Simpson had envisaged the export of surplus agricultural produce from a more diversified and civilized Oregon Country. He realized that the old order in the *pays sauvage*, centred on the fur trade, was fading with the depletion of fur bearers, the disaffection of fur buyers, and the encroachment of settlers. Activities like farming, fishing, lumbering, mining, and manufacturing, rudimentary under the fur trade, had to be developed further to sustain the company. To Simpson it was a question of sound business. The local market, however, was still small, so outside markets had to be found. Great Britain was a possibility. A connection already existed in the form of the London ship, which had to be continued in order to dispose of fur returns. The diminishing bales of furs did not fill the ship, and it could be ballasted more profitably with wool or tallow than with stones. Russian America, however, was a more promising customer. It was closer and more in need, having long suffered from inadequate supply, particularly of food.[3] Moreover, the Russians were to a large extent being supplied by American ships that competed with both the Hudson's Bay Company and the Russian-American Company for the peltry of

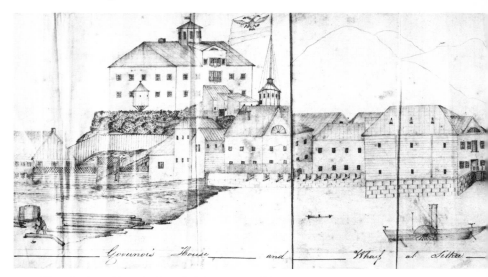

Plate 8. View of New Archangel (Sitka) circa 1840.

the Northwest Coast. Simpson reasoned that if his company were to outdo the Americans in supplying New Archangel (Sitka) (Plate 8), it would also outcompete them in trading on the coast, for much of the Yankee profit margin on the Nor'west trade was derived from the Sitka traffic. In 1824 he wrote:

> The Russian Settlements have hitherto been principally supplied with goods for their trade by the American adventurers on this coast payable by bills on St Petersburg or in Furs; but if we conduct our business with good management according to the present plan that channel will be shut up as we ought to be able to put down all competition on the Coast in which case 'tis probable we should be enabled to do business with the Russians on advantageous terms; it would be with a view to see what could be done in that way I should purpose visiting the Director Von Baranoff at their principal establishment of New Archangel in Norfolk Sound.[3]

With American competition lessened or eliminated, the high cost of company operations in the Columbia Department, Simpson's bugbear, would be lowered. McLoughlin asserted in 1840 that because of the presence of competition and the requirements of maritime activity, it was ten times more expensive for the company to obtain furs on the Northwest Coast than in the Mackenzie District or New Caledonia.[4]

Simpson was unaware in 1824 that Governor Alexander Baranov (1799-1818) had been replaced six years earlier, and Simpson himself did not visit Sitka until 1841. In the meantime he sent a cousin, Lieutenant Aemelius

Simpson, to Russian America's capital in 1829 with an offer to supply the Russian-American Company annually with fifty to one hundred tons of manufactures from the British Isles and four to five thousand bushels of grain and eight to ten thousand pounds of salted pork and beef from the Columbia Department in exchange for furs, bills of exchange payable at London or St. Petersburg, or specie.[5] George Simpson was confident that "we could . . . furnish them with Provisions, say Grain, Beef & Pork, as the Farm at Vancouver can be made to produce, much more than we require."[6] The Russians declined, however. They believed that although the Honourable Company could undoubtedly supply textiles more cheaply than the Americans and grain more cheaply than California's missions or Siberia's Okhotsk, it would not be willing to ship such a small amount of textiles (five tons). Besides, the Russian-American Company did not want to rely upon a rival firm for such a basic need as grain. Governor Ferdinand Von Wrangell (1830-35) even reckoned that British prices would be "much higher" than American prices for the same goods.[7]

But Governor Simpson was not deterred. Neither was McLoughlin, who now extended his superior's proposal by suggesting that a separate agricultural enterprise be established to serve the export trade, leaving the company's posts to concentrate on the fur trade. McLoughlin was impressed with the progress of fort farming in his department and the success of the hide and tallow trade from California. Like Simpson, McLoughlin was planning alternatives to the waning fur trade. A company physician, Dr. William Tolmie, wrote in 1833 that:

> He thinks that when the trade in furs is knocked up which at no very distant day must happen, the servants of Coy. may turn their attention to the rearing of cattle for the sake of the hides & tallow, in which he says business could be carried on to a greater amount, than that of the furs collected west of the Rocky Mountains.[8]

In March 1832 McLoughlin formally proposed to the Governor and Committee the formation of "The Oragon Beef & Tallow Company," a joint-stock association with an initial capital of £3,000 (divided into three hundred shares) "for the purpose of rearing (on a large scale), Cattle, with the view of opening from the Oragon Country an Export trade with England & elsewhere, in the articles of Tallow-Beef-Hides-Horns etc etc." It was proposed to buy from 700 to 800 cattle in California and drive them overland to Puget Sound, where, it was estimated, they would multiply to 12,405 head by 1842. Indeed, the venture's prospectus boasted that the lower Columbia could support 500,000 cattle.[9] In August 1834, Simpson "strongly" recommended to Beaver House that "cattle rearing on a large scale on the banks of the Wilhamet, on the Cowlitz Portage [between the Columbia River and Puget Sound], or elsewhere . . . be established." He believed that the soil and climate of the lower Columbia were well suited and that the venture could be launched with little expense and would soon yield great profits. The governor pointed out that the number of cattle on the Columbia had already increased to more than 600 from only about 30 head in 1834, and he suggested that 1,000 head be bought in California and driven to the Willamette. He insisted, however, that the scheme be undertaken "by the Honble. Company, as a branch of the Fur Trade," not by an association or individuals. (This insistence was characteristic of Simpson, whose passion for economy and profit was exceeded only by his concern for the company's monopoly.)[10] The Governor and Committee agreed, and at the end of 1834 they wrote to McLoughlin that they could not sanction a private concern because "it would be detrimental, if not dangerous to the Fur Trade . . . [which] has a right to the best exertions, and to the undivided time and attention of every Chief Factor and Chief Trader as well as Clerk & Servant." They added, however, that "it might be profitable if carried on by the Fur Trade, with all the Protection and facilities, which might be easily supplied by the Company's establishments." Accordingly, they sent McLoughlin £300 for the purchase of Californian cattle. He was also cautioned to establish the concern on the northern side of the Columbia, possibly Whidbey Island or the head of Puget Sound, wherever there were "most of the advantages of good Harbour and shipping Place, good soil and Climate, healthy situation and open Pasture ground." The northern side of the Columbia was also preferred because it was considered certain to become formal British territory.[11] At the beginning of 1835, Beaver House authorized McLoughlin to establish a "grazing farm" under the company's management between the Columbia River and Puget Sound with from 500 to 1,000 Californian cattle, plus sheep, for the export of hides, tallow, and wool. And in the spring, Simpson ordered McLoughlin to begin the venture "on a grand scale at once" with 5,000 cows and a "proportionate number" of bulls.[12]

Meanwhile, the Russians were becoming increasingly interested in Simpson's offer of a trade agreement. Their change of heart was prompted by the demise of several of their usual sources of supply. Since the middle 1810's, Russian America had relied heavily upon the cornucopian missions of Alta California for provisions, with one or two Russian-American Company ships annually exchanging cloth, utensils, tools, and other goods for wheat, beef, tallow, and lard at San Francisco, Monterey, and elsewhere. This trade was crippled by the secularization of the missions in the middle 1830's. The Indian workers left and the untended fields reverted to grass and brush. Already in 1831 a Russian agronomist observed that "at the missions . . . almost nothing is sown now," and in 1835 Governor Von Wrangell bemoaned "the utter decline of the prosperity of the missions in California."[13] California's ranchos produced surpluses, but they did not match mission output. In the summer of 1839, a Russian official admitted to Roderick Finlayson of the Hudson's Bay Company that the supply of wheat from California was "sometimes precarious."[14]

Grain for Alaska and Wool for England 81

By now another source of provisions for Sitka had proved even more uncertain. Fort Ross had been founded in 1812 on the coast of California some seventy-five miles north of the Golden Gate as a sea otter hunting base for the Russian-American Company. Within half a dozen years, however, the sea otters had been depleted by the Russians' proficient Aleut and Kodiak hunters, and following a brief experiment with shipbuilding, the post had concentrated on farming from the middle 1820's. But little surplus was produced for export to Sitka, owing to the insufficiency and inexperience of agricultural labour, the rugged terrain, and the cool, damp climate. Russian California went further and further into debt, and by the spring of 1839 the Russian-American Company admitted that its high hopes for farming there had been "completely dashed."[15]

By the late 1830's, too, American vessels had virtually abandoned the maritime fur trade of the Northwest Coast. Since the late 1780's they had competed with the Russians for the sea otters of the Gulf of Alaska, offering the natives more and better trade goods, including spirits and firearms. They had even incited the Tlingits against the Russians, who had already incurred the wrath of these most formidable of the Northwest Coast Indians by invading their territory. The Russians, undermanned and undersupplied, were unable to repulse the "Boston men." Besides, Sitka found the Americans to be a willing source of provisions and manufactures, particularly up to the late 1810's. After 1821, when California's ports were officially opened to foreign trade, ending the era of clandestine American traffic, the Yankee victualers were gradually replaced by the missions. And as sea otters disappeared from the coast, so did American vessels. By the late 1830's only one to three ships per year from the United States were trading at Sitka, compared with five to eight a decade earlier.[16]

The Americans were also discouraged by increasing British competition. Governor Simpson was determined to ensure his company's monopoly and oust the Yankee traders by the use of both posts and ships to trade superior goods or offer higher prices on the coast. Fort Simpson was erected at the mouth of the Nass River in 1831 and Fort McLoughlin on Milbanke Sound in 1833, and half a dozen vessels were acquired to service them. Soon the Bay men had replaced the Boston men as the rivals of the Russians for Tlingit and Haida furs. And again the Russians were outcompeted. Trade goods from St. Petersburg or Okhotsk cost the Russian-American Company twice as much as similar goods from London cost the Hudson's Bay Company. Also, the British were able to pay the coastal Indians two to three times as much for furs as the Russians.[17] And, complained the Russians, the British did not strictly observe the 1825 Anglo-Russian Convention's ban on the trading of liquor and ordnance to the Indians. (In fact nobody — Americans, British or Russians — did so.) Thus, in 1839 Governor Ivan Kupreyanov (1836-41) reported that "year by year I see the Tlingits drawn more and more from our places to the English frontier."[18]

This mounting rivalry reached a flashpoint with the *Dryad* or Stikine affair. In the spring of 1834, the Hudson's Bay Company's brig *Dryad* tried to ascend the Stikine River in order to found a post beyond the Russian-American Company's territorial ten-mile limit, but it was blocked at the mouth of the river by a Russian gunboat and the new Russian redoubt of St. Dionysius (whose supply of furs would be intercepted by a British port upstream). One indignant Bay man wrote: "I trust *John Bull* will answer these hairy beasts in their own way — and see justice is done to the *loyal and dutiful subjects* of *his Britanic* majesty."[19] The Hudson's Bay Company angrily claimed a breach of the Convention of 1825 and demanded £22,150 in damages. It also appealed to the Foreign Office for support.

In the process of trying to resolve this dispute, both sides came to realize the advantages of a general agreement that would allow them to share the coastal trade amicably at the expense of the pesky Americans. In September 1838 at St. Petersburg, Governor John Pelly of the Hudson's Bay Company proposed to Baron Von Wrangell, former governor of Russian America and now director of the Russian-American Company, that the latter's demand for manufactures and provisions be met by the Honourable Company, so that the American "Birds of passage" would be excluded from the coastal trade and consequently the sale of guns and spirits to the Indians would be reduced. This arrangement in turn would permit both companies to economize by decreasing the number of employees, posts, and ships that were needed for the trade. The Russian firm, however, hesitated because it was obtaining wares and victuals from Chile "on very moderate terms and by means of their own ships." Also, its second twenty-year charter was about to expire.[20] Then the tsarist government intervened. Perhaps under British pressure, the Russian Ministry of Finance ordered the Russian-American Company at the end of 1838 to reach a "peaceful agreement" with the Hudson's Bay Company. Von Wrangell contacted the Governor and Committee in order to secure "rapprochment with it, agreement on mutual interests, and avoidance of hostile conflict in the future."[21] The British company repeated its offer to supply Russian America annually with manufactures and provisions at moderate prices but suggested that they be provided in exchange for trading rights to the *lisière* or coastal strip of mainland extending north and south of the Stikine River. Such a lease would effectively eliminate both American and Russian competition from the coastal trade. At the beginning of 1839, the Russian-American Company, with the tsar's approval, sent Von Wrangell abroad to negotiate an agreement. He and Simpson met halfway at Hamburg, where on 6 February they signed an accord. Under its terms, which took effect on 1 June 1840 for a period of ten years, the Russian-American Company agreed to abandon St. Dionysius Redoubt at the mouth of the Stikine, to discourage American vessels from visiting Russian America to sell goods (except in emergencies), and to lease the *lisière* between Cross Sound and Chatham Sound (54°40' N.) for an annual rent of 2,000 land otters.[22] The

Hudson's Bay Company agreed to abandon its claim for damages resulting from the *Dryad* affair. It also agreed to deliver annually to Sitka 2,000 land otters from the western side of the Rocky Mountains at 23 shillings each and 3,000 land otters from the eastern side at 32 shillings each, as well as 8,400 bushels of wheat (4,200 bushes in 1840) at 6-2/3 shillings per bushel, 8 tons of wheat flour, 6-1/2 tons of barley groats, 6-1/2 tons of peas, 15 tons of salted beef, 1-1/2 tons of ham, and 8 tons of butter, plus sundry manufactures at £13 per ton.

The Hudson's Bay Company's farms at Forts Vancouver and Langley could not meet all of the food demands of the "Russian contract," so agriculture had to be expanded.[23] Thus was born the Puget's Sound Agricultural Company. More correctly, thus was McLoughlin's beef and tallow company finally realized, for the contract was to be met largely by the agricultural output of Forts Vancouver and Langley, with the Puget's Sound Agricultural Company catering mostly to the British market and, it was hoped, to the Sandwich, or Hawaiian, Islands as well.[24]

5

SUCCESS AND FAILURE: THE PERFORMANCE OF THE PUGET'S SOUND AGRICULTURAL COMPANY

> *It cannot fail of proving a highly profitable speculation.*
> ASSISTANT SURGEON SILAS HOLMES
> OF THE U.S. EXPLORING EXPEDITION, 1841

> *I wish I had Kept as Clear of the Puget Sound business, as I did of that of the Silver Plate to his Excy [Governor Sir George Simpson]. That Sheep Concern which I, with many others, joined in an evil hour, will I fear involve us all they have already 50£ of my money and it is said we are all in for 200£ more, without the slightest prospect so far of a farthing in return. . . . It has been instituted I believe for no other purpose but to glean the little from us which we may happen to save in the Fur Trade.*
> CHIEF TRADER JOHN TOD TO
> EDWARD ERMATINGER, 1842

After concluding the "Russian contract" with Director Von Wrangell in Hamburg in early February of 1839, Governor Simpson returned to London, where at the end of the same month, together with Governor John Pelly, Deputy Governor Andrew Colvile, and Chief Factor John McLoughlin, he attended the formative meeting of the Puget's Sound Agricultural Company. They approved the prospectus of a joint-stock association "for the rearing of Flocks and Herds on an extensive scale, with a view to the production of Wool, Hides and Tallow, and for the cultivation of agricultural produce."[1] The main object was to be the production of wool, hides, and tallow, not crops or dairy products, which were considered less profitable.[2] The association's capital was £200,000, divided into two thousand shares of £100 each. It was founded as a separate concern because the Governor and Committee believed that "agricultural pursuits on a large scale cannot conveniently be combined with the fur trade."[3] In other words, they did not want agriculture to interfere with the fur trade at the Hudson's Bay Company posts, whose servants were not necessarily proficient in farming.[4] Beaver House even expected that fort farming would decline as Puget's Sound Agricultural Company operations expanded, so that "the fur trade should depend more on that Association for its supplies in grain, beef, Pork etc. for the Coasting trade and the Sandwich Islands market than on its own productions."[5] The Governor and Committee elaborated its view to Simpson in 1841:

> we beg it may be understood, that it is not our desire that agricultural pursuits, say both tillage & grazing should be prosecuted at Fort Vancouver or any of the trading establishments on the West side the Mountains, beyond what may be necessary to meet the demands of the trade, that is the maintenance of the establishments, transport, provisions & supplies for our shipping; ... and it is our desire that all farm produce required for supplying contracts with the Russians, for meeting the demands of the Sandwich Islands and for an export trade to this country, in wool, hides, tallow etc be procured from the Pugets Sound Agricultural Company, as we are quite sure that the Fur trade and farming pursuits, branches of business so foreign to each other, will be done more justice to and be much more likely to prosper under distinct managements and separately attended to, than if combined, and conducted by persons, who from being totally inexperienced, can have little or no Knowledge of farming operations.[6]

Although *de jure* the two companies were separate, *de facto* the crop growing and stock raising enterprise was a subsidiary of its fur trading parent.[7] The Puget's Sound Agricultural Company's first directors in London were Pelly, Colvile, and Simpson, who had exclusive direction of its affairs, and its first superintendent or manager in the field was McLoughlin. Half of its shares were reserved for directors and stockholders of the Hudson's Bay Company and their friends, and half for servants of the same company. The Governor and Committee stressed that the fur trade was to remain paramount:

> we have made it an express condition as a protection to the Fur trade that the principal superintendence of its affairs, shall be vested in a Gentleman interested in the Fur Trade, and that neither the Agricultural Association nor any of the people who may be taken to the Country for its service or employed in its affairs, shall have any dealings, either directly or indirectly, with Indians or others at variance with the interests of the Fur trade.[8]

The Hudson's Bay Company had agreed to sell land, livestock, and seed to the Puget's Sound Agricultural Company and to recruit farmers. In March 1839 the new company's agents instructed Superintendent McLoughlin to begin operations at Cowlitz (just northeast of present-day Toledo, Washington) and Nisqually (just west of present-day Fort Lewis, Washington) at either end of the Cowlitz Portage, a locale which, they asserted, "appears well suited for the operations of the Company, inasmuch as it combines all that is necessary for pasture and Tillage farms, and is accessible by water, both by the Columbia River and the straits of De Fuco."[9] This location would strengthen the British claim to the right bank of the Columbia, which, it was confidently assumed, would eventually become the boundary between British and American territory.[10] Farming was to be commenced at Cowlitz, which had been established in the fall of 1838, when James Douglas sent "Mr Ross & eight men with a number of agricultural implements" and ninety-five cattle there from Fort Vancouver.[11] The property was sold to the Puget's Sound Agricultural Company in 1839. A "large Stock Farm" was to be established in the "best situation" on the banks of the Cowlitz River. Here, McLoughlin was told,

> it is desirable to break up and lay under crop as much land as convenient with the least possible delay, in order to maintain your people, to relieve the Hudson's Bay Company of a contract for agricultural produce which they have entered into with the Russian American Company, and to make provision for settlers that will be sent from England as early as you can conveniently receive them.[12]

Nisqually was not transferred to the Puget's Sound Agricultural Company until the winter of 1840-41.[13] Stock rearing was to predominate at Nisqually because its "light and dry" soil was more amenable to grazing than tilling. Also, its site was more accessible to shipping than that of Cowlitz, so that time would be saved and damage would be minimized in unloading sheep and loading wool.[14] The agents believed that

> the country in the neighbourhood of this establishment is better adapted for pastoral than tillage farming, as the soil is light and thin; we therefore think you should there confine your attention principally to the rearing of flocks and herds, cultivating no more ground than may be necessary to maintain the establishment and provide Mangel Wurzel etc for the sheep and cattle.[15]

Because the natural pasture at both farms was too coarse for sheep, McLoughlin was ordered to burn the "long dry grass" in the fall or spring and then reseed with clover and "other artificial grasses" in wet weather after the fields had been closely grazed to produce a "close sward" or a "finer ... grass." Reseeding was necessary "as the quality of the wool and condition of the sheep depend principally on the Pasturage."[16] McLoughlin was also instructed to build sheds for sheltering sheep and for storing fodder crops.

The Hudson's Bay Company promised to transfer to the farming firm "their entire stock of sheep, and as many of their stock of cattle as can be dispensed with, say by the 1st of June 1840 about 2,000 head of sheep, and about a thousand head of cattle with any Horses, Pigs and other live stock you may require."[17] Half of the cattle and all of the sheep, including any brought from California, were to be kept at Nisqually. "As early as possible" the London ship was to transport 2,000 to 2,400 sheep in three trips from California (taking care "to guard against suffocation in the hold"), and 1,000 cows were to be driven from the same source in 1841 by the returning Bonaventure (Sacramento) River trapping expedition.[18] The Californian ewes were to be bred to Merino and Leicester rams, and the crossbred lambs were to be taken to Cowlitz, where "none except improved breeds" were to be kept. It was expected that the livestock would multiply to 10,000 head of sheep and 2,000 head of cattle by the end of 1841.[19] The Puget's Sound Agricultural Company's surplus output of livestock and crops was to be sold to the Hudson's Bay Company.

At Cowlitz "the necessary complement of people ... for the purpose of simultaneously erecting buildings, fencing and breaking up land, and watching your flocks and herds," a total of twenty-four men, "with any additional assistance in Indian herdsmen you may require," were to be transferred temporarily from the Hudson's Bay to the Puget's Sound Agricultural Company "until they can be replaced by others that may be sent from this country from time to time by the annual ship."[20] At Nisqually, nine men, including four herdsmen, were to be maintained. In addition, when the sheep and cattle had increased to 12,000 head by the end of 1841, the agents intended "to send a few respectable farming families from this country, each family to be accompanied by two or three labouring servants under engagements, with the view of gradually forming an European Agricultural Settlement."[21] These colonists were to be settled on halves; each family was to be leased 1,000 acres of land (including 100 acres fenced and cultivated) with a house and a barn, livestock (500 sheep, 29 cattle, 6 horses, and "a few" pigs) at fixed prices, necessary implements, and one year's provisions (repayable) in return for one-half of the output. McLoughlin was urged to employ Indians, especially as herdsmen, "as a means of affording the benefit of cheap labour," and he was advised to keep Indian wages low.[22] It was hoped that eventually most agricultural operations would be performed by settlers on halves, with company activity being confined to Nisqually, which would serve as a model farm, exporting and marketing the output of the settlers.[23] Thus, the company was intended to specialize in livestock products for the British market, since Nisqually was to be primarily a pastoral operation. Accordingly, at the end of 1839, McLoughlin was instructed to meet the new company's current expenses from sales of flour, pork, and butter in Russian America, the Sandwich Islands, and elsewhere until its flocks and herds had multiplied to the point where returns could come from the export of wool, hides, and tallow.[24] And all of this new agricultural activity, the Hudson's Bay Company hoped, would enable its posts to confine their agricultural operations to the demands of the fur trade alone, so that they could concentrate on

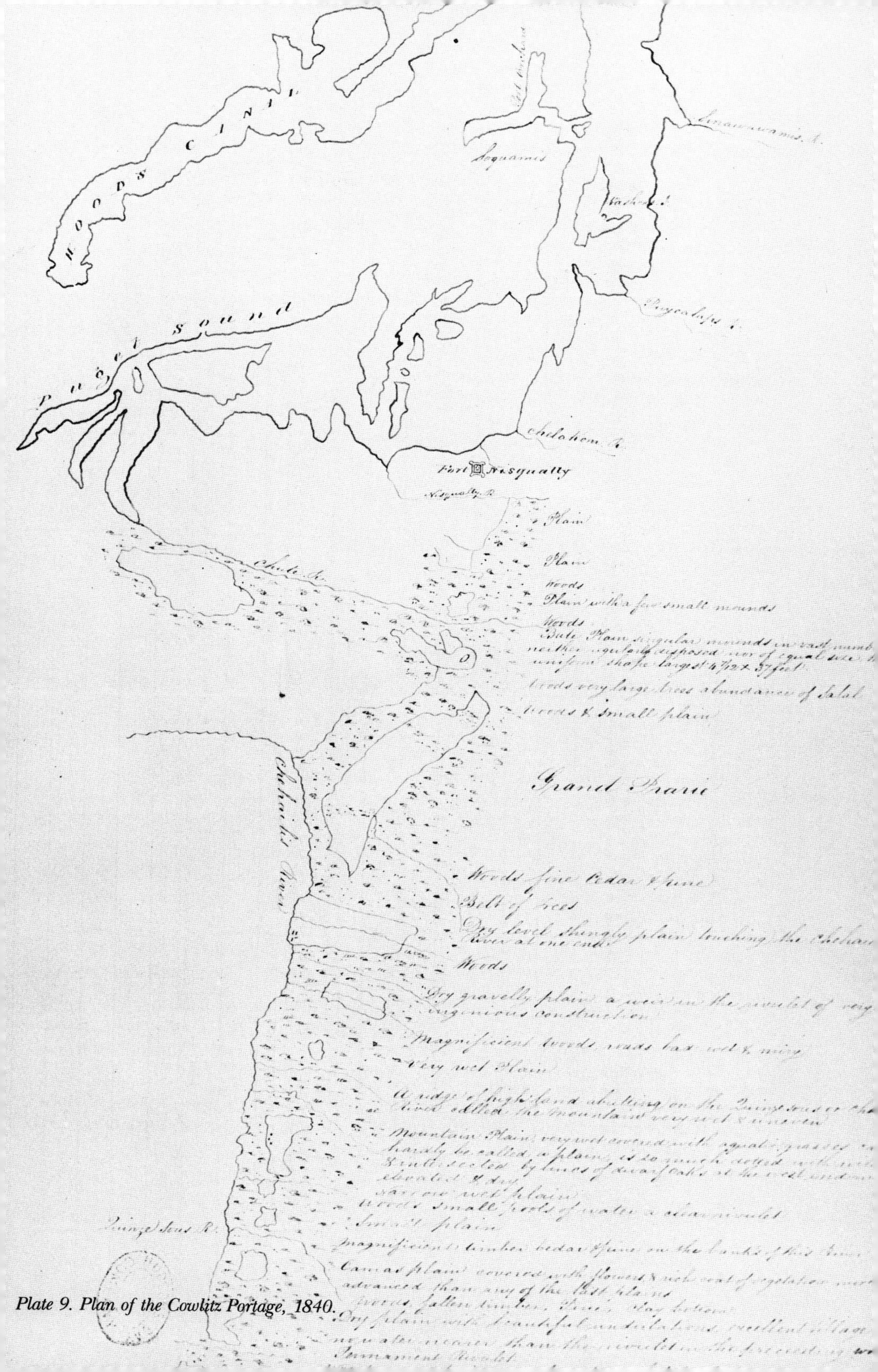

Plate 9. Plan of the Cowlitz Portage, 1840.

the latter in the face of declining returns. At the end of 1839 McLoughlin was ordered

> to reduce the tillage farming operations of the Fur Trade as early as possible, in measure as you can increase those of the Puget's Sound Company, as it is intended that the Russian American Compy's contract and the export trade to the Sandwich Islands shall be supplied by the Puget's Sound Company when you may be in a condition to meet those demands, and that the Hudson's Bay Company's farming operations shall be confined to the demands of the Fur Trade alone.[25]

This, then, was the scenario, and McLoughlin with his customary diligence undertook to make it a reality.[26] The locale was the Cowlitz or Puget Sound or Mountain Portage, a rolling interfluve between the head of Puget Sound and the estuary of the Columbia River and between the Coast Ranges and the Cascade Mountains (Plate 9). The coniferous forest cover was interspersed with grassy openings known as "plains" or "prairies." Some of these prairies were dotted with circular mounds (the "mima mounds" of uncertain geomorphological origin). Prairie de Bute, for example, which was located between the Nisqually River and Mud Mountain, was described by Paul Kane in 1846: "This is remarkable for having innumerable round elevations, touching each other like so many hemispheres, of ten or twelve yards in circumference, and four or five feet in height."[27] In 1839 Father François Blanchet, a Catholic missionary, counted eighteen prairies in the sixty miles between Cowlitz and Nisqually. The next year McLoughlin reported that the prairies between the Cowlitz River and the portage's water divide totalled five to six thousand acres and that those between the divide and Puget Sound each measured fifteen to twenty miles long and five to fifteen miles wide, with light, gravelly soil.[28] Here, in the words of William Tolmie, "the Puget Sound Company occupied 144 or 160 square miles of alternate Prairie and woodland, prairie predominating."[29]

Cowlitz Farm was situated on Cowlitz Prairie on the right bank of the Cowlitz River twenty-five to thirty-five miles upstream from the Columbia (Plate 10). The prairie lay five hundred yards from the river and a mile from the landing.[30] It was at least four miles in length in a northeast-southwest direction and nearly one mile in width, and two-thirds of it were "very level."[31] Cowlitz Prairie encompassed three thousand acres of unforested land; half belonged to the company and half to a Catholic mission and some private settlers. Douglas described the site in early 1839:

> The site chosen for the Grazing Farm, is 36 miles north of the Columbia River, at the east end of the Cowelitz Portage, in the midst of a rather extensive Plain, and the buildings are erected upon an angular projection of the second level [terrace] above the Cowelitz River, commanding a

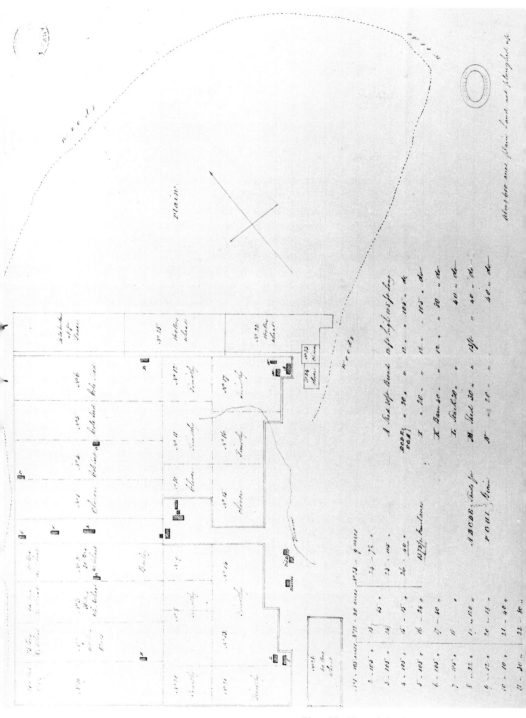

Plate 10. Plan of Cowlitz Farm, 1844-45.

distinct view of every acre of arable land in the neighbourhood, and, in every respect decidedly the most eligible situation that could be found: although not free from the disadvantages common to all the Cowelitz Plains, of being separated from the River by a steep, rugged, hill impracticable in its present state, to wheeled carriages: and the excavation of a convenient road, will be an enterprise attended with great labour and expense.

The entire Plain contains a surface of about 3000 acres of clear land; the half of which has been laid out in small farms, and ceded to the Catholic Missionaries and our retired servants, leaving a reserve of 1500 acres for our own use. The soil is of the best quality for growing wheat, consisting of a fine rich loam, running at the depth of 15 inches, into a subsoil of stiff clay.

Wood and water for every purpose of utility are found at a few hundred yards from the door, and there is no stream of sufficient power, within a convenient distance, to propel machinery.[32]

Douglas rated the prairie "probably the finest tract of tillable land in the Indian Country," a surprising assessment in view of the obvious superiority of the Willamette Valley.[33]

Farming had been initiated at Cowlitz in 1833 by two retired Canadian engagés, François Faignant (Failland) and Simon Plamondeau (Plamondon), first-rate axe men who had erected the company's wooden forts in the Columbia Department.[34] Their "two small farms" were still there in 1839, by which time they had been joined by at least two other Canadian farmers, Michel Cognoir and Joseph Rocbrune.[35] Company operations began in the autumn of 1838; a year later Chief Trader John Tod had cultivated 200 acres, "broken up" another 135 acres, and planted 275 bushels of wheat. By the end of 1839, 230 acres had been fenced and seeded (mostly to wheat), and another 320 acres had been ploughed and fallowed. By the spring of 1840, six hundred acres had been "broken up," houses had been built, and a "considerable quantity" of livestock had been brought from Fort Vancouver and California.[36] A year later a large dairy had been completed and a gristmill and a sawmill were under construction.[37] The cultivated acreage expanded rapidly from six hundred acres in the spring of 1841 to one thousand acres in the fall of the same year and fourteen hundred in the fall of 1845 and spring of 1846.[38] Speaking before a joint British-American committee formed in the middle 1860's to adjudicate Hudson's Bay and Puget's Sound Agricultural Company claims in Washington and Oregon Territories, a onetime employee recalled that:

At Cowlitz, the Company had, in 1846, and for many years previously, a large, arable farm, with pasture land on three sides of it; I suppose it consisted of about four thousand acres of land, but it has not, to my knowledge ever been accurately measured. In 1846 there were about twelve hundred acres enclosed and subdivided by fences and ditches, into fields of convenient size, say from fifty to one hundred acres. Portions of this land were laid down under cultivated grasses, and the pastures were fully stocked; indeed, in 1846, the Company had flocks of sheep on the public lands to the north of their claim, and in 1846, as well as for several years previously, they regularly pastured their horses on a prairie, on the south or left bank of the Cowlitz River, opposite their farm. In 1846 they had, at Cowlitz, about eight hundred head of cattle; about one thousand sheep; about a hundred and twenty head of horses; about three hundred large and half grown hogs, besides the young of that year. They had a comfortable, commodious dwelling house; a large two-story granary, with barns and sheds, conveniently distributed at various points over the farm. They had a wagon road to the bank of the Cowlitz River, made at considerable cost. They had a gang saw-mill on the timbered land, near the Cowlitz River, within their claim; also dwelling houses for their servants, at various points on their farm, stables, cow sheds and piggeries.[39]

Even allowing for exaggeration (intentional or otherwise) on the part of this witness for the claimants, his picture is that of a large-scale and well-equipped enterprise. It employed twenty-four men, who were supervised by Charles Forrest (1841-47).[40] A "large number" of Indian men and women were hired at harvest time; in July 1840, for example, forty Indian cradlers were employed to cut grain.[41]

Cowlitz Farm specialized in grain growing. Its clayey loam, "a mixture of Clay, Sand and decayed vegetable matter," was more suited to cultivation than the thin, stoney soil at Nisqually Farm. Cowlitz's primary function was to fulfil the grain requirement of the 1839 agreement.[42] Spring wheat, barley, and oats and winter wheat were grown, as well as buckwheat, peas, potatoes, turnips, colewort, flax, clover, and timothy. In order to avoid grain loss from "shake" (the dislodging of the kernels from the ears through overripeness) the different grains had to ripen in succession rather than simultaneously. Thus, peas were planted before 20 March (and matured after 120 days), oats before 10 April on "low moist land" (and ripened after 137 days), spring wheat before 10 May on "rich and well tilled soil" (and matured after 127 days), barley before 20 June (and ripened after 103 days), and fall wheat before 20 October.[43] In 1845 wheat occupied one-half of the farm's cropland.[44] Generally about 10,000 bushels of grain, mostly wheat, 1,000 to 2,000 of peas, and 1,000 to 2,500 of potatoes were produced annually (Table 14). Wheat yielded ten bushels per acre in 1840 and from fifteen to twenty in 1841, when it weighed sixty-three to sixty-eight pounds per bushel and was deemed "excellent" in quality. In 1843 grain yielded sevenfold, peas nearly fourfold, and potatoes nearly elevenfold.[45] These yields enabled the Puget's Sound Agricultural Company to sell 4,000 to 6,000 bushels of wheat annually to the Hudson's Bay Company (Table 15). In none of these years, however, was output sufficient to meet the "Russian contract," which stipulated 8,400 bushels yearly. In 1845 McLoughlin admitted that the agricultural company did not produce enough wheat to satisfy the terms of the contract, so that the parent company had to meet the deficiency.[46] The shortfall was obtained from Fort Vancouver's farm, settlers in the Willamette Valley, and even California. In 1840, for example, McLoughlin had to buy four thousand bushels of wheat in California in order to satisfy all of his needs.[47] The Willamette purchases were crucial, as Simpson acknowledged in 1844, when he informed the Governor and Committee that grain production at Cowlitz Farm and Fort Vancouver together was such that "without purchasing, we should not have sufficient for our own demands."[48] Company harvests and purchases combined provided ample grain for export. In late 1843 McLoughlin estimated his wheat stock at 25,300 bushels, comprising 7,300 bushels on hand, a crop of 3,000 bushels from Fort Vancouver, a crop of 5,000 bushels from Cowlitz Farm, and an expected purchase of 10,000 bushels from the Willamette settlers; of this total, 15,300 bushels were needed locally and for Russian America, and the remaining 10,000 bushels were to be sold in the Sandwich Islands.[49] And in late 1845 McLoughlin reported that the Hudson's Bay Company had 30,000 bushels of wheat in store at Fort Vancouver and in the Willamette Valley and 10,000 bushels at Cowlitz Farm; after supplying the Russians, 6,000 barrels of flour would remain.[50] From 1841 the Hudson's Bay Company even had enough grain to be able to offer to provision Kamchatka via the Russian-American Company. The Russian firm accepted, for under the terms of its third charter of 1842 it was obliged to victual the peninsula. Therefore, from 1842 the Honourable Company supplied Sitka with an extra ten tons of flour (Table 16).

Table 14

OUTPUT OF GRAINS, PEAS, AND POTATOES AT COWLITZ FARM, 1840-44 AND 1846

Year	Grains	Peas	Potatoes
1840	8,000 bus.[a]	? bus.	? bus.
1841	11,000-12,000[b]		?
1842	4,800	464	1,900
1843	9,174	900	2,574
1844	10,370[c]	1,000	?
1846	9,000	2,000	1,000

For sources see *Sources for Tables*.
[a]All wheat.
[b]Including from 7,000 to 8,000 bushels of wheat (Dunn, *History of the Oregon Territory*, p. 214; HBCA, D.4/59: 85, D.4/110: 23; Schafer, "Letters of Sir George Simpson," p. 78; [Wilkes], *Diary*, p. 26; Wilkes, "Oregon Territory, p. 19).
[c]Including 7,000 bushels of wheat (HBCA, D.4/67: 85v.).

Table 15

SALES OF WHEAT FROM COWLITZ FARM BY THE PUGET'S SOUND AGRICULTURAL COMPANY TO THE HUDSON'S BAY COMPANY, 1840-44 AND 1846

Year	Wheat	Year	Wheat
1840	2,500 bus.	1843	5,591[a]
1841	4,500	1844	7,000-8,000
1842	4,161	1846	5,349[b]

For sources see *Sources for Tables*.
[a] Plus 360 bushels of oats and 302 bushels of peas (HBCA, F.12/2: 33).
[b] Plus 807 bushels of oats (HBCA, F.8/1: 63).

Table 16

SHIPMENTS OF PROVISIONS TO SITKA BY THE HUDSON'S BAY COMPANY, 1840 AND 1843-49

Year	Wheat	Flour	Biscuit	Peas	Pork	Beef	Butter
1840	3,206 bus.	0 lbs.	0 lbs.	0 lbs.	0 lbs.	0 lbs.	2,664 lbs.
1843	9,314	41,200	0	17,584	0	33,456	5,448
1844[a]	15,600	39,200	0	17,300	6,000	33,725	4,592
1845[b]	8,016	39,200	0	17,300	0	33,600	4,584
1846[c]	17,021	79,968	0	16,349	0	32,400	3,360
1847	13,068	51,548	336	16,240	3,000	33,400	3,416
1848	8,446	51,352	336	19,775	4,500	32,400	3,360
1849	7,875	51,352	336	19,110	0	29,600	3,360

For sources see *Sources for Tables*.
[a] Plus 2 kegs of pickles, 269 pounds of dried apples, and 1,878 gallons of molasses (HBCA, B.223/d/158: 76d).
[b] In fact, 20,000 bushels of wheat may have been delivered in 1845 (HBCA, B.223/z/4: 208; "Papers Relative to the Expedition of Lieuts. Warre and Vavasour, p. 82v).
[c] Plus 3,154 pounds of mutton (HBCA, B.223/d/166: 9d).

Although Cowlitz Farm produced mostly grain, peas, and potatoes, some livestock were kept (Table 17). The cattle in particular were probably more than sufficient for local needs, so that many of them may have been sold annually to the Hudson's Bay Company for beef, tallow, and hides. In the spring of 1841, Lieutenant Wilkes was told by Superintendent Charles Forrest that because of the mild climate the cattle were only "partially" housed and "little or no" hay was made, the grass being "sufficient" year round, although during dry spells they had to be driven "a distance" for pasture.[51] In late 1841 the agents instructed McLoughlin how to turn the virgin land at Cowlitz and Nisqually Farms into productive pasture by rotation: first one or two crops of oats, then one crop of coleseed (to be grazed and manured by sheep), then one crop of wheat, then another crop of coleseed or turnips, then another crop of wheat, then yet another crop of coleseed or turnips, then a crop of barley and grass together, followed by grazing for three or four years. McLoughlin implemented this rotation system at Cowlitz Farm, where he hoped to treble pasture yields.[52]

Table 17
NUMBER OF LIVESTOCK AT COWLITZ FARM, 1839-42[1]

Year	Cattle	Horses	Pigs
1839	174	24	0
1840	122	40	1
1841	206	?	?
1842[a]	290	148	36

For sources see *Sources for Tables*.
[1]These figures represent spring populations, except for autumn 1841.
[a]Plus 2 mules (HBCA, F.8/1: 33, F.12/1: 564).

Stock rearing loomed much larger at Nisqually Farm (Plates 11 and 12), which had existed since 1833 and was transferred to the Puget's Sound Agricultural Company in the winter of 1840-41. The farm occupied a "boundless & picturesque prairie" on the right bank of the Coe, or Nisqually, River.[53] Dotted with islands of forest, this Grand Prairie or Nisqually Plain extended thirty miles east-west and twenty miles north-south.[54] Such prairies along Puget Sound were the result of infertile shingly soil and excessive drainage, plus annual midsummer burning by Indians, anxious to maintain both camas fields and game pastures. Nisqually Farm's sandy to gravelly soil was correctly reckoned more suitable for grazing than tilling. After reconnoitering the locale, Simpson reported to the Governor and Committee in 1841 that there was "a large extent of fine pastoral land . . . covered with a tufty, nutritious grass," although the soil, "being light and shingly, is not so well adapted for tillage."[55] Douglas praised the dry underfooting and the nutritious pasture for livestock, although he admitted that the grass was short, thin, and tufty and yielded no more than one-sixth that of ordinary cultivated land in England.[56] A member of the U.S. Exploring Expedition nevertheless asserted that "this country . . . is perhaps one of the best for grazing in the world." He added:

> Every where in this part of the country the Prairies, open wide, covered with a low grass of a most nutritious Kind which remains good throughout the year. In September there are slight rains, at which time the grass starts; and in Oct. and November there are a good Coat of green grass, which remains so until the ensuing summer; and about June is ripe in the lower

plains, and drying without being wet is like made hay; in this state it remains until the autumn rains begin to survive it.[57]

He was unaware that in summer climatic and edaphic aridity could sear the grassland.

The Hudson's Bay and Puget's Sound Agricultural Companies occupied roughly 140 square miles of this seaboard by the end of 1846. Their property ran from the mouth of the Nisqually River up the coast to a point two miles north of the mouth of the Steilicoom River, from there twelve miles northeastward and then southward to the Nisqually River twelve miles from its mouth, and from there downstream to the sound.[58] In 1865 the "sheep concern" described its Nisqually establishment glowingly to the British and American Joint Commission:

> At Nisqually, the Company had about eight thousand sheep in 1846, divided into flocks of about five hundred each, herded by white men, Indians and Sandwich Islanders, and regularly moved from one pasture to another, so as to prevent destruction of the grasses by over-feeding. They had also in the year 1846, very nearly three thousand head of horned cattle, which were herded by white men, with Indian assistance, and which cattle were then tame and easily driveable from place to place, and from two hundred and fifty to three hundred horses, also driveable and easily managed. At this time they had full and undisturbed possession of their claim. The Company had several small farms, in charge of their head shepherds and cattle herdsmen, at convenient points; they had at the main station, known as Fort Nisqually, a farm of several hundred acres enclosed, and on the clearing and draining of the swampy portion of which considerable outlay had been incurred prior to 1846. The Company's business was, in 1846, in a flourishing condition — the early difficulties, incident to its commencement, having by that time been overcome. They had, by 1846, made at considerable expense, wagon roads through the different belts of timber existing on the claim, and two wagon roads to the sea beach, involving much grading, one at which is now known as Nisqually Landing, and the other at the mouth of Steilacoom Creek. They had cleared many of the swamps scattered over the claim, for the sake of the hay to be obtained from them. At Fort Nisqually they had a dwelling house, for the officer in charge, and several dwelling houses for the laborers, bastions for defence, and two large stores, one at the landing and the other at the fort — each a story and a half in height, and 60 x 40 feet in dimensions — barns and outhouses of various kinds; also a large house, with convenient parks surrounding it, for shearing sheep, and a dam on the Segwalitchew Creek, for washing the sheep in. All the dwelling houses, stores, bastions, and the

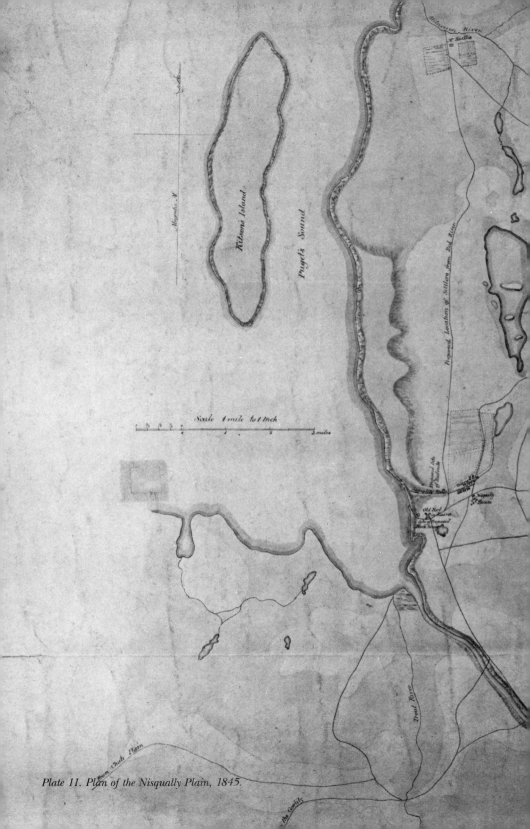

Plate 11. Plan of the Nisqually Plain, 1845.

Eye Sketch of the Plains &c about Nisqually
at the Head of Puget's Sound

Woods Colored Green.
The Roads are merely Indian Paths

Thick Woods commence here

To Walla Walla

Sheep Hdd.

Douglass River

Nisqually River

M Vavasour
Lieut Royal Eng.rs

Plate 12. View of Nisqually Farm, 1845.

sheep shearing house were constructed of square timber — French Canadian fashion. At the out-stations already mentioned, the Company had dwelling houses for their overseers and other employés, and from twenty to forty acres enclosed at each.[59]

Seven men were employed in the fur trade and four in farming as shepherds and herders in 1841, when Alexander Anderson was superintendent. (He was succeeded by William Tolmie in 1842 at the same time that Forrest was put in charge of Cowlitz Farm.)[60] But from one-half to two-thirds of the farm work was done by Indian labourers. They were paid from £4 to £8 per year, plus provisions; the British shepherds received £35 per annum.[61]

The farm hands were preoccupied with livestock, for Nisqually Farm was the pastoral arm of the Puget's Sound Agricultural Company, just as Cowlitz Farm was the arable arm. Stock were obtained from Hudson's Bay Company posts in the Columbia Department, Mexican ranchos in California, and farms in England. The first sheep came from California in the summer of 1838, when the barque *Nereide* reached Nisqually Farm from San Francisco with 634 head, having lost 166 in passage. In 1840 Alexander Simpson, the governor's cousin, bought 700 Californian ewes for the "sheep concern," but half of them died soon after arriving by sea.[62] Finally, in the winter of 1840-41, Douglas purchased 3,670 ewes and 661 cows in California (the former at $2 each and the latter at from $5 to $6 each, payable half in money and half in goods), and in the spring they were driven to the Columbia by thirty drovers. After three arduous months they reached Fort Vancouver with a loss of 470 sheep and 110 cattle, suffered mostly in crossing rivers and ravines. (One Hudson's Bay Company report stated that 700 sheep and 80 cows were lost en route.) The surviving 3,200 ewes and 551 cows were forwarded to Nisqually Farm.[63]

The first cattle came from Fort Vancouver in the summer of 1839, when seventy head were driven over the Cowlitz Portage. Two years later 111 cattle were sent from Fort Colvile. And in early 1842 all spare Hudson's Bay Company cattle at Forts Colvile, Okanagan and Nez Percés, amounting to 283 head, were driven in a body to Nisqually Farm via the Yakima Valley and White Pass.[64]

Breeding stock were brought from England on the annual London ship to improve Nisqually Farm's flocks (and obviate the costly drives from California). At the end of 1839 twelve purebred Leicester and Merino sheep (six rams and six ewes) with two Scottish shepherds were sent on the barque *Columbia*, and in 1840 twenty-four more of the same breeds were shipped on the barque *Cowlitz* and the barque *Vancouver*.[65]

The Californian ewes (degenerated Merinos) were crossed with the English rams. They were kept in several flocks, which were shifted daily from pasture to pasture. The flocks were folded nightly in "parks" (pens), which were moved periodically to manure as much land as possible and to make the sheep more comfortable.[66] The entire population was moved quarterly between the southern or Nisqually range and the northern or Puyallup range until the summer of 1844; thereafter, it was divided into two groups of five flocks and three flocks that were alternated semi-annually between the two ranges.[67] The shepherds, who were under the direction of Mr. Lewis, one of the Scottish hirelings, lived with their flocks in temporary huts and a house on wheels.[68] Lambing began in late March or early April and lasted six weeks. Shearing, which started in late May or early June right after lambing, likewise lasted six weeks. The wool was "picked" (combed) and sent to Fort Vancouver for shipment to England.

All of the cattle, except the seventy milch cows, were penned at night for protection from wolves and were moved every week to manure the soil and accustom the animals to being driven and herded; the cowpens were made of pine and formed in the manner of Virginia rail fences. Mounted Indian herders guarded and mustered the cattle. Two-thirds of the milch cows were left with their suckling calves; the rest were milked at a dairy some five miles from the fort. According to Wilkes, Nisqually Farm's livestock "thrive uncommonly well."[69] They increased nearly sixfold in number between 1840 and 1842 (Table 18).

Even more impressive was the propagation of all Puget's Sound Agricultural Company livestock (Table 19). They doubled between 1841 and 1842 and again between 1842 and 1846. Every year saw a sizable increase, except between 1842 and 1843, when five hundred sheep succumbed to an unknown disease during the wet season (autumn-spring) and two hundred of the cattle kept at Fort Vancouver strayed en route to Cowlitz and Nisqually Farms.[70] All of the company's sheep and most of its cattle were held at Nisqually, and all of its horses, pigs, and mules and the rest of its cattle at Cowlitz (except in 1841 and 1842, when some sheep and cattle were kept at Fort Vancouver). All of Nisqually Farm's commercial output consisted of animal products: wool, hides and skins, and horn for England and butter for Russian America. London was the main market (Table 20).

Table 18

NUMBER OF LIVESTOCK AT NISQUALLY FARM, 1840-42[1]

Year	Sheep	Cattle
1840	820	82
1841	2,342	?
1842	4,194	924

For sources see *Sources for Tables*.
[1]These figure represent spring populations.

A "large supply" of butter for the Russian-American Company was churned at Nisqually's "filthy" dairy.[71] The butter clause was the only stipulation of the 1839 contract that worried McLoughlin because, in his words, "our cows are very bad milkers," some of them not giving more than one pint of milk per day. He tried to solve the problem by milking as many cows as possible at as many posts as possible. Five dairies were established at Fort Vancouver, two at Fort Langley, and one each at Nisqually Farm and Cowlitz Farm.[72] McLoughlin's concern was well founded, for usually less than half the required amount of eight tons was shipped to Sitka (see Table 16). From 1843 Nisqually Farm also supplied the new fort at Victoria with beef and mutton.[73]

From at least 1843 it was not intended to cultivate any more land at Nisqually Farm than was necessary to feed the establishment's personnel and livestock, since the soil was "not so well adapted for tillage." It had been Douglas's hope in 1839 that the farm would become the "grannary of the north" through the improvement of its soil by grazing and manuring. That hope was not realized, however. In 1841, 100 acres were ploughed and winter and spring wheat, oats, barley, peas, and potatoes were planted. By 1846 the cultivated acreage had more than doubled to 220 acres.[74] Not surprisingly, yields were low (Table 21). They averaged twelve and one-half bushels per acre in 1841-42. (Wilkes asserted in 1841 that there were 200 cultivated acres that yielded fifteen bushels

Table 19
NUMBER OF PUGET'S SOUND AGRICULTURAL COMPANY LIVESTOCK, 1841-46[1]

Year	Sheep	Cattle	Horses	Pigs	Mules	Total
1841	2,342	649	12	0	0	3,003
1842	5,295	1,414	148	36-42	2	6,895-6,902
1843	5,888	1,501	176	80	3	7,648
1844	6,996	1,921	186	136	2	9,241
1845	8,833	2,436	299	182	2	11,752
1846	10,578	3,063	343	?	2	>13,986

For sources see *Sources for Tables*.
[1]These figures represent spring populations.

Table 20
SHIPMENTS OF PUGET'S SOUND AGRICULTURAL COMPANY WOOL, SHEEPSKINS, CATTLE HIDES, AND CATTLE HORNS TO LONDON, 1842-47

Year	Wool	Sheepskins	Cowhides and Oxhides	Cowhorns and Oxhorns
1842	7,270 lbs.	0 pieces	0 pieces	0 pieces
1843	6,263	0	0	0
1844	7,682-8,621	608	0	470
1845	10,842-11,845	160	75	133
1846	14,624	137	95	24
1847	16,569	1,926	158	431

For sources see *Sources for Tables*.

per acre. He possibly deliberately overstated the agricultural potential of the Nisqually area in order to maximize its worth to Washington during the boundary talks. Certainly he stressed the naval and mercantile importance of American access to Puget Sound's harbours.)[75] According to Douglas, until 1839, when the ploughland was first manured, crop yields seldom exceeded twofold "through the unproductive character of the soil, which is hardly fit for tillage, in its natural state." McLoughlin declared that wheat never yielded more than threefold and grain usually from threefold to fourfold (in fact, between fourfold and fivefold). He added that no grain at all would be grown at Nisqually Farm were it not needed for sheep fodder in wet or cold weather, owing to the scarcity of meadow grass and the failure of timothy and clover plantings.[76]

Table 21

ACREAGES, SOWINGS, AND HARVESTS OF GRAINS, PEAS, AND POTATOES AT NISQUALLY FARM, 1841-44

Year	acreage	Sowing (bus.)			Harvest (bus.)		
		grains[a]	peas	potatoes	grains[a]	peas	potatoes
1841	100	117	44	105	738	0	492
1842	126	158	42	109	837	90	708
1843	?	136½	42½	100	384	61	808
1844	?	111	60-66	130-164	489	450-451	1,040

For sources see *Sources for Tables*.
[a]Including autumn (of the previous year) and spring sowings.

At first the Puget's Sound Agricultural Company did not make a profit. That was to be expected, given heavy initial outlays for land, personnel, livestock, seed, implements, and the like. Both farms showed losses in 1838 and 1839. In 1841 the company owed its parent firm £4,108 "for Stock transferred & supplies furnished."[77] The next year showed a loss of £732 to £1,322, incurred mostly by Cowlitz Farm, and 1843 registered a loss of £1,285.[78] During the first half-dozen years of the company' existence, any profits were added to its capital in order to bolster financial security. Some investors, like John Tod, expecting an early and large return, were disappointed. And McLoughlin himself complained in 1845 that the venture had not met the expectations of shareholders, thanks to "so many contrarities."[79] But in 1844 the company saw its first profit — £1,313 to £1,577, and in 1845 it netted £2,220.[80] There was an upsurge in sales in 1846 to British warships on the coast, and in the winter of 1845-46 the company's agents paid the stockholders their first dividend (5 per cent), just when the latter were beginning to consider the concern a "dead loss." Dividends of 5 to 10 per cent followed annually through 1854.[81] Thereafter losses ensued, partly because of encroachments by American squatters, who allegedly deprived the company of £50,000.[82] The losses were absorbed by the Hudson's Bay Company, which was owed £37,400 by its offshoot by 1870.[83] This debt was repaid after the settlement of Puget's Sound Agricultural Company claims against the United States in 1868, and the company was again solvent.

The company's success or failure ought to be measured in terms of its purposes. One was to meet the provision terms of the 1839 contract. This it failed to do by itself. It has already been demonstrated that the company produced only about one-half of the amount of wheat contracted by the Russian-American Company. The rest had to be furnished by the Hudson's Bay Company's Fort Vancouver farm and purchased from the Willamette settlers. This defeated the Honourable Company's intention that the success of the "sheep concern" would allow its Columbian posts to reduce farming in favour of fur trading (or at least deferred it until after 1845, when the Puget's Sound Agricultural Company's 1841 goal of twelve thousand sheep and cattle was finally reached [see Table 19]). At the foremost agricultural post of Fort Vancouver, for instance, cultivated acreage, wheat output, and cattle numbers generally increased rather than decreased during the first half of the 1840's (see Tables 3, 4 and 5). Nevertheless, the wheat was still sold at a profit. It was bought on the Willamette for 3 shillings per bushel, paid in goods, and sold at Sitka for 5 shillings, 4½ pence, less freight costs; this profit amounted to £750 per year.[84]

Also, it was the original aim of the Puget's Sound Agricultural Company not so much to produce wheat, beef, and butter for Alaska as to yield wool, skins and hides, and tallow for England. Few skins and hides and no tallow were shipped, but by 1846 more than seven tons of wool (twice as much as in 1842) were exported (see Table 20), and from 1843, when its quality improved markedly, it fetched moderate prices or better.

In addition, the Hudson's Bay Company profited from the transport of English manufactures under the 1839 agreement. The company charged the Russians £10 per ton, which was more than double the amount billed the company by chartered ships.[85] This freight profit totalled nearly £4,000 in 1844 alone.[86] (Little wonder that in the 1850's the Russian-American Company began chartering supply vessels itself.)

Moreover, the Hudson's Bay Company reaped the fur harvest of the *lisière*. This source of income was more important to the company than provisioning or freightage, especially since the fur returns of the interior of the Columbia Department had declined sharply. As McLoughlin admitted, "I never expected we would make any thing from the prices at which we sell our wheat, flour, and butter; but I looked for a remuneration to the fur trade of the coast." In 1843 the company collected 12,343 pelts from the *lisière*, worth about £8,000.[87]

Furthermore, by selling provisions and manufactures to the Russians so close to cost, the Hudson's Bay Company undercut Sitka's traditional American suppliers. That was a crucial blow, since not infrequently the profits of Yankee fur trading voyages were determined largely or solely by the Sitka market for flour, liquor, sugar, tobacco, textiles, hardware, and other goods. To squeeze the American shipmasters even further, the company offered the coastal Indians more for their furs, at least whenever American vessels were on the coast. The Americans could not compete, and by 1842 they had abandoned the coast trade, just as American mountain men had earlier been driven from the transmontane West by the company's trapping expeditions.[88] The British finally had the coast trade to themselves (except that of the Alexander Archipelago); in their own words, the trade was now "tranquil."[89] With American competition removed and Russian cooperation secured, Governor Simpson was able to reorganize the coast trade by reducing the number of posts, and therefore expenses, in favour of shipping.[90] Forts McLoughlin, Stikine, and Taku were dismantled. Simpson's strategy had won. With much satisfaction he reported to the Governor and Committee in mid-1841 that "it is highly gratifying to be enabled to say that all opposition from Citizens of the United States is now at an end, both in the interior & on the Coast." In 1841 and 1842 Simpson could point to profits of £1,805/5/- and £1,460/17/9 on the company's transactions with the Russians. These figures represented just under 9 per cent and just over 6 per cent of the Columbia Department's profits for those years. A company servant, Dugald MacTavish, testified before the British and American Joint Commission in 1866 that until 1846 his employers had netted from $8,000 to $10,000 annually on the "Russian contract."[91]

Overall, then, historian John Galbraith's argument notwithstanding, the Puget's Sound Agricultural Company was largely an economic success, particularly in terms of the all-important fur trade.[92] But the "sheep concern" also had a political purpose, which was to substantiate the British claim to the right bank of the Columbia River with agricultural settlement, which commonly provided perhaps the most solid basis for territorial claims, apart from military occupation. In a letter to Chief Trader James Hargrave at York Factory, Douglas implied that the political was more important than the economic aim:

> How shall I reply to your enquiries respecting our prospects in the Pugets Sound business? You are of course perfectly informed of its political aim and bearings, and not one of those who expected an immediate harvest from it, consequently you will feel no disappointment in learning that it is not calculated to yield an immediate cash returns.[93]

He added that "in the course of time it will become a business worth pursuing; though that time may be yet remote."[94] In fact, within three years the company was turning a profit. Its political role, however, was frustrated by the preference of British settlers for the better farmland of the Willamette Valley. This preference gave England a claim to the left bank of the Columbia as well, but after 1841 British settlers in the Willamette were greatly outnumbered by Americans. And in 1846 the right bank of the Columbia was lost to the United States, thanks to the firm persistence of the American government and the willingness of the British government to relinquish an area in which they had little vested interest. That demarcation, as historian Frederick Merk has demonstrated, had little or nothing to do with agricultural occupation, since no more than half a dozen American settlers had ventured north of the Columbia (and even then not until 1844), and they were clearly outnumbered by Puget's Sound Agricultural Company and Hudson's Bay Company farmers.[95] After the boundary settlement, the Puget's Sound Agricultural Company's Cowlitz and Nisqually properties were subject to piecemeal seizures by American squatters, whose actions were generally overlooked by American officials, despite the fact that the boundary award had confirmed the possessory rights of both companies below 49° (it had also given the Hudson's Bay Company free navigation of the Columbia, but U.S. revenue officers imposed duties anyway). The harassed Puget's Sound Agricultural Company finally moved its operations to Vancouver Island on the outskirts of Fort Victoria. It was not until 1934, nearly a century after its founding, that the "sheep concern" was formally dissolved.

The 1839 contract and the Puget's Sound Agricultural Company were more successful from the standpoint of the Russian-American Company. At the outset the Russians believed that the agreement would prove advantageous because the Hudson's Bay Company would supply provisions and manufactures at "very moderate" prices and in a "reliable" fashion, and they would be payable in bills of exchange on St. Petersburg rather than in scarce cash or kind at Sitka. Moreover, in case of a rift between Russia and England (relations between the two countries were strained at that time) the contract would remain in force. For its part the Russian-American Company did not begrudge the loss of the fur trade of the *lisière* because it was "very poor" in sea otters, the mainstay of its operations. The company had merely to promise not to encourage American vessels to visit Russian America to sell goods, except in emergencies.[96]

The Russians were not disappointed. Columbian wheat proved no more expensive than Californian wheat and of higher quality; it was also more readily obtainable, entailing less waiting.[97] At the same time, Governor Michael Tebenkov (1845-50) acknowledged that two sources were better than one, for they made supply more certain and less costly.[98] The Russian-American Company received so much grain from the Hudson's Bay Company that by 1842 Governor Adolph Etholen (1840-45) reckoned that 361,130 pounds of flour could "easily" be spared for sale in Kamchatka.[99] Similarly, the English manufactures delivered by the Hudson's Bay Company proved "incomparably cheaper" than those brought by American traders. They were also of the "best quality."[100] Furthermore, they were better and cheaper than Russian wares received overseas via Cronstadt or overland via Okhotsk.[101]

Despite the equivocal economic success of the Puget's Sound Agricultural Company and British and Russian satisfaction with the 1839 contract, when it was renewed in 1849 the provision clause was deleted. By then neither the Hudson's Bay Company nor the Puget's Sound Agricultural Company could grow enough grain. The boundary settlement of 1846 had isolated the farms at Fort Vancouver, Cowlitz, and Nisqually. Output had been reduced by a crop failure in 1847. Even before the expiration of the contract, its provision clause was being met largely by farms at Forts Victoria and Langley.[102] Then the California gold rush erupted and drained the Oregon Country of farm labour. Governor Tebenkov reported to the Russian-American Company's Head Office in St. Petersburg in 1849 that "everything has been abandoned in the

Columbia or Oregon, where they formerly cultivated wheat; everyone has gone to look for gold in California, where grain is now imported from Chile and sold for up to 20 silver rubles per pood [thirty-six pounds] of ordinary wheat flour!"[103] Douglas told Tebenkov that about two-thirds of the Euroamerican population of the Columbia Department had abandoned their farms for the goldfields of California, so that it would be "very difficult" for the Hudson's Bay Company to provision Sitka at any price.[104] Similarly, an American overlander of 1845 wrote to her sister in 1849 that "every thing was prosperous [in Oregon] until the breaking out of the gold fever in California, when the men, most of them left plow and ax in search of the glittering dust."[105] The Russian-American Company's Head Office told Tebenkov that "the refusal of the English to supply the [Russian American] Colonies with provisions is based upon quite valid reasons."[106] Then in 1851 the Head Office was informed by the Honourable Company that California's gold fever was abating and more and more settlers were entering the Oregon Territory to farm. It therefore instructed Governor Nicholas Rosenberg (1850-53) to determine whether Oregon could again meet Russian America's grain needs and thereby obviate lengthy and costly round-the-world shipments from European Russia.[107] The very same settlers, however, dealt the Puget's Sound Agricultural Company a final blow by plundering its land, livestock, and crops, and the settlers' own output was absorbed by the growing cities of the American West Coast.[108]

6

HALF SHARES AND MEAN LANDS: THE PROBLEMS OF THE PUGET'S SOUND AGRICULTURAL COMPANY

From what I have been able to learn from the best informed is that whenever the [boundary] question does come under Consideration that [our] occupancy will have great weight & whatever we may do exclusively on the N of the Columbia it will be very desirable to have the do. occupied by Friends of the Company such as half breed retired servants or British subjects.

GOVERNOR JOHN PELLY TO
GOVERNOR GEORGE SIMPSON, 1841

The settlers make a very poor figure, all except five have migrated to the south side of the Columbia, and I fear the others will soon follow, as they do not work like men that have made up their minds to live and die here.

CHIEF FACTOR JAMES DOUGLAS TO
GOVERNOR GEORGE SIMPSON, 1843

The political failure of the Puget's Sound Agricultural Company — its failure to lend weight to the British claim to the northern side of the Columbia River by peopling it with agricultural settlers — as well as the inability of the company's operations at Cowlitz and Nisqually to produce enough provisions for Sitka without supplements from Forts Vancouver and Langley and the Willamette Settlement, stemmed from the reluctance of settlers to migrate to, and remain on, the company's holdings. There were certainly enough employees. The company's agents estimated at the end of 1842 that no more than fifteen men were needed at Nisqually and fourteen at Cowlitz; two years later there were fifty-six hands at both places.[1] But employees had less choice than settlers. And the latter chose to go elsewhere.

Efforts to recruit settlers began right after the formation of the Puget's Sound Agricultural Company. Originally the agents intended to persuade skilled and loyal British colonists to settle on halves, that is, land in return for half their yearly output. But it was decided to encourage immigration from Assiniboia too because, the agents believed, Red River settlers, although less experienced farmers, were more used to the rigours of pioneering and to dealing with Indians, especially since many of the settlers themselves were Métis.² The first Red River colonists to arrive on the lower Columbia were four who returned with John McLoughlin from his furlough in 1839.³ Two years later Lieutenant Wilkes found them "quite flourishing and apparently happy and enjoying plenty" at Cowlitz with "log huts & young orchards."⁴ But more British settlers were needed, and a note of urgency was sounded by reports that two hundred homesteaders and thirty missionaries were about to leave New England for the Willamette Valley. Alarmed, Governor Simpson instructed Duncan Finlayson, Governor of Assiniboia, in the autumn of 1839 to persuade "some good steady respectable halfbreed Scotch or Canadian families" to migrate to Puget Sound in 1840 in order to relieve population pressure along the Red River and enhance the British presence on the northern side of the Columbia.⁵ Simpson added:

> I would recommend that you look out for such heads of families as may be steady and industrious, with young growing lads and as few useless hangers-on and children as possible. These people may be engaged as servants at the usual wages or as settlers on the Pugets Sound portage, where they will be established on such lands as they can bring into cultivation, and we shall provide them with stock, agricultural implements etc, they holding their farms and stock on halves. To such parties advances may be made of £10 to each head of a family, payable either by servitude to the Company or in Farm produce as C.F. McLoughlin may determine. It is desirable such migration should commence next summer and be followed up from year to year afterwards.⁶

At the end of 1839, the Governor and Committee informed McLoughlin that Finlayson was recruiting Red River settlers for the Columbia in the hope that these "attached and useful" Canadians would strengthen the Puget's Sound Agricultural Company and protect the fur trade of the Northwest Coast against "miscellaneous and restless" Yankees. Beaver House hoped that Finlayson could annually recruit from ten to twenty migrant families "in the prime of life, of good character and industrious habits, with small families." It took Finlayson some time to "beat up for recruits to proceed across the plains."⁷ In the spring of 1840 he wrote to McLoughlin from Norway House that he had not had time to recruit any migrants; he added that although the scheme had been "favourably received" by most settlers, more would move if the terms were "more encouraging." In a letter to the Governor and Committee, Finlayson was more explicit, warning that the Assiniboians were "averse to taking farms on halves" and suggesting that land be sold to them.⁸ This warning was the first acknowledgement that the settlers did not like the terms, which amounted to indenture. They were freemen and wanted to remain so. Although it had been decided at the time of the formation of the Puget's Sound Agricultural Company to allow settlers farms on halves, the agents themselves admitted that "a business on an extensive scale is more likely to be well conducted where so many people have an interest in it, than if left to servants acting for a public company."[9] In fact, settlers on halves would be bondsmen, having to pay interest on their advances, to sell their own share of output solely through the company, to forfeit any improvements upon expiration of their leases, and to be supervised by company officials. These restrictions were to prove one of the critical weaknesses of Simpson's resettlement scheme.

In the summer of 1840, Simpson told Finlayson that his project had aroused "much interest & attention" on the part of the Governor and Committee. They hoped that fifty families would migrate in 1841.[10] But the main purpose of Simpson's letter was to reassure Finlayson and any migrants that the terms of resettlement were reasonable and even favourable. He wrote:

> You say the people are averse to taking Farms on halves, and suggest that Land should be sold to them, as in the United States and Canada. In the present unsettled state of the Boundary question it is impossible for us to effect sales, the sovereignty of the country is not even, as yet, determined, and altho' we have every reason to believe the Columbia River will be the boundary, and that the country situated on the Northern bank of that river will become British territory, still we have no assurance that such will be the case, nor can the Company, in any shape, effect sales of Land on the west side the Mountains. But even if sales could be effected, we think it would be preferable for the people to take farms on halves, as already suggested, as, by so doing they would be put in possession of certain parcels of Land, part of which will be broken up, houses will be erected for them, stock, such as cattle, sheep, horses etc. provided, likewise agricultural implements, without any advance being required from them, in fact the Company is willing to provide them with capital, their proportion of the capital being labour, and the Company looking to be repaid for their advances in the shape of produce, say half the increase of stock and produce of every kind. These terms are more favorable to industrious settlers than any we have yet heard of, either in the United States, Canada, or any of the new colonies. You may therefore inform such people that allotments will be made to each family, of at least 100 acres of land, besides the use of common or pasture lands, part thereof broken up, with the necessary buildings erected for them, and live stock advanced to each family, of a Bull and ten or more cows, 50 to 100 Ewes, with a sufficient number of rams, hogs, Oxen for agricultural purposes, and a few horses; in short, as many of those different stocks as they may be equal to the management of; all valued at low money prices, the expences of erecting the buildings being a charge upon the farm; the cattle

valued at £2 a head, the sheep at 10/- a head, horses at 40/-each, and other stock in proportion; a credit given to them from year to year for their increase, produce or returns, at such fair prices as the state of the markets may afford. To every family who may feel disposed to accept this offer, an advance may be made of £10, to make the necessary prepatory arrangements for their journey. By crossing the plains, they will be able to provide their own subsistence in most cases by hunting, but at each establishment where they may touch, they will be furnished with the necessary supplies, provisions etc etc, and by starting from the settlement as early in the season as it is possible to travel, say in the early part of April, and taking the route to Edmonton, thence by the Coutonais Portage, they may reach Fort Vancouver before the close of the season. You will understand, however, that none are to be sent except people of industrious habits and good character, & no family should exceed four in number, the whole party not being more than 200 souls.

Simpson stressed that "none except good subjects, i.e. tractable well disposed industrious people, can be allowed to migrate to that new and promising country," and that they should depart "very early in the season" in order not to have to winter at a post en route.[12] Typically, the governor was determined to effect his scheme as economically as possible and with as much company control as possible. The Governor and Committee were somewhat more liberal. They wrote to Simpson:

we think it is very desirable that some regular plan should be formed for drafting from 15 to 20 families annually across the Mountains to the Cowlitz Settlement, as the circumstance of a British agricultural population being settled there, would greatly strengthen our claims to the Territory in any adjustment of the Boundary Line; while the presence of a respectable British force will be a great protection in the event of our Wallamatte neighbours, who have lately crossed the Mountains, attempting to carry their threats of lawless aggression into effect. We are unprepared to say what outlay should be incurred in effecting this, whatever it may be however, we feel, it ought to be borne by the Fur trade, as the relieving of Red River Settlement of its surplus population and the increasing the British population on the West side the Mountains, we consider matters of much importance at present to both the immediate and permanent interests of that Concern.[13]

It was not until the spring of 1841 that Finlayson was able to report from Fort Garry that he had finally found seventeen migrant families (eighty-five persons) and might be able to recruit others, but no more than twenty or twenty-two families altogether (ninety to one hundred persons). However, in order to convince them, he had been obliged to adjust the terms. Firstly, he had had to take some families with more than four members because most families contained more than two children, so that the migrants would average five persons per family. Secondly, because farm tenure on halves was "by no means a popular measure," he had had to promise them that as soon as the Columbia Department became British territory that form of tenure would be rescinded and the land would be sold to them ("otherwise not a single family of respectability would embrace the terms offered"). Although there was a "strong desire to remove," the settlers were reluctant to leave "old quarters" for unknown territory, and they dreaded the long journey.[14] Finlayson had also encountered

> a great deal of under hand opposition to this measure — the most ridiculous stories were industriously circulated to throw cold water upon it — even the Doctor's [McLoughlin's] reputation as a disciplinarian was not forgotten, & had it not been well managed, not one family would be found to embark thereon.[15]

The migrants refused to leave before 1 June, when there would be enough grass for their cattle. Early that month twenty-three families (121 persons, including 77 children) in fifty carts with sixty horses, seven oxen, and two cows left White Horse Plain under James Sinclair, a free trader.[16] Simpson predicted that they would be increased to 200 persons by freemen and others by the time they reached Fort Vancouver, and he asserted that at least 1,000 Red River settlers could be persuaded to migrate to Puget Sound. The Hudson's Bay Company furnished a guide, horses, provisions, and goods en route, and houses, barns, and fenced fields with fifteen cows, one bull, fifty ewes, one ram, and oxen or horses plus seed and implements on arrival. The settlers agreed to relinquish half of their crops for five years and half of their increase of livestock after five years.[17]

While Finlayson and Simpson were disagreeing over the terms of resettlement, McLoughlin and Simpson were disputing the locale. In the winter of 1840-41, McLoughlin inspected the Cowlitz Portage and reported that "the country does not appear well adapted for an extended settlement, although . . . favorable for the operations of the Puget's Sound Company alone," and he recommended that arriving migrants and retiring servants settle in the Willamette Valley.[18] McLoughlin told the Governor and Committee:

> If there was more Prairie Land at the Cowlitz it would be possible to encourage emigration to that place but the Puget Sound Association requires all there is and though the soil is equally as good as that of the

Wallamette the larger extent of the Prairies of the Wallamette and the great abundance of Deer on them and their more beautiful scenery causes them to be preferred to the Cowelitz and settlers will never settle on it till the Wallamette is settled or till the wood at the Cowelitz comes in demand.[19]

Simpson demurred. He believed that the shores of Puget Sound with their arable plains, spacious and secure harbours, and absence of "intermittent fever" (malaria or influenza) held more promise than the lower Columbia-Willamette district with its navigational obstacles and unhealthfulness.[20] He declared:

> The Straits of de Fuca afford a safe and ready access at all seasons to these districts of country, where there are many safe & commodious harbours; & as the climate is healthy, the Intermittent Fever being unknown in that quarter, there is no doubt that that country will in due time become important as regards settlement & commerce, while the country in the vicinity of the Coast, bordering on the Columbia & Willamette Rivers, so much spoken of in the United States as the El dorado of the shores of the Northern Pacific, must from the dangers of the bar & the impediments of the navigation, together with its unhealthiness, sink in public estimation.[21]

Simpson reckoned that there was plenty of land suitable for agricultural colonization in the form of "a chain of plains" between Cowlitz and Nisqually, along the shore of Puget Sound, in the valleys of the Chehalis and Black Rivers, and on Whidbey Island and the southern tip of Vancouver Island.[22] He was likewise certain that the Columbia River would become the boundary between the United States and British North America, so that any settlement in the Willamette Valley would eventually be lost to the Crown. He wrote:

> I am decidedly opposed to C.F. McLoughlin's opinion as to the policy or expediency of allowing any of our people to settle on the South side of the Columbia, as I have no doubt that river will become the boundary between British & United States territory in that country, & in that case we should in removing them thither, be raising opposition to ourselves, by strengthening the American interests in that quarter.[23]

As usual, the governor had his way, and the settlers were directed to the Cowlitz Portage. Prophetically, McLoughlin warned that they would soon desert.[24]

Sinclair's party of overlanders, twenty-two families or 116 persons (three babies were born en route and one family apparently turned back), reached Fort Vancouver in early autumn, 1841. Fourteen families, totalling seventy-seven persons, mostly "English half-breeds," were located on farms at Nisqually on halves under the Puget's Sound Agricultural Company. They were joined by three Englishmen, retired servants of the Hudson's Bay Company.[25] The remaining seven families, thirty-eight persons or one-third of the migrants, mostly "Canadians and [French] half breeds," were "disinclined" to do likewise and took farms "on their own accounts" at Cowlitz outside the Puget's Sound Agricultural Company but with advances of seed and implements from the Hudson's Bay Company. The official reason for their change of heart was their hunting and trapping background, which allegedly made them less suitable for sedentary stockbreeding.[26] The real reason, however, may have been Simpson's threat upon their arrival at Fort Vancouver that the Puget's Sound Agricultural Company would fulfil its obligations to only those who settled at Nisqually. Moreover, no houses, ploughs, or cattle were provided; the result was "much discontent and loud murmurings."[27]

It was not long before the settlers on halves also withdrew. They had not arrived in time to grow a crop, so the first winter found them short of food, if not shelter as well. Captain William McNeill of the Honourable Company's "naval department" told James Douglas at the end of 1841 that "the Settlers in general wish some person to come from Vancouver to regulate things so far as concerns themselves they complain greatly about their food." By the spring of 1842 thirteen families of migrants and four of retired servants were still at Nisqually on halves. According to Douglas, they were well established and appeared pleased with the locale and satisfied with their condition.[28] Chief Trader Dugald MacTavish was less sanguine:

> The Red River Emigrants are another set, the Canadian part of them with the exception of one Man, at their own desire, had the Engagements they made at Red River cancelled, and are now at the Cowelitz — on the same footing as the other settlers in the Columbia, The other part have kept to their Agreements, and are accordingly settled at Nisqually — where from the nature of the Country which though beautiful as regards scenery etc. etc. convenient as to the sea shore, is barren & sterile in the extreme, and has more the appearance of a rough shingley Beach than *Soil* — I am afraid they will never do much — they are besides a most extravagant set, and if we may judge by our Wallamette Farms, it will take each of them their returns for 3 years — to pay their expenses for one.[29]

Drought, climatic aggravated by edaphic, struck in the summer and nullified their hard work. Disheartened by the crop failure and the "very indifferent" soil and attracted by the advantages of American jurisdiction, including homesteading rights, and the superior land of the Willamette Valley, for which they had always had a "strong inclination" anyway, five families quit Nisqually in late summer for the Palatine (Tualatin) Plains on the lower Willamette's left bank. Another six families followed in early autumn. The remaining six families persisted another year, and in the fall of 1843 they likewise headed south, since "the Wallamette is a finer country for tillage than Nisqually."[30] As McLoughlin admitted, "the fact is the soil of the Wallamette is so much better than that of Nisqually, that no man who can get a lot of land by squatting in the Wallamette will remain at Nisqually." McLoughlin also contended that the Red River settlers were extravagant and negligent.[31] Superintendent Tolmie concurred:

> The last five of the Red River settlers have left for the Wallamette this Autumn, and I do not at all regret their departure as being indolent, and thriftless they would have been an ever increasing burthen on the Association which can itself more profitably occupy the Nisqually plains.[32]

The seven Red River families who had gone to Cowlitz joined the two families of retired engagés who had been there since 1837. This community numbered thirteen families (eleven French Canadian and two Indian) or sixty persons in the autumn of 1842, sixty-four persons in the spring of 1843, and from nineteen to twenty families in 1845. Each family had fifty acres of land.[33] By not moving to the Willamette, these settlers bolstered the British claim to the Columbia's right bank, where they were not outnumbered by American migrants until 1847.

The departure of most of the overlanders prompted the Puget's Sound Agricultural Company to forego migrants from Assiniboia and to rely instead on a few experienced English farmers and retired servants, who were likewise allotted farms on halves. Indeed, it had been intended in 1839 to send twenty "respectable industrious agriculturalists" (Scots) on the London ship in 1840, but the plan had been deferred by the prospect of migration from Red River.[34] When that prospect temporarily faded in 1840, Simpson tried to recruit from ten to twenty farming families in Scotland for the company either as "farm servants" on halves or as "farm laborers" on salaries (in both cases for five-year terms) but in vain. Eventually, several English shepherds and cowherds did go, but most of them proved as unsatisfactory as the *Métis* migrants and soon left. In 1843 McLoughlin reported that at first they sometimes refused to undertake certain tasks, such as erecting pens, washing sheep, and sleeping outside with the sheep, apparently considering such chores to be too menial or undignified. In the same year, Douglas complained that two of the departed English herdsmen, including Mr. Steel, the head shepherd, had worked "diffidently." Another English settler, Joseph Heath, who arrived in 1844, worked conscientiously but remained in debt until his death in 1849. Even Superintendent Tolmie was not above reproach. In 1845 Simpson urged Peter Ogden, McLoughlin's successor at Fort Vancouver, to replace Tolmie because both his practical knowledge and his authority over the local Indians were wanting.[35] Tolmie, a physician by training, lacked not only expertise but evidently initiative and diligence as well.

Indian manpower was also a problem. It was desirable to employ natives not only because they offset the labour shortage but also because they were economical. In 1839 the Governor and Committee suggested to Douglas, McLoughlin's proxy, that he recruit "a few docile young native Indians, say emancipated Slaves" as herdsmen, carters, and weeders, for their labor would be found cheaper than that of regular servants."[36] From one-half to two-thirds of the company's work at Nisqually Farm was done by Indians, who were paid £4 to £8 per year, plus provisions. (The British shepherds earned £35 per year.) But in Tolmie's opinion the Indian labour still did not pay. Unless closely supervised, he said, native workers were lazy, troublesome, and unprofitable. Their low productivity was at least partly attributable to a low level of mechanization. In 1843, for example, threshing at Nisqually was done by hand by Indians and Kanakas, who barely averaged one bushel each per day. And the lack of a winnowing machine was also a "great drawback."[37]

Not surprisingly, the Puget's Sound Agricultural Company was even less successful in attracting retired Hudson's Bay Company servants from the Willamette Valley to the Cowlitz Portage. In 1838 Douglas and Father Blanchet exerted their "utmost influence" on the Willamette settlers in a vain attempt to persuade them to move.[38] Douglas reported to Beaver House:

> I acquainted the Wallamette Settlers with your intention of colonizing the Cowelitz River, but have not succeeded in inducing any of them to remove from their present habitations, into the new colony. They have incurred considerable expense in improving their present possessions, which, altho' desirous of seconding our views, they are naturally reluctant to abandon, without receiving a full equivalent in return. The expense of a removal, is also a very serious consideration to persons, who have no property besides their Farms; it would, I fear, reduce most of them to poverty, and the poorer farmers could not accomplish the object without our assistance. The influx of American Settlers may probably enable the Canadians to sell out to advantage, and their natural dislike of Jonathan [the United States], a feeling that absence from their Native Province has not blunted, may probably incite them to emigrate into some quarter, where American influence will not predominate. In the mean time, I will endeavour to raise the character of the Cowelitz Country, which was never popular, and permit none of our people, except a few, to whom the Company are pledged, to settle on the Wallamette. The arrival of a Priest and the erection of agricultural machinery will, I am certain, decide the point in favour of the Cowelitz; The Settlers there, are put to serious inconvenience from the want of a grist Mill, and I have sent them, a small hand Mill, as a temporary substitute.[39]

Nevertheless, the following year the Governor and Committee instructed McLoughlin to encourage retired servants to shift from the Willamette to the Cowlitz

> with the least possible delay, as the presence of those people would not only be important in affording us protection and support, but the fact of a numerous body of British subjects having formed themselves into an agricultural settlement on the North side of the Columbia River would, we conceive, strengthen the claims of Great Britain to the Country so occupied: and as a means of preventing any Citizens of the United States from establishing themselves as squatters on the Cowlitz and on other advantageous positions on the North side of the Columbia, we think it would be advisable to take immediate possession of those positions, by blazing the trees, cutting down timber, cultivating small patches of land etc. etc. as is usual in taking possession of wild districts of Country in the United States and Canada.[40]

The chief factor was also ordered not to assist American migrants, except in emergencies. Again, in 1840 Simpson ordered McLoughlin to try to dislodge the Willamette Settlement "either by getting the people removed to the Sandwich Islands, California, or other parts, or getting our own retired servants removed to the Cowlitz, where they would be more immediately under our guidance & control."[41] But the Willamette settlers were unmoved. They were not about to surrender their hard-won holdings for indentures on the shingly prairies of the Cowlitz Portage.

Lacking settlers, the Puget's Sound Agricultural Company was left with its own hired hands, and the company was either unwilling or unable to employ enough of them. At the beginning of 1845 Simpson advised Tolmie that not all ewes should be bred at the same time; rather, they should be divided into three or four flocks, and each flock should be bred at one-week intervals; otherwise, they would all lamb at the same time, for which there were not enough workers. He further advised Tolmie to hire "careful Indians" if sufficient Euroamericans were unavailable. The Indians, however, were neither used to, nor fond of, agricultural labour, particularly on a regular basis. And the Euroamerican hands had little or no vested interest in their work. In 1845 the company's agents were reported by Simpson to be considering allowing the chief shepherds to share in the "increase of the flocks" in order "to induce a greater degree of interest," but this belated and halfhearted measure does not seem to have been implemented.[42]

The shortage of labour was compounded by the infertility of the land. Douglas even asserted that the sandy and pebbly soil of the Nisqually Plain was the chief obstacle to farming there. The soil was "unproductive" in its natural state, but with manuring and rotation wheat yields doubled.⁴³ Similarly, on the Cowlitz Prairie Lieutenant Wilkes found that the soil was "not so deep" and the pasture "not so good" as in the Willamette Valley.⁴⁴ McLoughlin, however, felt that the soil of the Cowlitz was as good as that of the Willamette, although the latter's "larger extent of the Prairies," "great abundance of Deer," and "more beautiful Scenery" made it preferable to settlers, who would not choose the Cowlitz until the Willamette was completely settled and the Cowlitz's timber was in demand. The woodland of the Cowlitz Portage was so dense, he added, that it was difficult to clear, with tree roots taking several years to rot.⁴⁵

What little prime land that the Puget's Sound Agricultural Company did enjoy was eventually diminished even further by American squatters. By 1845 "a few" American families had settled between Cowlitz and Nisqually, but they were not numerous enough to exert pressure on company property. By the fourth article of the 1846 boundary treaty, all land below 49° N. became American, but the company's land title was confirmed, although the U.S. government reserved the right to buy the company's holdings at evaluated prices.⁴⁶ Nevertheless, some American homesteaders regarded the company as a foreign, monopolistic, anti-republican trespasser with too much land. At Nisqually encroachments began in January 1847, one month after word of the boundary settlement reached the lower Columbia. In 1851 alone nine squatters claimed 2,720 acres.⁴⁷ The squatters not only seized the company's land but also tracked and shot its livestock. These depredations, which were tolerated by the American authorities, were partly responsible for the Hudson's Bay Company's refusal in 1849 to renew the provision clause of the Russian contract.⁴⁸ Costly and unsuccessful litigation in the courts of Oregon Territory spurred the company's departure from the United States. Already in 1847 the agents ordered the superintendent to prepare to move operations to Esquimalt Harbour on Vancouver Island (McLoughlin had contemplated such a move as early as 1842 because of the disadvantages of Nisqually). Four farms were founded on Vancouver Island in the early 1850's, and they increased considerably in value with the Fraser River gold rush in 1856. Meanwhile, from the early 1850's Cowlitz Farm, and from the middle 1850's Nisqually Farm, were operated at a loss. The former was abandoned in 1855.⁴⁹ A prospective settlement with the U.S. government for up to $500,000 was postponed by the disintegration of President James Buchanan's administration and then by the Civil War. The Puget's Sound Agricultural Company went deeper and deeper into debt to its parent for supplies. A settlement of claims was finally reached in 1869 for only $200,000.

Indians, as well as American squatters, caused additional problems for the Puget's Sound Agricultural Company by rustling company livestock. The "beef-eating propensity" of starving natives was especially felt at Nisqually in the winters of 1841-42 and 1844-45.⁵⁰ Wolves and "panthers," or cougars, likewise preyed on livestock. A member of the U.S. Exploring Expedition reported in 1842 from Nisqually that "wolves are very numerous in this part of the Oregon Territory and are very destructive to the sheep when they get a start among them." He added that "these animals when shoved with hunger often attack horses."[51] In 1842 there were "extraordinary losses" of livestock to predators and disease, particularly at Fort Vancouver. In 1844 at least ten sheep, eight pigs, and two horses were killed by wolves at Nisqually.[52] A settler reported in the middle of January 1845 that wolves "are becoming very common and daring on the [Nisqually] Plains, being driven from the mountains by the snow." And on one night in 1845 a single cougar killed twelve sheep at Nisqually; on another night two cougars killed twenty more.[53] Indian dogs also harassed company sheep, and lambs fell prey to eagles as well as to spring cold and wet. McLoughlin received strychnine from London for use in wolf bait, and "high prices" were set on wolf skins to encourage the Indians to hunt the beasts. In eighteen months in 1841-42 a shepherd at Nisqually Farm killed more than 100 wolves.[54]

Sheep in particular suffered from disease. Scab struck Nisqually Farm's sheep in 1842, killing 140 head in March alone. It was particularly prevalent during the rainy season, as in the winters of 1842-43, when more than five hundred sheep perished, and 1844-45. Scab reduced fleeces and exposed sheep to the cold and wet. Scabby animals were dressed with mercurial ointment or a solution of corrosive sublimate in tobacco water. Douglas recommended that the sheep be tarred at the beginning of every winter; although he admitted that the resultant tarry wool brought lower prices, he thought that the additional weight would offset the poorer quality![55] Braxy, or intestinal inflammation, also struck in 1842, causing a "serious loss" of sheep at Fort Vancouver. In 1845 it killed ninety-nine head at Nisqually Farm. It may have arisen partly from the dirty condition and crowded quarters of the animals. Occasionally livestock succumbed to outbreaks of severe cold, as on 9 March 1843, when an eighteen-hour snow storm left eight inches of snow on the ground at Nisqually Farm.[56]

Most livestock losses, however, were a result of insufficient tending. Many animals strayed, fell over precipices, drowned, or collapsed from exhaustion during drives from California to Fort Vancouver, Cowlitz, and Nisqually, and some of those shipped from California and England died at sea. For example, nearly half of the three hundred cows being driven from Fort Vancouver to Cowlitz Farm in early 1842 were lost. In the winter of 1842-43 two hundred cattle escaped from a herd en route to Nisqually Farm from Fort Vancouver. And in 1845 twenty-three of one hundred cattle strayed from a drive between Cowlitz and Nisqually.[57] California cattle, being half-wild, were especially unmanageable, so much so that in 1842 a herd could not be driven from Fort Vancouver to Nisqually Farm, the drovers being unable to restrain them. By the end of 1841 two bands of such cattle, totalling 102 head, were roaming wild in the woods around Plamondeau's Plain on the banks of the Cowlitz River, having bolted in the autumn from a drive from Fort Vancouver to Nisqually Farm. In the spring of 1842, 200 of the Puget's Sound Agricultural Company's 1,414 cattle were "running wild" along the Cattlepoodle (Lewis) River. These feral animals even threatened travellers. The cattle that strayed in 1841 and 1842 were not recaptured until 1844.[58]

At first the cattle at Nisqually Farm, like the sheep, were penned every night in order to manure the grudging soil. The confined animals became so nervous, however, that pregnant cows often aborted. So in 1843 Tolmie ended cowpenning, with the result that the animals became calmer and fewer calves were lost. As an observer noted, "since the practice of soiling the land with cattle has been discontinued, the cattle are becoming much tamer, as they are not harassed by the daily driving, necessitated by this mode of fertilizing land, which is also open to the more weighty objection of causing a loss of calves."[59] Thereafter the cattle were collected only occasionally. In winter at Nisqually Farm they were allowed to range thirty miles "in circuit" in order to find better pasture. In ranging so freely, however, they became quite wild. In 1845 Simpson complained to Tolmie of the "difficulty of collecting the cattle," adding that "while they are in so wild a state that, it is impossible to collect them, they are of no other benefit to the concern than as an article of food, the produce of the chase, in like manner as Buffalo are in the Saskatchewan."[60]

Running at large, the cattle strayed and fell easy victim to predators and Indians. In addition, their wild calves were caught with difficulty. Nisqually Farm's increase of 430 to 432 calves in 1843 was less than Simpson expected, and he assumed that the cause was "their wild and unmanageable state rendering it difficult to collect them, & many, no doubt, have gone irrevocably astray." He therefore advised that the cattle be assembled at particular places at least once a week, and the agents recommended the adoption of the Californian style of rounding up cattle by enticing them with salt and blood. In order to optimize calving and beeving, Simpson instructed Tolmie not to slaughter cattle until they were at least seven years old, and then only those in prime condition, to sell the meat locally and render the inside fat into tallow, and to stretch, dry, and salt the hides and ship both the tallow and hides to England.[61] Simpson was pleased to learn in 1846 that, apparently because of these measures, the cattle were becoming "more manageable" and multiplying "very rapidly."[62]

Nisqually Farm's sheep were more tractable and less vulnerable, since they were penned nightly in flocks of five hundred with a guard. Losses of sheep, particularly lambs, were nevertheless high. Early 1843 was "memorable" for Nisqually's sheep because of, in Tolmie's words, "the concentration of disasters which then assailed them," including predation, cold and wet, and infanticide. Of three thousand lambs dropped at Nisqually that spring, only 1,739 remained by 21 May for a loss of 42 per cent. And by 1 October 1844 twelve hundred of three to four thousand spring lambs had died, a loss of one-half to two-thirds. If the lambs were born too early in spring, they could die of exposure and hunger. In late 1842 the agents warned McLoughlin that the ewes should not be "tupped" (bred) too early in the autumn lest they lamb before the spring grass was sufficiently advanced to support nursing.[63]

Superintendent Tolmie complained of infanticide in early 1844, when he told McLoughlin that there had been "an extraordinary loss of lambs" at Nisqually Farm (3,307 ewes bore only 1,575 lambs). He ascribed the toll to "a strange and unnatural propensity" on the part of the ewes to nibble their young, a result, he thought, of a deficiency of salt in their diet. So he pastured some of the flock on salt marshes astride the river mouths along Puget Sound, where the old ewes and weak lambs improved.[64] And McLoughlin moved some ewes to Cowlitz Farm, where they likewise improved, but the pasture was limited. Simpson finally wrote to Tolmie that English experts had advised him that the infanticide probably arose from the failure of the shepherds to separate the parturient ewes from the others during lambing, but just in case he promised to send rock salt for the sheep to lick at pasture. Whatever the cause, the high loss of lambs meant lower production from older ewes. In 1845 and 1846, 573 ewes at Nisqually and Cowlitz, 3 per cent of the entire stock, died of old age.[65] In 1845 Tolmie began the system recommended by Simpson of breeding the ewes later in the autumn in small groups at intervals of several days and carefully tending the ewes during lambing. "I hope to learn," Simpson wrote, "that your flocks become more healthy & productive from year to year." And they eventually did, although as late as the middle of 1846 Simpson complained to Tolmie that the increase of sheep was "still miserably small," probably because of the high proportion of old ewes resulting from the previous high loss of lambs.[66]

Similarly, there was a high loss of foals. Because of the shortage of geldings, Tolmie explained, brood mares were worked too much. Consequently, they suffered miscarriages and became barren, and Tolmie was ordered by Simpson not to use brood mares as work horses.[67]

The quantity and quality of livestock products, particularly wool, were also problematical. All of the company's wool was intended for export to England, and the first batch arrived in London in 1839. But it was found to be so coarse, dirty, and uneven that its value was lowered "very much," and McLoughlin was instructed at the end of that year to ensure that the sheep were thoroughly washed and dried before shearing and that the fleece was completely sorted and tightly baled before shipping.[68] To further improve the wool, twelve high-grade Merino and Leicester sheep from the farm of Lord Western were shipped to the Puget's Sound Agricultural Company in 1840 for crossbreeding with the degenerated Californian sheep.[69] The first crossbreeding occurred in the fall of 1841, and that year saw some improvement. The wool received in London in 1841, although small in quantity, was rated "good and sound";

however, because the sheep still had not been washed before shearing it was unclean and brought the low price of from 3½ to 6 pence per pound.[70] The London buyers estimated that at least 75 per cent of the wool's weight was dirt. The 1842 shipment was also small and "exceedingly dirty and nearly unmarketable," apparently because until 1843 or 1844 some of the sheep were clipped at Fort Vancouver, where the Columbia's banks and water were muddied by spring run-off and rains at shearing time.[71] It was not until 1843 that the quality of wool shipments became satisfactory; being clean, it fetched "very fair prices" that year. The company's wool broker in London, Mr. Hazard, reported to the agents that the 1843 shipment was "very much improved in condition," being "particularly soft and kind in the staple & very sound," as well as free of the burrs and grass seed that depreciated wool from the colonies of New South Wales and Cape of Good Hope. Tolmie informed Simpson that the main problem had not been washing the sheep but keeping them clean "in the period intervening between washing and clipping while the wool is drying" because "dust abounds during the dry season in the Nisqually plains."[72]

The yield per sheep, however, was still low, averaging 1 pound, 2 ounces at Nisqually Farm and 2 pounds, 6½ ounces at Fort Vancouver.[73] Although the wool was of "fair quality" and the mutton was in "good condition," Simpson was disappointed in the low yield, ascribing it to either underbrush on the pasture or late shearing. He urged that the sheep be grazed on grass free of brush and that they be shorn earlier, after having been well washed in clean water and thoroughly dried. Simpson also complained that the wool was not properly sorted during baling, so that it could not be sampled fairly during the London sale; he therefore urged that it be uniformly sorted and tightly packed.[74]

The 1844 shipment was the largest yet, as much as 8,621 pounds, and was better in quality, fetching an average of 8 pence per pound, but there was still "much room for improvement" in the weight and quality of the wool. The 1845 shipment was even larger and better; the 10,842-11,845 pounds of wool brought an average of 10¼ pence per pound.[75] But Simpson, a perfectionist, still complained that the amount was "very small" in relation to the size of the flock. Finally, in 1846, 14,624 pounds of wool, at least a 23 per cent increase over 1845, were sold in London at an average of 8 pence per pound, the reduction in price from 1845 resulting not from a decline in quality but from the depression of wool prices. Nevertheless, the 1846 sale had to have realized nearly £500, virtually the same as in 1845.[76]

Crops at Colvile suffered less than livestock at Nisqually, but they were not immune to drought and frost. Summer drought merely drove the cattle "a distance" for pasture, whereas dry weather and bad ploughing wrought a "very inferior" harvest in 1839, and the 1842 crop was affected "very much" by "severe drought."[77] Drought could also generate forest and grass fires, as in the early summer of 1833, when most of the prairie around Fort Nisqually burned. Hard frost killed most of Nisqually Farm's planting of winter wheat in the early spring of 1843. And mild blight reduced Cowlitz Farm's wheat crop in 1840.[78] Generally, however, the company's crops had sufficient moisture and heat, and meagre harvests were no more common than bumper ones, while ample outcomes were the rule.

Similarly, storage and transport of output presented few problems. The Cowlitz River was a challenge, as Father Blanchet noted in 1839: "Cowlitz River is excessively tortuous. Its course is filled with trunks of trees, which renders its navigation difficult and often dangerous, even for small craft. Numerous rapids are found there, very difficult to ascend; and its steep banks present an aspect gloomy and wild."[79] In 1844 the *Cowlitz* waited six weeks at the mouth of the river while wheat was lightered downstream from Cowlitz Farm. A "large quantity" was damaged aboard ship by heating, having been damp when loaded, all for want of a granary at the mouth of the Cowlitz where the wheat could be dried and stored.[80] One was built soon afterwards. Transport between Nisqually and Cowlitz was not crucial because farm output was not conveyed over the Cowlitz Portage; besides, the Fort Vancouver-Fort Nisqually trail was improved during the summer and autumn of 1840.

Thus, by the middle of the 1840's, the Puget's Sound Agricultural Company had successfully adjusted to the physical peculiarities of the Cowlitz-Nisqually tract. However, the firm did not flourish because it failed to attract many British settlers. They opted instead for the freer and richer land of the Willamette Valley, which also became the lodestone for American migrants on the Oregon Trail.

Part Three

HOMESTEAD FARMING
Pioneer Agriculture in the Willamette Valley

The apsect of the [Willamette] country, in its natural state is strikingly beautiful. The intermixture of woods & fertile plains, peculiarly adapts it for the residence of civilized man, affording lands easily tilled, excellent pasture, fuel and building materials of the best quality.

Chief Trader James Douglas to the Governor and Committee of the Hudson's Bay Company, 1838

7

THE PROMISED LAND: THE FORMATION OF THE WILLAMETTE SETTLEMENT

> *There appears to be an Oregon fever in the [United] States.*
> REVEREND HENRY SPALDING TO
> REVEREND ELKANAH WALKER, 1839

> *The people of the west are crouding into the [Willamette] country, by sea and land as fast as they can come.*
> CHIEF FACTOR JAMES DOUGLAS TO
> GOVERNOR GEORGE SIMPSON, 1845

Most of the Red River migrants who quit the Puget's Sound Agricultural Company's farms, as well as many of the servants who retired from the Hudson's Bay Company's posts in the Columbia Department, settled in the Willamette Valley (Plate 13), which a retired Bay man pronounced "the most fertile district in all Columbia."[1] This structural lowland, stretching about one hundred miles south of the Columbia and twenty to forty miles west to east, was accurately described by a physician with the Hudson's Bay Company, Dr. Meredith Gairdner, "as consisting of a series of extensive plains or prairies covered with grass, interspersed with belts of fir and oak, and bounded on the east by the [Cascade] mountain chain of snowy peaks, and on the west by the [Coast] ranges immediately skirting the Pacific."[2] The temperate climate, fertile alluvial soil, lush grassland, rolling parkland, gentle hills, and majestic peaks drew almost unanimous praise. An American settler of the mid-1830's, Philip Edwards, who wrote a guide for immigrants, stated that the climate was "mild and equable," thanks to the high Cascades, which kept cold continental air masses at bay (except at the Columbia gap, where they occasionally chilled Fort Vancouver), and the low coastal mountains, which only partially blocked the tepid and moist Westerlies, resulting in few extremes of temperature and moderate precipitation from autumn through spring. As an overland migrant of 1845 noted, "the climate is mild and pleasant . . . it scarcely ever snows, and if any snow falls at all it melts quickly. Men can work in thin shirt sleeves all winter."[3] The Valley's soil impressed Edwards even more:

Plate 13. View of the Willamette Valley, 1845.

The soil is generally of a silicious nature, and bears little resemblance to the dark vegetable mould which we of the west are used to prefer. It produces well without the application of manure; but I have never known any country in which its happy effects are so palpable. Even the ashes deposited from the burning of stubble or other remains of the previous year's produce, effect a marked improvement in the crops. The soil is deep and its productive qualities durable, but little if any deterioration being yet perceptible in the oldest fields. Capt. Wyeth considers "the soil equal to that of any part of New York." It is adapted to the culture of wheat, rye, oats, barley, and generally all sorts of small grain; all varieties of peas and beans, Irish potatoes, and nearly all sorts of roots cultivated in the United States . . . In no country in the world, may be the husbandman look forward with more assurance to the reward of his toil.[4]

Perhaps the most striking and attractive feature of the Willamette was the tall grassland of the valley bottoms and the oak parkland of the low interfluves. The grass, "as high as your saddle sometimes," was annually regenerated by Indian burning, which was intended to facilitate the hunting of deer (by restricting their grazing grounds and by encircling them with a decreasing ring of fire), the gathering of grasshoppers, wild honey, sunflower seeds, and "wild wheat" (tarweed), and the sighting of enemies.[5] With the decimation of the native population by disease and warfare, the grassland reverted to woodland. But the earliest settlers found numerous and extensive "prairies of good land with handsome islands of timber interspersed here & there."[6] The Hudson's Bay Company was well aware of the valley's appeal, and so were the British authorities. In 1845 Lieutenant William Peel, son of the British prime minister, reconnoitered the Willamette and reported:

> The land in the Wallamette is rich and well watered, generally clear of Timber, and covered with a fine luxurious grass.... The face of the Country is very picturesque, the eye resting upon large Prairies, divided by belts of timber running along the Banks of the Rivers and Streams, and separated from each other by a soft outline of hilly land, whilst the snowy Mountains to the Eastward, forming a chain parallel to the Coast, are distinctly visible and make a beautiful background.[7]

The Willamette Valley offered obvious advantages to agricultural settlers. The open grasslands precluded labourious clearing, and — à la New Zealand — winter housing and feeding of livestock were unnecessary in the mild climate (although it was prudent to shelter animals from the cool drizzle). For pigs there were plenty of roots (wappatoos) and white oak mast. Nathaniel Wyeth found in 1832 that above Willamette Falls the "oak prairies" of one to thirty miles in extent with their "extremely rich" soil were very suitable for cultivation (below the falls seasonal flooding endangered farming). He asserted that the valley had two major agricultural advantages: because of the mild climate ploughing could be done eleven months of the year (even year round as often as not), and livestock could graze all year.[8] He might have added that few Indian occupants remained to impede settlement. And in 1843 the provisional government of the Willamette settlers confirmed pre-emption rights. Marcus Whitman, the leader of the Congregational-Presbyterian missionaries, wrote to his brother in 1846 that "there are many and great advantages offered to those who come at once. A mile square, or 640 acres of land such as you may select and that of the best of land, and in a near proximity to a vast ocean and in a mild climate where stock feed out all winter, is not a small boon."[9] To Whitman the only drawback of the valley was its raininess:

> The greatest objection to the country west of the Cascade range is the rains in winter. But that is more than overbalanced by the exemption from the care and labour of feeding stock. It is not that so much rain falls, but that it

rains a great many days from November to April or May. People that are settled do not find it so rainy as to be much of an objection. It is a climate much like England in that respect.[10]

Most migrants were not deterred (although some did return or proceed to California or Hawaii), and the Canadians at least would undoubtedly have agreed with the assessment of John McLoughlin after he visited the valley for the first time in 1832: "certainly it is deserving all the praises Bestowed on it as it is the finest country I have seen and certainly a far finer country than Red River for Indian traders to retire to."[11]

In fact, by the 1830's "Indian traders" had been retiring to the Willamette for a decade or more. The first to do so were the "Willamette freemen," McLoughlin's term for retired company servants who homesteaded the valley, including former Astorians and Nor'westers; and nearly all were Canadians. Astor's Pacific Fur Company, manned mostly by Canadians, had established the first post in the valley, Wallace House, near Salem in 1812-13. The Northwest Company countered with Willamette Post near Champoeg in 1813. Both companies used the valley as an abundant source of beaver and venison; thus, from the beginning the valley was exploited for food, initially game and later grain. Perhaps as early as the mid-1810's and at least by the early 1820's, ex-company and ex-private trappers (*lachés* or freemen) were settling and farming the Willamette. Certainly by the middle 1820's few fur bearers and little game remained in the valley to support *lachés*. In 1827 McLoughlin recollected that "formerly ... there were many Freemen in the Willamette" who had supplied Fort George with a "great quantity" of "leather" (dressed elk hides).[12]

The Astorians and Nor'westers were succeeded in 1821 by the Bay men, for whom the Willamette provided deer and elk hides and pasture for cattle and horses. Soon Hudson's Bay Company servants were retiring there, partly because they were used to, and fond of, the lower Columbia, partly because they were aware of the agricultural advantages of the valley, and partly because they knew that their "country" wives and children would be ostracized in the Canadas.[13] The first may have been Etienne Lucier and three other retired *engagés*, who as early as 1830 established farms just above Willamette Falls (the future Oregon City) on French Prairie (between the Willamette and Pudding rivers), which became the centre of the "Canadian Settlement" in the valley. They were joined by several other retirees in 1831, 1832, and 1833, including seven in 1832 alone. By 1833 there were eight or nine farms on the Willamette.[14] That year an active fur trader reported that "the Willamette is getting much in vogue."[15]

At first McLoughlin, acting on Governor Simpson's orders, refused to allow any retired servants to settle in the valley. "It is true I Know," he wrote, "and Every One Knows who is acquainted with the Fur trade that as the country becomes settled the Fur trade Must Diminish and I therefore Discouraged our people from settling as long as I could without exciting ill Will towards the Company." Indeed, when Lucier applied in 1828 for implements for starting a Willamette farm, McLoughlin "did not give them and Dissuaded him from it and to get Rid of him Granted him and his family a passage to Canada." But Lucier failed to cross the Rockies because of the lateness of the season and returned to join the California trapping expedition, which was directed by McLoughlin to hunt towards the Bonaventure (Sacramento) River in hopes of finding "a place Where we could Employ our Willamette freemen so as to remove them from a place where they were Anxious to begin to farm."[16] That tactic also failed, however, and McLoughlin feared that Lucier and others like him would join the "opposition" (American fur trading vessels on the coast and mountain men in the interior) or become freelance competitors or renegade allies of the Indians. So McLoughlin decided to let them open farms in the Willamette Valley, provided they had families and at least £50 in capital for the purchase of implements and clothing.[17] Otherwise, he reasoned, "It would have Disaffected them to the Company, Excited their ill Will towards us, and Encouraged our opponents to persist in their Endeavours to get a footing in the country"; besides, he added, the company might profit from the marketing of their output of grain. McLoughlin lent each of the settlers seed, flour, and two cows and sold implements to them at 50 per cent of prime cost. But because their former contracts with the company stipulated that they were not to be discharged in the Indian Country, McLoughlin kept them on the firm's books as servants but without duties or wages.[18]

Under such conditions, the "Canadian Settlement" grew slowly. By 1834 it comprised only a dozen or so families on French Prairie stretching fifteen miles along the right bank of the Willamette River above Campment du Sable (Champoeg). Two years later there were seventeen to eighteen families and eighty-three to ninety-five individuals, including fifty-nine children. In early 1837 an American agent, Lieutenant William Slacum, toured the settlement and took a census that listed about one hundred souls and just over a section enclosed and just under a section cultivated (Table 22).[19]

By the mid-1830's, the Canadians had been joined by American settlers, who remained a trickle of mostly retired mountain men until the flood of migrant farmers of the mid-1840's. The first American homesteaders were dropouts from the two abortive trapping, fishing, and trading ventures of Nathaniel Wyeth in 1832-33 and 1834-35. Altogether, twenty-two of Wyeth's men remained, including Wyeth himself, who at the end of the summer of 1834 "laid out" a farm on a one hundred-square-mile prairie on the middle reaches of the Multnomah (Willamette) River. Wyeth also erected a post, Fort William, on the southwestern side of Multnomah (Wappatoo) Island, which was overgrown with forest and prairie and abounding in "considerable" deer (its Indian population having just been annihilated by disease). Here in the spring of 1835 Wyeth

Table 22
THE CANADIAN SETTLEMENT IN THE WILLAMETTE VALLEY, 1836

Name of Settler	Year of Settlement	Number of Children[a]	Number of Houses	Fenced Acres	Tilled Acres	Wheat Crop (bus.)	Horses	Pigs
Jean Baptiste Desportes McKay [Dupaté]	1831	3	3	69	35	556	33	22
Joseph De Lor [Délard]	1832	5	2	28	28	280	11	28
Joseph Gervais	1832	7	3	125	65	1,000	19	55
Etienne Lucier	1832	6	4	70	45	740	21	45
Jean Baptiste Perrault	1832	2	3	80	60	500	4	20
L. Arquette [Amable Arcouët?]	1833	3	2	80	50	600	5	31
Pierre Bellique [Bélèque]	1833	3	2	50	45	700	9	28
Pierre Depeau [Dépot]	1833	1	2	40	35	500	8	39
[François] Xavier Ladéroute	1834	1	2	36	36	350	11	35
William [James?] Johnson	1834	2	2	45	25	300	2	14
Louis Fournier [Forcier]	1835	3	1	34	34	540	9	10
André Longtain [Lonetain]	1835	4	2	45	24	400	3	33
Charles Plante	1835	4	2	60	60	800	12	14
Charles Rondeau	1836	3	1	24	24	200	9	10
André Picard	?	?	?	?	?	?	?	?
Charles Chateau [Chartier?]	?	?	?	?	?	?	?	?
William McCarty	?	?	?	?	?	?	?	?
Louis Labonté	?	?	?	?	?	?	?	?
A. Laferté [Lafortier?]	?	?	?	?	?	?	?	?

For sources see *Sources for Tables*.
[a] In March, 1836.

pastured cattle, sheep, goats, and pigs that he had obtained from Indians, California, and the Sandwich Islands and planted wheat, corn, potatoes, peas, beans, and turnips and apple and other fruit trees.[20] With the decline of the Rocky Mountain fur trade, as a result of overtrapping, changing fashion, and Hudson's Bay Company pressure, more and more mountain men retired to the arcadian Willamette with its combination of promising arable and plentiful game. (The best known settler was perhaps Joe Meek, who was to become a sheriff under the provisional government.)[21] Their hub was the Rocky Mountain Retreat or Tualatin Settlement on the Tualatin Plains, where in the summer of 1841 one ex-trapper had from thirty to forty acres fenced and ploughed.[22]

By then "Oregon fever," a burning desire on the part of many Americans east of the Mississippi to reach what they regarded as the El Dorado of the New Northwest, had ignited. It was fanned by the jingoistic and bellicose exhortations of expansionists like Senators Lewis Linn of Missouri and Caleb Cushing of Massachusetts and inflamed by the intemperate charges of British oppression and conspiracy by publicists like Hall Kelley and Senator Thomas Benton of Missouri, all of whom petitioned and appealed for the extension of American sovereignty and settlement to the Pacific. Informed by this propaganda as well as by travel accounts, public lectures, and personal contacts, hundreds of mostly young male farmers from the Middle West states, particularly Missouri, headed for Oregon. The route of the migrants was the Oregon Trail, which began at a rendezvous on the Missouri River, commonly around Independence, and wound westward across the Great Plains, Rocky Mountains, and Great Basin via Fort Laramie, South Pass, Fort Bridger, Fort Hall, Fort Boise, Fort Nez Percés, and Fort Vancouver to the Willamette. The wagon trains crossed the Missouri right after the spring freshet and reached the Columbia in the autumn just before the onset of the severe interior winter after five arduous months and two thousand interminable miles, with occasional Indian attacks, rough and dusty terrain, and high "mountain prices."[23]

The motives of the migrants were identified by Peter Burnett, one of their 1843 number: to assert an American presence on the West Coast and overcome the British presence; to improve their livelihood and avoid the economic uncertainty and natural hazards, particularly floods, of the Middle West; and to regain their health and escape the morbidity of the eastern states, especially malaria in the Mississippi Valley.[24] The patriotic motive was expressed by the chauvinistic Marcus Whitman in a letter to his brother: "It is now decided in my mind that Oregon will be occupied by American citizens."[25] A year later, after accompanying and assisting the "great migration" of 1843, he declared: "As I hold the settlement of this Country by Americans rather than by an English Colony most important I am happy to have been the means of landing so large an Imigration onto the shores of the Columbia with their Waggons Families & stock, all in Calefty [Safety]."[26] Patriotism notwithstanding, economic motives were paramount. The chief goal of most migrants was a donation claim of 640 free acres of arable land in the salubrious Willamette. Lieutenant Peel summarized the main motives of the Oregon pioneers on the eve of the boundary settlement. "Some are induced to come over from not finding a market for their produce in that Country; others come merely from speculation and a restless disposition, and some either to recover or get rid of their debts, or to escape justice."[27]

The first wagon train left Missouri in 1840; it began with four families in four wagons, plus forty to fifty American Fur Company trappers in thirty to fifty carts (each pulled by a pair of draught mules) with sixty pack mules.[28] Annually thereafter hundreds of American overlanders poured into the Oregon Country via the Oregon Trail (Table 23).

Table 23

AMERICAN MIGRATION TO THE OREGON COUNTRY VIA THE OREGON TRAIL, 1840-46

Year	Departures from Missouri	Arrivals in Oregon	Wagons dep.	Wagons arr.	Cattle dep.	Cattle arr.
1840	?	60 fams.	?	?	?	?
1841	54 inds.	24-136 inds.	?	?	?	?
1842	105 inds.	100-152 inds.[a]	16	?	36	?
1843	990-1,000 inds.	500-1,000 inds. or 100 fams.[b]	120-121	100-150	973[c]	1,300
1844	800 inds.	475-1,500 inds.[d]	?	160-200	?	1,000
1845	?	2,000-5,000 inds.[e]	?	570-600	8,000	2,500
1846	?	1,000-1,350 inds.	?	250	?	?

For sources see *Sources for Tables*.
[a]The modal figure is 137.
[b]The modal figure is 875.
[c]Plus 698 oxen and 296 horses (Anonymous, "Documentary," pp. 190-191; *Oregon Spectator*, 21 March 1846).
[d]The modal figures are 1,000, 1,200, 1,475, and 1,500.
[e]The modal figure is 3,000.

Not every migrant on the Oregon Trail homesteaded the Willamette or even reached the valley. Some died on the way, some turned back, some turned off to California, and some went on to Hawaii. In 1841 thirty of the fifty-four who left Missouri went to California rather than Oregon. "Most" of those who reached Oregon in 1842 were disappointed, and "near half" of them departed for California.[29] In 1843, 293 adult males (over the age of sixteen years) left Missouri for Oregon; six died en route, five turned back, and fifteen went to California, leaving 267 to make the Willamette. Livestock losses were higher, as in 1844, when the migrants lost from one-half to two-thirds of their cattle in crossing the snowbound Cascades.[30]

Most overlanders, however, persevered and soon filled the valley. Following the migration of 1842, before which Canadians were still a majority, Archibald McKinlay reported from Fort Nez Percés that "Americans are getting as thick as Mosquettoes in this part of the world." The "great migration" of 1843 overwhelmed the "Canadian Settlement"; it "more than doubled the resident civilized population" of the Willamette Valley, adding 267 adult American males to the 157 already there.[31] McLoughlin reported in 1844 that "the Country is settling fast." He added that the Americans "have hitherto settled on the South side of the Columbia" but warned that they might start to populate the opposite bank too, because the Willamette was "filling up rapidly."[32]

The Willamette Settlement grew apace (Table 24), the number of Euroamerican inhabitants increasing tenfold between the autumn of 1838 and the autumn of 1841 and again by the autumn of 1845, when up to six thousand occupied the valley. The settlers occupied the margins of the open grasslands of the lower Willamette. There they were closer to the woodland, which provided timber, nuts, and berries, and the spring water of the foothills; they were also farther from the flood waters, heavier soils, and malarial mosquitoes of the river bottomlands. Every free man or widow could claim 640 acres (square or oblong in form), which had to be surveyed, marked, and recorded as well as occupied and improved. It took a settler up to three winter months to erect a one-room cabin of fir logs with a roof of cedar shakes. In early spring he began a vegetable garden and broke the grassland with an ox-drawn wooden plough. Then he had to enclose the ploughland with a Virginia rail fence and his garden with a picket fence, make implements, and build facilities like carts, pens, troughs, and threshing floors. He also raised barns and sheds to protect his livestock from the elements and predators, mainly wolves and coyotes.[33]

For some time agriculture predominated. By the autumn of 1839 there were fifty to sixty "fine" farms on the Willamette. In the summer of 1841 the settlement comprised a string of farms along the river above Campment du Sable between a Catholic mission on the north and a Methodist mission on the south, plus outlying farms in the Yamhill district to the west. According to Lieutenant Wilkes, the Canadian settlers exhibited "cheerfulness and industry" and the Americans "neglect & discontent."[34] One of Wilkes's officers, however, found the Canadians "cultivating small farms from 10 to 40 acres in extent" and wanting "both means and energy ever to become very formidable competitors in the farming line."[35] Governor Simpson toured the valley in the autumn of the same year and was "surprised at the prosperous & promising condition of that infant settlement." Although the settlers had little capital, they were "industrious

Table 24

EUROAMERICAN POPULATION OF THE WILLAMETTE SETTLEMENT, 1838-45

Date	Canadians	Americans	Missionaries	Total
March, 1838	?	?	?	55 inds.
October, 1838	23 inds.	18 inds.	16 inds.	57 inds.[a]
autumn, 1839	24 inds.	20 inds.	10 inds.	54 inds. or 75 fams.
spring, 1840	50-60 inds.	?	?	100-115 inds.
summer, 1840	?	?	?	60-70 fams.
February, 1841	75-80 inds.[b]	50 inds.[b]	?	?
June, 1841	?	?	?	60 fams.
autumn, 1841	up to 80 fams.	?	?	500 inds. or 100 fams.
November, 1841	350 inds. or 61 fams.	150 inds. or 65 fams.	15-16 fams.[c]	500 inds. or 141-142 fams.[d]
autumn, 1842	360 inds. or 96 fams.	?	?	?
December, 1842	?	?	?	500 inds.
early 1843	460 inds. or 96 fams.	84 fams.	20 fams.	809 inds. or 200 fams.
mid-1844	1,000 inds.	2,000 inds.	?	3,000 inds.
late 1844	?	?	?	4,000 inds.
March, 1845[e]	?	?	?	4,000 inds.
autumn, 1845	1,200 inds.	?	?	5,000-6,000 inds.

For sources see *Sources for Tables*.
[a] About 100 Euroamerican and Indian families (HBCA, D.4/106: 28).
[b] All married to Indian women (National Intelligencer, "From Oregon," p. 1).
[c] Comprising from 14 to 15 Protestant families and 1 Catholic priest.
[d] Plus 1,000 Indian hired hands (Schafer, "Letters of Sir George Simpson," p. 82).
[e] Joe Meek's census of the spring of 1845 listed 2,110 individuals (1,259 males and 851 females) (Meek, "Census of Oregon," nos. 12193-194). His count was obviously incomplete.

& enterprising." It was reported in 1845 that the Canadians were "more lavish and less enterprising" than the Americans, but that their farms, being older with perhaps choicer land, were in "better order."[36] It seems that the post-1842 American settlers were more industrious than their predecessors. By the autumn of 1845 the Willamette Settlement extended sixty miles upstream from Willamette Falls along both sides of the river.[37] The "upper settlement" centred on Willamette Station and the "lower settlement" on Campment du Sable and Willamette Falls; another nucleus was the Tualatin Plains.

Towns began to arise as centres for trading, servicing, and processing. The largest was Willamette Falls, which became Oregon City (Plate 14). The site was

Plate 14. *View of Oregon City, 1845.*

the Indian village of Walaml beside the twenty-foot-high cascade, where the local natives caught salmon and trout and traded them to the Cowlitz Indians for camas root. McLoughlin was quick to appreciate the site's water power and transshipment potential and established a claim there in 1829. The settlement grew from one building in July 1840 to five or six in May 1843 and seventy-five in December 1843 (the value of land increased 300 per cent in eight months in 1843). By the autumn of 1845 there were three hundred residents and ninety to one hundred houses.[39] The town's services and trades (tailoring, gristmilling, sawmilling, distilling, tanning, blacksmithing) made the Willamette settlers less dependent upon the Hudson's Bay Company at Fort Vancouver. But until the boundary settlement the Honourable Company remained the chief market for the settlers' main crop — wheat. Especially wheat but oats and potatoes too, as well as livestock, flourished in the "garden of the Columbia."

8

THE "GARDEN OF THE COLUMBIA": THE SUCCESS OF HOMESTEAD FARMING

The Hudson Bay Company was a great element of order & law in this community. It was here, & organized under its own system when these people [pioneers] came here. It had its peculiarities; & of course it was not intended to permit the settlement of this country. It was a fur company; its interest was in favor of keeping the country in the condition of a fur producing country. But it had a man at its head who was very much of a philanthropist, a noble man in many respects; & while he did not invite the American people he treated them very well when they came, & facilitated settlement by giving them what they wanted to enable them to make farms & live.

MATTHEW DEADY, PIONEER, 1878

Unlike the retired Canadian servants of the Hudson's Bay Company, who knew the Willamette well and did not have to trek across half a continent to reach it, the American migrants were often exhausted and impoverished by the time they began to homestead the valley. For some time, however, both groups shared a dependency upon the company as a source of goods and as a market for their crops. In particular, those migrants who had lost their possessions on the Oregon Trail were obliged to take work with the company in exchange for supplies or with established settlers, often relatives or friends, in exchange for food and seed, or to sharecrop until they had accumulated enough capital to preempt or purchase their own land and develop commercial farms.[1] (Newcomers thus provided a cheap labour pool and a sizable wheat market, game being scarce by the middle 1840's.) Dependence upon the Hudson's Bay Company especially galled many American settlers, who disliked and opposed the concern anyway because it represented John Bull, colonialism, monopolism, and non-agrarianism. But, at least in the beginning, they had little choice. Most settlers needed supplies and implements, and Fort Vancouver was the nearest source. In addition, in order to rise above subsistence farming, the settlers required a market for their surplus produce, and the company and its

Russian contract was the closest outlet; indeed, without the Sitka market the company would have bought little, if any, wheat from the settlers.[2] Much to the consternation of the Governor and Committee, the "courteous and kindly" John McLoughlin loaned supplies to needy American migrants, who after enduring the hardships of the Oregon Trail, arrived to face the onset of the rainy and cool Willamette winter. They repaid initially in labour and eventually in kind (wheat or shingles), as did indebted retired servants. In autumn 1843 the American explorer Captain John Frémont noted the company's assistance to American overlanders:

> I found many American emigrants at the fort . . . and all of them had been furnished with shelter, so far as it could be afforded by the buildings connected with the establishment. Necessary clothing and provisions (the latter to be afterwards returned in kind from the produce of their labor) were also furnished. This friendly assistance was of very great value to the emigrants, whose families were otherwise exposed to much suffering in the winter rains, which had now commenced, at the same time that they were in want of all the common necessaries of life.[3]

The Willamette settlers also paid in wheat or shingles for company goods that were otherwise unobtainable (in 1845 they bartered their wheat for company goods at sixty cents a bushel and their shingles at four dollars per thousand). This system of commodity exchange lasted until 1848; by then Fort Vancouver had declined in the wake of the British withdrawal from the lower Columbia, and American manufacturers and retailers had emerged on the Willamette.[4]

Some settlers did not honour their debts to the Hudson's Bay Company. Presumably they were those 1842 arrivals whom an American missionary found to be "not generally very calculated to make wholesome citizens."[5] Similarly, a company clerk declared that "upon the whole, I have no taste for a new country, and if the Americans who come here are a fair sample of their law and order, neither should I desire to live in what they call a free country." Most settlers, however, seem to have been law-abiding and hard-working and to have repaid their debts. Governor Simpson stated that the Americans had a "strong feeling of nationality" but noted that there were "few cases of outrage or atrocity"; he added that generally they were "industrious and enterprising."[6] James Douglas, who knew the settlers even better, concurred:

> No sort of manufacture is yet introduced, but the restless Americans are brooding over a thousand projects, for improving the navigation, building steam Boats, erecting machinery and other schemes that would excite a smile, if entertained by a less enterprising people, with the same slender means. After, however, having witnessed the perfect indifference, with

which an American embarks his last shilling in more unpromising speculations, I really think, that very slight encouragement would give the necessary impulse.[7]

Such industry quickly transformed the landscape of the Willamette Valley. Already in 1836 the Canadian settlers reported to the Bishop of Juliopolis that "the farms are All in a very thriving state and produces fine Crops."[8] And in 1841 a member of the U.S. Exploring Expedition described the valley just above the falls: "the land as far as I could see on both sides cultivated with wheat and everything bore the same appearance of the ordinary farms in New England."[9]

Farming was considerably eased by the mildness of winter and the availability of readymade grassland. A *New York Herald Tribune* reporter wrote in early 1841 that "these people ... live very much the same, in all respects, as our farmers at home, with the exception of not being obliged to labor half as much .... The people here cut no hay and make no pastures."[10] Later that same year, Lieutenant Wilkes estimated that a man would have to work three times as hard in the United States as in the Willamette Valley to achieve an equal result, mainly because on the Willamette it was not necessary to house livestock in winter.[11] Winters were damp as well as mild. The rainy season, which lasted from mid-October until mid-April, impressed Peter Burnett:

> But during most of the rainy season the rains are almost continuous. Sometimes the sun would not be seen for twenty days in succession. It would generally rain about three days and nights without intermission, then cease for about the same period (still remaining cloudy), and then begin again. These rains were not very heavy, but cold and steady, accompanied with a brisk, driving wind from the south.[12]

Perhaps the winter dankness helps to explain Burnett's eventual departure for sunnier and drier California, where he became that state's first governor. Be that as it may, the temperate winters permitted autumn (November-December) as well as spring (February-March) sowings. Most grains, including wheat, were planted in autumn; oats, peas, and potatoes were seeded in spring. Harvesting began in the third week of July; the summer heat sometimes reduced barley and potato yields.[13] In response to the demands of the Fort Vancouver market, wheat was the dominant commercial as well as subsistence crop, even being recognized by the provincial government as legal tender (one bushel = one dollar).[14] Wheat growing was recalled by an American settler:

> The farmer enclosed as much land as he desired to cultivate with a rail fence. When the September showers came to moisten the soil, he ploughed and sowed his grain, which grew slowly during the winter and ripened for

the harvest in July. This was cut with scythe and cradle, bound into bundles, and drawn to a point in the field where a small space of ground was smoothed to a level surface. There a circular enclosure was made from forty to fifty feet in diameter, the grain was spread on the ground, and a bank of six to ten horses was turned in, and a man or boy standing in the centre with a long whip urged them around until their hoofs had shelled the dry grain, when the straw would be thrown over the fence, the grain floor again covered and threshed until the crop was finished. Then the grain and chaff were shoveled into one great heap, a fanning mill turned by a crank by hand separated the grain from the chaff, and the gathering of the crop was complete.[15]

The main types of wheat were two anonymous varieties called "spring red" and "[winter] white seed."[16] The mild winters allowed the settlers to grow winter wheat, which yielded twice as much as the spring planting. Spring wheat yielded fifteen to thirty bushels per acre and averaged twenty bushels; winter wheat yielded thirty to fifty bushels per acre (and oats, thirty, barley thirty-five, and peas fifteen bushels per acre).[17] The low wheat yields resulted from crude ploughing, seeding, and threshing. Lieutenant Wilkes nevertheless asserted that Willamette wheat was better and heavier, by nearly four pounds per bushel, than that grown in the United States. Indeed, the Willamette Valley was extolled by one American migrant as "one of the best wheat countries in the world. You can sow wheat any time of the year, and you are sure of a good crop."[18]

The first wheat crop was grown in 1832. In 1833 one Canadian settler on French Prairie harvested 500 bushels; in 1842 some Canadian settlers each owned 100 or more horses and reaped 1,660 bushels of wheat.[19] The *New York Herald Tribune* reported in early 1842 that each of them reaped 300 to 500 and sometimes 1,000 to 1,200 bushels of wheat and owned 50 to 100 horses, 50 to 100 pigs, and 25 to 50 cows. In the spring of 1843, the 809 Willamette settlers possessed 6,284 acres of improved land, 3,519 cattle, 2,683 horses, 1,733 pigs, and 135 sheep and had harvested 31,698 bushels of wheat in 1842.[20] Clearly the settlers had prospered. Their wheat output increased from 3,000 bushels in 1835 to 5,000 in 1836, 5,000 again in 1837, 8,000 to 10,000 in 1841, and up to 52,000 in 1846.[21] They produced 27,675 to 35,000 bushels of all kinds of grain in 1841 and 47,725 bushels in 1842.[22]

The surplus wheat was sold to the Hudson's Bay Company: 1,000 bushels in 1836, 4,000 in 1841, 20,000 to 26,000 in 1845 (about half of the settlers' surplus), and at least 12,000 in 1846.[23] McLoughlin paid liberally for wheat — three shillings or $1.10 per bushel in 1841 — but he paid half in cash or scrip, redeemable only at the company's stores at Fort Vancouver and Willamette Falls, and half in the form of a 50 per cent discount on purchases of company goods from the same stores. According to McLoughlin, the Canadian settlers furnished three-quarters of the wheat bought by the company.[24] Either they were more indebted to the company than the American settlers or they were simply favoured as suppliers.

The company erected "receiving stations," or depots, at Campment du Sable and Willamette Falls for storing the wheat bought from the settlers. It was ground at three mills: Fort Vancouver, Campment du Sable, and Chemeketa (North Salem). George Roberts, a company clerk at Fort Vancouver, recollected that this traffic was "not very profitable or desirable," but "it enabled the company to gather in a good deal of old debt."[25] In addition, of course, it enabled the company to meet the terms of the Russian contract and reap its benefits.

Although less calorific and more perishable than wheat, potatoes were also a staple, tolerating a wide array of environments and requiring little attention. The Canadian settlers, in addition, grew oats for their horses, as well as peas and beans, and had mature apple, pear, and peach trees.[26]

With its year-round grazing and readymade grassland, the Willamette Valley invited stock rearing, and the settlers were quick to respond. An early settler noted in the winter of 1841-42 that "they have large droves of horses and cattle, who graze on the green grass all winter; and there is no other cost or trouble to raise stock, then to keep them from going wild."[27] The Canadian settlers used horses as draught animals, but the Americans preferred oxen because they grazed better, strayed less, endured more than either horses or mules, and furnished more edible meat and more potable milk. Indian, Spanish, and American stock were reared. The horses brought by American migrants were gentler and more fine blooded than the Cayuse ponies, but the Spanish horses from California, big, strong, tough, spirited, and fleet, were better yet.[28] The black Spanish cattle, however, were inferior to the American stock (Shorthorn and Durham), which were smaller but gentler and gave more milk. American cows cost up to four times more, partly because of their scarcity. That fact prompted the driving of livestock from California. Moreover, the Hudson's Bay Company would not sell cattle to the settlers, although it would lend them "as many as any person wished to use."[29] In 1837 twenty to twenty-five men of the Willamette Cattle Company under Philip Edwards drove 700 to 800 neat cattle and 40 horses from San Francisco Bay to Campment du Sable, arriving with 630 head of cattle, mostly heifers, and all of the horses. In 1841 two more drives brought 3,500 cattle and sheep.[30] And in the spring of 1843, 1,200 cattle, 600 sheep, and 200 horses left California for Oregon; "probably" 1,500 head of cattle and sheep reached the Willamette that summer.[31] These drives swelled the livestock population of the Willamette Valley and eventually made the settlers independent of the Hudson's Bay Company for cattle, horses, and sheep. As immigration increased, the Spanish longhorns were moved southwards to the more extensive and less settled grasslands, where their independent and combative nature was better suited to less tending and more predation. Pigs were also well suited to frontier conditions, being hardy animals that required little care. "Grunters" were versatile feeders, heavy breeders, and collectively spirited fighters.[32]

In 1838 the Willamette Settlement had 600 cattle and "a good stock of swine & horses in sufficient numbers for the purposes of tillage." In the autumn of 1841 there were 3,000 cattle, 3,000 pigs, and 2,000 horses, and in the spring of 1843, 3,519 cattle, 2,683 horses, 1,733 pigs, and 135 sheep.[33] The latter year saw the arrival of the "great migration," which tripled the population of the Willamette Settlement. Thereafter the Canadians were in a negligible minority, and the Hudson's Bay Company was on the defensive. American acquisition of at least the left bank of the lower Columbia was a foregone conclusion.

The success of the Willamette Settlement was unequivocal. Its population grew steadily and, from the early 1840's, rapidly, and its farms raised abundant crops and reared sizable herds. Any problems, such as the poisoning of cattle by water hemlock, were minor. Two factors account for this success: the munificence of the Willamette Valley and the liberality of the Hudson's Bay Company at Fort Vancouver. Physically, the valley was a "soft" land, with few, if any, extremes or insurmountable obstacles. Dry summers and hard winters, for example, were rare, and the large extent of readymade grassland minimized clearing and facilitated grazing. Indian resistance was almost non-existent because the Indians themselves had been nearly exterminated by the time the first Euroamericans settled the land.

The settlers faced only one major problem — a shortage of capital — and that was overcome by the Hudson's Bay Company, or, more specifically, by Chief Factor John McLoughlin, for the attitude of his superiors to the settlers was hostile. The Governor and Committee realized that hunting-trapping and large-scale farming were incompatible, and they feared that their monopoly would be broken. In 1838 Douglas warned Beaver House that "the Wallamatte Settlement is annually growing in importance, and threatens to exercise, in course of time, a greater influence, than desirable over our affairs."[34] He added:

> The interests of the Colony and Fur trade will never harmonize, the former can flourish only through the protection of equal laws, the influence of free trade, the accession of respectable inhabitants; in short by establishing a new order of things, while the fur Trade, must suffer by each innovation.
>
> The only perceptible effect yet produced on our affairs, by the existence of the Settlement, is a restless desire in the Companys servants, to escape from our service to the Colony; but in the present state of the country when the Settlers are so entirely dependent on us, that every man must go down to Vancouver, to sharpen his share, his coulter, and his mattock, we have no reason to fear desertion; however, when the introduction of foreign capital

terminates this dependence such events may be expected; and our general influence will decline as the wants of the Settlement find a provision in other sources.[35]

Douglas was right, of course. His fears were realized even before the signing of the boundary treaty in 1846. In 1844, for example, Chief Trader Paul Fraser reported from Fort Umpqua at the southern end of the Willamette Valley that because of the influx of American settlers, "many of the Indians had shifted their location, hunting was neglected and our business very poor."[36] Simpson was even more anxious, writing to McLoughlin in 1840:

> The Governor & Committee notice with much alarm the increase of population at the Wallamette, and from the crafty and designing character of the more enlightened residents there, such as Bailey, Farnham, and I may even add, the United States Missionaries, it is quite evident that they are likely to become very dangerous neighbours; no pains or expence ought, therefore, to be spared in endeavouring to break up that settlement, either by getting the people removed to the Sandwich Islands, California, or other parts, or getting our own retired servants removed to the Cowlitz, where they would be more immediately under our guidance & control. Unless that settlement be broken up, it will become the resort of the worthless & dissaffected outcasts from the United States and Sandwich Islands, and may prove not only highly injurious to our own views in regard to agricultural & pastoral pursuits, but to the best interests of the Fur trade.[37]

Simpson always feared that the Willamette Settlement would attract desperate and unscrupulous men who would oppose the company's interests, and given the sizable number of ignorant and truculent Southern uplanders from the "Western states," such as Missouri and its neighbours, among the migrants, his fears were not groundless.

At Fort Vancouver the powerful but aging McLoughlin, the "king of Oregon" to the settlers, was caught in the middle. He was torn between feelings of political and commercial loyalty to his British employer and compassion for the needy American immigrants. Moreover, McLoughlin had a vested interest in the Willamette, where he had a disputed claim at Willamette Falls. And there was a growing rift between him and Simpson. At any rate, McLoughlin chose to aid the settlers, and that aid was crucial in giving them a start. At least he was able to show the "great folk" that most of the settlers paid their debts in wheat that was needed to honour the Russian contract.

McLoughlin's assistance was appreciated and acknowledged. Silas Holmes, a surgeon with the U.S. Exploring Expedition, wrote: "he has uniformly furnished to settlers every possible facility: allowing them the use of his horses, cattle and utensils of every kind till such time as they are able to provide them for themselves; their crops are purchased by him for the Co. at prices which if not large, at least furnish a fair remuneration."[38] The Methodist missionary Elijah White declared in 1843 that "the gentlemen of this company have been fathers and fosterers of the [Willamette] colony ever encouraging peace industry and good order and have sustained a character for hospitality and integrity too well established easily to be shaken."[39] And the 1843 migrant Peter Burnett asserted that "had it not been for the generous kindness of the gentlemen in charge of the business of the Hudson's Bay Company, we should have suffered much greater privations. The Company furnished many of our immigrants with provisions, clothing, seed, and other necessaries on credit."[40]

McLoughlin's aid may have been motivated by more than just altruism. Henry Warre and Mervyn Vavasour, the two British military officers who inspected the lower Columbia in 1845, reported to the Secretary of State for the Colonies that the company had calculatingly assisted the Willamette Settlement for commercial as well as humanitarian reasons:

> In conclusion, We must beg to be allowed to observe, with an unbiassed opinion, that what ever may have been the Orders, or the Motives of the Gentlemen in charge of the Hudsons Bay Company's Posts on the West of the Rocky Mountains, their policy has tended to the introduction of the American Settlers into the Country.
>
> We are convinced, that with out their assistance, not 80 American Families, would now have been in the Settlement.
>
> The first Immigrations in 1841 or 1842, arrived in so miserable a condition, that had it not been for the Trading Posts of the Hudsons Bay Company, they must have starved, or been cut off by the Indians.
>
> Through motives of humanity, We are willing to believe, and from the anticipation of obtaining their Exports of Wheat & Flour to the Russian Settlements & to the Sandwich Islands, at a cheaper rate; The Agents of the Hudsons Bay Company gave every encouragement to their Settlement, and Goods were forwarded to the Willamette Falls, and retailed to these Citizens of the United States at even a more advantageous rate than to the British Subjects.
>
> Thus encouraged, Emigration left the United States in 1843, 1844 and 1845, and were received in the same cordial Manner.
>
> Their Numbers have increased so rapidly, that the British party are now in the Minority, and the Gentlemen of the Hudsons Bay Company, have been obliged to join the Organization [provisional government] with out any reserve, except the mere form of the Oath of Office; Their lands are invaded, Themselves insulted, and They now require the protection of the

British Government, against the very People, to the introduction of whom, they have been more than accessory.[41]

But by then the question of motives was academic. The boundary had finally been drawn (and farther to the north than either McLoughlin or Douglas had expected), and the Old Doctor had cast his lot with the Willamette Settlement, retiring there and eventually becoming an American citizen. And the settlement itself was well on the way to becoming the heart of the new Oregon Territory.

Part Four

MISSION FARMING
Protestant and Catholic Husbandry on the Lower Columbia

Missionaries. Little has yet been effected by them in christianizing the natives. They are principally engaged in the cultivation of the Mission farms and in the care of their own stock in order to obtain flocks and herds for themselves, most of them having selected lands. As far as my personal observations went, in the part of the country where the Missionaries reside there are very few Indians, and they seemed more occupied with the settlement of the country and in agricultural pursuits, than Missionary labors.

>*Lieutenant Charles Wilkes, commander of the U.S. Exploring Expedition, 1841*

9

THE "MACEDONIAN CRY": THE ADVENT OF MISSIONARIES

> *The missionaries are faithfully and successfully laboring to scatter the Bread of Life among the poor benighted children of the forest in those ends of the earth.*
>
> ANNUAL REPORT OF THE BOARD OF FOREIGN MISSIONS OF THE METHODIST EPISCOPAL CHURCH, 1842

Missionaries were the last group to undertake agriculture on the Oregon frontier, lagging behind even the pagan Indians whom they came to rescue from damnation. Methodism, an evangelistic, revivalistic creed which held that all men can be saved, was first in the field. An evangelical revival in North America and Western Europe had awakened religious zeal. Missionary societies were formed, particularly in England and the United States, and young zealots, recalling Jesus Christ's behest to "go ye into all the world, and preach the Gospel to every creature," were dispatched to distant lands to convert the heathen by means of preaching, teaching, and publishing. Evangelism provided the impulse; commerce revealed the prospects and furnished the transportation or at least the guides. In 1834 the Reverend Jason Lee of the Mission Society of the Methodist Episcopal Church and his nephew, the Reverend Daniel Lee, were dispatched to the Oregon Country by the Missionary Society of New York City to Christianize the Indians. They reached Oregon in company with the returning Nathanial Wyeth and were warmly welcomed and generously assisted by John McLoughlin, for he saw the missionaries as agents of civilization bent on saving souls rather than trading pelts. The "gentlemen of [Fort] Vancouver" advised the Lees to locate in the Willamette.[1] The presence of arable land, fresh water, workable timber, and at least some Indians decided Jason Lee in favour of the Willamette Valley over the remote Flathead Country, his original destination. Personal comfort and economic potential may also have swayed Lee, since the Willamette had lost most of its Indians by then to warfare and disease, and its climate was balmier and its agricultural prospects brighter than those of the upper reaches of the Columbia. Besides, the Flatheads were neither numerous nor sedentary.[2] The Methodist brothers were supposed to be itinerant but seldom seem to have left their stations and farms except to tour the country or visit one another. Indeed, the increasingly secular nature of the Oregon Mission was to lead to Superintendent Lee's recall by the Board of Managers of the Missionary Society of the Methodist Episcopal Church in 1843 and the reduction and eventual closure of his mission.

The first Methodist mission was the Willamette Station or Mission Place, founded by Jason Lee in the autumn of 1834 at Mission Bottom on the right bank of the Willamette (opposite present-day Wheatland). It was established with a particular view to farming; McLoughlin loaned eight cows, eight calves, seven oxen, and one bull.[3] In 1838 the mission contained from sixteen to twenty-five persons, mostly farmers, mechanics, and teachers rather than clerics. In the fall of that year, James Douglas reported that "the Mission is, at present, the life and soul of the [Willamette] Settlement, dispensing its bounties with a liberal hand." "Last winter," he added, "its Members laid out upwards of £500 in various improvements, purchases of Land & farm stock, which gave an extraordinary impulse to industry and greatly inhanced the price of labour."[4] By the summer of 1840 there were sixty-eight souls at the mission.

In 1837 the Willamette Station was augmented by spring and autumn overland parties (thirteen and seven persons, respectively) headed by Dr. Elijah White and David Leslie, so that it was possible to open a new mission, Wascopam, in the spring of the following year under Daniel Lee on the southern side of the Columbia four miles below the Dalles. Thomas Farnham, one of the few survivors of the disastrous "Peoria party" that trekked westward in 1838 with the motto "Oregon or the grave," described the mission:

> The buildings of the mission, are a dwelling-house, a house for worship and for school purposes, and a workshop, etc. The first is a log structure thirty by twenty feet, one and a half floors high, shingle roofs, and floors made of plank cut with a whipsaw from the pines of the hills .... Their premises are situated on elevated ground, about a mile south-west from the river. Immediately back is a grove of small white oaks and yellow pines; a little north, is a sweet spring bursting from a ledge of rock which supplies water for house use, and moistens about an acre of rich soil. About a mile to the south, are two or three hundred acres of fine land, with groves of oak around, and an abundant supply of excellent water.[5]

The ambitious Jason Lee believed that the Oregon Mission had to be expanded to succeed. In 1838 he returned to the United States and undertook a lecture tour to arouse support for more American settlement, both clerical and secular. One result was the "great reinforcement" of 1840, when fifty-one recruits reached Oregon aboard the *Lausanne* from New York. This addition spawned three more missions in the spring of that year: Clatsop Station under

John Frost at the mouth of the Columbia, Nisqually Station under William Wilson at Fort Nisqually, and Willamette Falls under Alvan Waller. (The last conflicted with McLoughlin's land claim at the falls and generated a prolonged dispute.)

At the same time, mission headquarters were moved ten miles up the Willamette from Mission Bottom to Chemeketa, "place of rest," where, it was hoped, disease would be less prevalent and the Indians less indifferent and indolent. New farmland was immediately opened, and a gristmill and a sawmill were added in 1842.[6]

Thereafter, the Oregon Mission declined as the Mission Board, alienated by Jason Lee's publicist activities and "unheard of amount of temporal business," withdrew support. Jason Lee was removed in 1843, and Daniel Lee in 1844, after an expenditure of more than $100,000, and personnel and operations were drastically reduced.[7] Nisqually Station was abandoned in the summer of 1842. By the spring of 1843 twenty-nine adult missionaries remained at four missions with forty-seven Euroamerican and eight hundred Indian members. In the summer of 1844, when the mission was dissolved, there were sixty-five whites at the four stations, two-thirds of them at Chemeketa.[8] The mission had been hampered by weak leadership and personality conflicts. Jason Lee himself had not been sending detailed, regular reports to the Mission Board, and some mission members had resigned and criticized the superintendent's management. (Particularly disaffected was Dr. White, who was expelled in 1840 and became a United States Indian Agent.) All but two of the missionaries had deserted their posts. Between 7 November 1834 and 20 June 1838 only twenty Indians had been "admitted" to the mission, and twelve of them died or left.[9] Lee's successor as superintendent, George Gary, conducted an investigation and concluded that "there were more persons connected with the mission than could be profitably employed; and that more property was held by it than was for its advantage either temporally or spiritually."[10] So Gary discharged all of the lay employees save one, the farmer Henry Brewer at Wascopam, and sold all of the property except the churches, parsonages, and Wascopam farm for $30,000 (which, incidentally, was guaranteed by the new provisional governor, George Abernethy, a former mission steward). The Mission Board even admitted that the "great reinforcement" of 1840, which entailed an outlay of $5,000 on the purchase of goods and the erection of a sawmill, had been unnecessary. This retrenchment was prompted by both economics and politics, for the claimed but unoccupied lands (thirty-six sections) of the mission had generated jealousy and even claim-jumping on the part of American migrants.[11] As Gary reported:

> the *secular character* of the mission had already excited suspicions and heart burning among the newly arrived emigrants, which threatened an almost entire loss of confidence in the purity of our motives in its establishment

and prosecution. This would have been a loss for which no amount of money could compensate. The hopes of the mission, for the future, depend principally upon the success of the Gospel among the emigrants. The Indians are comparatively few in number and rapidly wasting away. The Territory, however, is fast filling up with whites from the States, and the future character of this colony must depend greatly upon the impress it may receive in its infancy.[12]

Another Oregon Mission was launched in 1836 by the American Board of Commissioners for Foreign Missions (1810-60), a corporate creation of the Congregational, Presbyterian, and Dutch Reformed denominations of New England and the Middle States. These Calvinistic elements united to combat the spread of religious liberalism exemplified generally by rationalism and infidelity and specifically by Unitarianism. This missionary movement was an outgrowth of the Second Great Awakening, a revival of American evangelism in the late eighteenth and early nineteenth centuries that also found expression in camp meetings, Bible and tract societies, and evangelical magazines. In New England humanitarian and missionary zeal were particularly aroused by Samuel Hopkins's doctrine of "disinterested benevolence." Millenarianism, which demanded worldwide conversion in time for the Second Coming, also fueled missionary fervour. This chiliastic spirit lent a sense of urgency to the missionary movement. Dozens of young men and women, mainly residents of rural New England and graduates of Andover Theological Seminary, answered the call of Christian duty, and romantic adventure, to go and proclaim the Gospel in the "moral wastes" of the world. Their own country was expanding westward, and beyond the Mississippi lay uncounted thousands of heathen Indians whose salvation tweaked the evangelical conscience.[13]

The Oregon Mission was led by the determined Dr. Marcus Whitman, a physician and catechist whose love of country was exceeded only by his hatred of Catholicism. In 1836 he and his wife, Narcissa, founded Waiilatpu, "the place of rye grass" (Plate 15), on the Walla Walla River among the Cayuse Indians and close to Fort Nez Percés, a source of supplies and, if necessary, a refuge. Like the Lees, the Whitmans were helped by the Hudson's Bay Company. Narcissa referred to McLoughlin as "our Excellent friend & kind benefactor," and another missionary wife, Sarah Smith, declared that "I often think this company outdo the Americans in kindness & hospitality."[14] Mrs. Whitman described the mission's attractive site:

> It is indeed a lovely situation, we are on a beautiful level, a peninsular, formed by the branches of the WW River, upon the base of which, our house stands, on the S.E. corner near the shore of the main River. To run a fence across to the opposite river on the North, from our house this with

the rivers would enclose 300 acres of good land for cultivation, all directly under the eye. The rivers, are barely skirted with timber, this is all the wood land we can see, beyond them as far as the eye can reach plains & mountains appear. On the east a few rods from the house is a range of small hills covered with bunch grass a very excellent food for animals & upon which they subsist during winter, even digging it from under the snow.[15]

Plate 15. *View of Waiilatpu Mission, 1846.*

By 1842 there were two one-storey adobe houses, a small sawmill, and more than one gristmill.[16] Another missionary found that the acreage of arable land was "not very great," but he extolled the climate ("one of the finest climates in the world") with its short, dry winters and the "excellent" soil with its capacity to retain moisture. Consequently, he added: "It is not necessary to cut any hay for winter. Animals find grass the whole year. This is an immense saving of labor."[17]

Apart from farming, Marcus Whitman's chief aims were to convert the Cayuses, defeat the Catholic missionaries, and further American sovereignty (he could justifiably be labelled a conscious agent of American imperialism). He told his parents in 1844 that "I have no doubt our greatest work is to be to aid the white settlement of this country and help to found its religious institutions."[18] And two years later Whitman elaborated his goals to his brother:

> it is quite important that such a country as Oregon should not on one hand fall into the exclusive hands of the Jesuits, nor on the other under the English government ... the Jesuit Papists ... were fast fixing themselves here, and had we missionaries had no American population to come in to

hold on and give stability, it would have been but a small work for them and the friends of English interests, which they had also fully avowed, to have routed us, and then the country might have slept in their hands forever.[19]

Also in 1836, Henry Spalding, a sometimes impetuous and highhanded man, and his wife, Eliza, established the Nez Percés or Clearwater Mission of Lapwai ("the valley of butterflies") on the Clearwater River just above its junction with the Snake. It was found closer to timber and had access to more numerous and more receptive Indians (Nez Percés) than Waiilatpu. Its site was additionally advantageous in that the breeze off the river moderated the summer heat and discouraged the mosquitoes, especially at night. Lapwai was described by Narcissa Whitman right after its founding: "The land is very good but not very extensive but sufficient for the establishment, & most of the Indians. Enough may be found near on other streams for the remainder. Plenty of good timber, stone clay & water that is, fine spring, more timber on this location than on ours."[20]

Plate 16. *View of Tshimakain Mission, 1843.*

In 1837 one of the brethren, William Gray, an ill-humoured cabinetmaker and erstwhile missionary and physician, journeyed to the United States and returned in 1838 with a modest reinforcement, including his own bride and three missionaries and their wives: the Eellses, Smiths, and Walkers. Late that year Cushing Eells and Elkanah Walker were sent northward to found a mission among the Spokane Indians. With the assistance of Chief Trader McDonald of nearby Fort Colvile, they launched Tshimakain ("the place of springs") (Plate

16), several miles north of the Spokane River and seventy miles southeast of the Hudson's Bay Company's post. The rather dreary and lonely site was described by Mary Walker shortly after her arrival:

> our cabin which is sixteen feet squair without door window chimney or roof except a few boughs & grass stands at the western end of a plaine on the verge of a pine wood. The plaine is 6 or 8 mls. long elipticle or somewhat crescent form. It is intersected by a creek half as large as Breakneck Brook [Maine] which empties into a beautiful stream a few miles below called Spokane R. which is half the size perhaps of Laco [Maine]. Except one snowcaped mountain at the NE the entire horizon is met by an unbroken forest of pine.[21]

In 1839, after briefly helping the Spaldings at Lapwai, the Smiths moved sixty miles up the Clearwater and opened another mission among the Nez Percés. Kamiah's site was a "small circular plain of about 30 acres" on the right bank of the river.[22] Asa Smith's linguistic ability was offset by the hostility of the local Indians, and his work was hampered by his wife's chronic illness. In 1840 the aspiring Gray finally received permission to establish his own mission at Shimnap at the junction of the Yakima and Columbia Rivers. And in 1847 the station at Wascopam was purchased from the Methodists.

Most of the Congregational Presbyterians were not well suited by either temperament or training to cope with the isolation, discomfort, danger, and other rigours of frontier life. Few lasting converts were made among the unpredictable Indians. Personality clashes divided the missionaries and nearly led to their recall in 1842 (only Marcus Whitman's desperate winter ride to Boston persuaded the American Board otherwise). The Smiths soon abandoned Kamiah for the more salubrious Sandwich Islands. Gray quit Shimnap to homestead in the Willamette Valley (indeed, the mission may not even have been started at all). The most decisive setback was the Whitman Massacre of November 1847, when fourteen people, including the entire Whitman family and some overlanders, were killed by disillusioned and disaffected Indians. (The survivors were rescued and the murderers captured by a Hudson's Bay Company party.) That disaster spelled the end of the Oregon Mission of the American Board. Lapwai and Tshimakain were abandoned in 1848.

However, the Jesuit "black robes" were not far behind the Protestant "Bible colporteurs." The Roman Catholic Church had a long tradition of mission work, and in the New World the Jesuits, Dominicans, and Franciscans in New Spain and the Jesuits in New France had been particularly active, despite the Jesuit Suppression of 1773-1814. The Oregon Country offered two objects: the godless Indians and the priestless French-Canadian fur traders. The latter in 1821 and again in 1834 and 1835 petitioned the prelate of Red River for spiritual ministration. Finally, two priests, François Blanchet (later the Arch-

Plate 17. View of St. Paul Mission, 1846.

bishop of Oregon) and Modeste Demers (later the Bishop of Vancouver Island) trekked overland in 1838 from Québec to Fort Vancouver "to cultivate the vineyard of the Lord." At the end of that year Blanchet established St. Francis Xavier Mission at Cowlitz.[23] At the beginning of 1839, St. Paul Mission (Plate 17) was opened on the Willamette in French Prairie in a log church that had been built by settlers in 1836; in the autumn mission headquarters were moved from the Cowlitz to the Willamette, where most of the Canadian faithful lived. Blanchet assumed control of St. Paul, and Demers was put in charge of St. Francis. In 1844 St. Paul Mission was reinforced by a party of eleven. In addition, there were temporary "missions" at Fort Vancouver (Ste. Marie) and Willamette Falls (St. Louis).

Meanwhile, some interior Indian proselytes had lost patience. Four times, in 1831, 1834, 1837, and 1839, a mixed party of Nez Percés and Flatheads journeyed to St. Louis (the 1837 delegation was annihilated by Sioux en route), supposedly to find the "white man's book of heaven." Finally in 1839 in response to this "Macedonian cry" for spiritual aid, Father Pierre De Smet was sent west to preach and baptize among the Indians of the "Stonies." He did so in 1839 and 1840, returning to St. Louis for funds, supplies, and helpers. In 1841 a missionary party of seventeen, led by De Smet and including Father Gregory Mengarini, joined a group of sixty-four migrants bound for Oregon. At Fort Hall the missionaries left the wagon train, and at summer's end they reached the Flathead Country between the Bitterroot Range and the Rockies astride the Clark Fork. That fall St. Mary's [Marie] Mission was founded on the right bank of the Bitterroot River (near present-day Stevensville, Montana). Within two months one-third of the Flatheads had been baptized.[24]

De Smet energetically expanded the Flathead or Rocky Mountain Mission. In late 1842 Sacred Heart Mission was established among the Cour d'Alenes by Father Nicholas Point under De Smet's direction (in 1846 it was shifted to present-day Cataldo, Idaho). St. Joseph, St. Michael, and St. Pierre were opened in 1843. St. Ignatius Mission was founded by De Smet in 1844 among the Pend Oreilles (near present-day Newport, Washington); its first resident priest was Father Adrian Hoecken. And in 1845 De Smet launched St. Paul's Mission at Fort Colvile; Father Peter Vos was its first resident priest. Before long, however, the Catholic missions, like their Protestant rivals, were engulfed by the flood tide of Euroamerican settlement. Indian backsliding in the face of white hypocrisy and prejudice; Blackfoot hostility; the decline of financial support from Europe; and the shortage of priests and supplies — all resulted in the closure of St. Mary's in 1850 and the end of the Flathead Mission.

While the Oregon missions were active, however, all of the missionaries, Protestant and Catholic alike, divided their time among a daunting array of activities, ranging from studying Indian languages to making household implements and farming. Generally, they regarded crop and animal husbandry as virtuous (if not quite divine) occupations and lessons of Christianity, whereas hunting and trapping were rather wicked (if not quite satanic) pursuits commonly associated with heathen savages. But there were more practical reasons for mission agriculture as well. The more grains, vegetables, and fruits they grew and the more livestock they reared, the less dependent they were upon costly imported provisions and equally costly purchases from the Hudson's Bay Company, which to the American missionaries represented repugnant British interests and to the Protestant brothers also represented equally or even more repugnant Catholicism. Furthermore, transport of freight to the scattered missions was prohibitively expensive. At the annual meeting of the missionaries of the American Board in September 1839, "it was found that the expense of packing the provisions to the Stations would cost more than to raise it on the ground."[25] Missions near Hudson's Bay Company posts were less vulnerable, but for the Smiths at isolated Kamiah it was a matter of survival. "You may think," Asa told his father in 1838, "from the manner I have written that I think more about farming than I do of any thing else. It is true. I am under the necessity of thinking much about such things." "But I Hope," he added, "I shall not forget the spiritual good of these perishing souls around us."[26] Since mission funds were low and Hudson's Bay Company prices were high, a Methodist missionary on the Willamette pointed out that his station had to feed itself in order to sustain its principal goal of conversion:

> Though this outfit will be expensive, and for a time will require much to keep the mission in operation, yet if success crown our efforts by a prudent

management, the expenditure to the [Mission] Society will be diminished by the income from the cultivation of the farms, etc. And this mode of conducting the mission is considered essential to its successful operation, as there is no other way to furnish the mission family with provisions and other necessaries of life. The supply thus afforded therefore is considered only as subsidiary to the main object of the mission, which is to convert the natives to the knowledge of the truth as it is in Jesus.[27]

Furthermore, by farming, the missionaries hoped to set not only a Christian but also a sedentary example to the semi-nomadic Indians. If the natives could be induced by agriculture to abandon hunting, fishing, and gathering and to settle permanently, the missionaries reasoned, then they would be more accessible and more teachable. In the words of Joseph Williams, a Protestant clergyman:

> The people [Indians] are so scattered up and down in the mountains and valleys, that it seems hard to make much progress; and in the summer, they are all out digging roots and hunting. And in this scattered situation it is hard to keep up any kind of a society, until the young ones are informed by schools, and get to farming, and become a more settled people; and until our missionaries succeed in this work, I fear there will be but little good done towards religion; for while they live in their old Indian habits, they will not live up to any kind of discipline.[28]

Similarly, Father Nicholas Point believed that it was necessary to anchor his Flathead neophytes because their nomadic life was "full of perils" (it took them into enemy Blackfeet territory and imposed the hardships of travel on the very old and very young), brought them into contact with "strangers unsympathetic to religion," and exposed them to a dangerous "rapid transition from dire poverty to great abundance." "Therefore," he added, "it was our first concern to introduce them, little by little, to a much more sedentary existence. This could be done only by substituting the fruits of agriculture for those of the chase, the innocent pleasures of the fireside for those offered by the varied life of a hunter."[29] Besides, following the Indians to their seasonal fishing and bison hunting grounds was accompanied by too much hardship and too many dangers for the missionaries themselves, even for the more peripatetic Catholic fathers; for the married Protestant brothers it was nearly impossible. Henry Spalding, the superintendent of the Lapwai Station, pinpointed the motives behind mission agriculture when he replied to the admonition of the American Board "not to pay much attention to farming." He retorted that his mission had to farm because: there was no other reliable source of food in the area; farming was cheaper than importing; and it was necessary as an example to the Indians to ensure their sustenance.[30] Similarly, Elkanah Walker at Tshimakain advocated mission farming because: "it would diminish the expense" by eliminating costly imports; it would avoid a "precarious" dependence upon the Hudson's Bay Company's output, since "our being here is against the interest of the company in as much as the Indians do not hunt as much as formerly," and there was always the possibility "of their crops failing"; and it would bring "benefit to the Indians," for "while they retain the habit of roving they are but a part of the year under religious instruction of any kind," and "game is becoming scarce on their hunting grounds & unless they are prepared to gain a livelyhood in some other way the sad consequence of their starvation must take place." Walker believed that "we must use the plough as well as the Bible if we would do any thing to benifit the Indians," and he concluded that "to benifit the Indians at present the chief work of the Missionary must be tilling the soil."[31]

The Protestant missionaries were more successful farmers than their Catholic counterparts because they were less itinerant and more settled (as well as because they were more secular and occupied better land), but the "black robes" were more successful proselytizers (at least nominally), and partly for the same reason. The celibate fathers were able to live very near or even among the Indians either in camp or en route; the brothers with wives and children were more sensitive to the noise, dogs, and vermin of Indian camps, to the need for privacy, and to the extra burden of travel. Moreover, the more ceremonial Catholic liturgy seems to have had greater appeal to the Indians than the more austere fundamentalist rites. As Asa Smith reported, "One thing is certain, the natural heart loves such kind of instruction as the Catholics usually give."[32]

10

THE FRUIT OF THE FAITHFUL: THE OUTCOME OF MISSION FARMING

> *The missionaries in this field as well as all Indian missions have not only the spiritual wants of the people to attend to but are obliged to provide for their own sustenance & comfort by cultivating land building houses mills etc — & school houses — etc for the people.*
> NARCISSA WHITMAN TO MR. AND MRS. DUDLEY ALLEN, 1842

The most successful missionary farmers were the Methodists of the Willamette Valley. Few Indians had survived there for conversion, so that Jason Lee and his associates could spend more time and money on agriculture. Moreover, the Methodists were more entrepreneurial than their rivals; indeed, Lee's penchant for secular activities was to prove his undoing. Also, the Willamette was physically more conducive to agriculture than the rest of the Oregon Country; the Wesleyan efforts were therefore more bountifully rewarded.

The principal Methodist mission farm was at Willamette Station. It was rated "a good farm in order" in the spring of 1837 by Lee's future wife, Anna Pittman. The farm was opened in 1834, when forty acres were enclosed.[1] The first crop was harvested in 1835 (Table 25). At the end of 1836 the mission record book stated that "by the blessing of God we have an ample supply of wheat, peas, barley, oats, potatoes, pumpkins, squashes, carrots, beets, turnips, cabbages, onions, and a little corn, some beans a supply of butter a little cheese and four hogs fattening." By 1841 two hundred acres were under cultivation.[2]

In 1843 the "old" (Willamette Station) and "new" (Chemeketa) missions together held 286 acres of improved land; they produced 1,600 bushels of wheat and 1,440 bushels of coarse grains in 1842 and "probably" 5,000 of wheat in 1843.[3] The latter crop undoubtedly rendered the two missions self-sufficient for the first time. The "mission mill" at Chemeketa ground grain for both of the stations and for the Willamette settlers.

Willamette Station's first livestock came from Fort Vancouver, but in 1837 the mission obtained eighty head of cattle in California. They multiplied to 168 head in 1838 (plus 65 pigs and 63 horses and mules), 300 in 1842, and 600 in 1843.[4]

The only other Methodist mission with appreciable farming was Wascopam. Five acres were cultivated in 1839 and produced 25 bushels of small grains, 75 bushels of potatoes, and "considerable quantities" of vegetables.[5] The "small farm" yielded 120 bushels of wheat and 400 to 500 of potatoes in 1841, 50 bushels of wheat in 1843, and "probably" 500 in 1845. The wheat was ground at Fort Vancouver's mill; in the fall of 1841, for example, nearly 50 bushels were boated downriver for milling.[6]

Table 25

CULTIVATED ACREAGE AND CROP OUTPUT AT WILLAMETTE STATION, 1835-39

Year	Ploughland	Wheat	Barley	Oats	Peas	Potatoes
1835	30 acres	120-150 bus.[a]	56 bus.	35 bus.	60-87 bus.	225-250 bus.
1836[b]	45	500-750[c]	30	40	200	300-319
1837	150	600	?	?	?	?
1838	122	?	?	?	?	?
1839	?	1,100	?	?	250	840

For sources see *Sources for Tables*.
[a]Forty bushels per acre (Lee, "Articles on the Oregon Mission," pt. 2, p. 2).
[b]Plus 4½ bushels of corn and 3½ bushels of beans ([Slacum], "Mr. Slacum's Report," p. 194).
[c]From eighteenfold to twenty-eightfold the seed (Lee, "Articles on the Oregon Mission," pt. 5, p. 3).

Farming failed at the Clatsop Mission, which was provisioned overland from the Willamette; for example, fifty head of cattle and horses were driven from Mission Bottom in the summer of 1841.[7]

The principal American Board mission farms were at Waiilatpu and Lapwai. As at Willamette Station and Wascopam, Indian labour was hired. At Waiilatpu mostly native women and children were used to harvest corn; they were paid in kind. Besides garden vegetables (peas, beans, onions, tomatoes, beets, carrots, turnips, cucumbers, pumpkins, and melons), corn, wheat, and potatoes were the major crops at this "very good farm" (Table 26). By 1847 there were forty to fifty acres under cultivation.[8]

Waiilatpu's cattle herd grew slowly. By 1840 there were only twenty head, including ten cows, plus seventeen horses. Consequently, horsemeat replaced beef; five horses were butchered for the autumn of 1837 and the winter of 1837-38 for the mission's nine members. By the summer of 1841, when the first

Table 26
CROP OUTPUT AT WAIILATPU, 1837-41

Year	Corn	Wheat	Potatoes
1837	200 bus.	? bus.	250 bus.
1838	300[a]	60-100[b]	1,000[c]
1839	100	?	1,000
1840	130	250	?
1841	?	600	?

For sources see *Sources for Tables*.
[a] From thirty to forty bushels per acre.
[b] Forty-six bushels per acre.
[c] One hundred and sixty-seven bushels per acre.

cattle were beeved, thirty horses had been slaughtered for meat. It was not until the winter of 1843-44 that the Whitmans were able to substitute beef for horsemeat. In addition, sheep were kept — eighty by the spring of 1845, all derived from three ewes brought from Hawaii in 1838 and 1839.[9]

These results soon made Waiilatpu self-sufficient in food. Asa Smith reported in 1838 that although the Whitmans' house "is rather a rough one, but one part of it of logs, which was built the first year, the part added to it of dobies or clay dried in the sun ... at this station we have an abundance of the principal necessaries of life." The only disadvantage, he added, was the lack of a gristmill for grinding their grain, so that "much time is consumed in pounding it in a mortar."[10] In 1839 the Whitmans produced enough food for their family for one year plus some to spare for the Indians as payment for their labour. By 1840 Marcus Whitman was able to report confidently that "with a little arrangement, this Station may be made to support nearly all the others at a small expense."[11] Instead, Waiilatpu found itself feeding and doctoring not a few of the hungry, ailing migrants on the Oregon Trail. By the summer of 1842, the mission had a threshing machine and a flour mill, both powered by water, and was "farming to a considerable extent."[12]

Mission farming was even more successful at Lapwai, possibly because Henry Spalding had more agricultural experience than Whitman and probably because irrigation provided a more reliable water supply for the fields than at Waiilatpu. Spalding described his operations in 1842:

> The buildings consist of a saw-mill capable of cutting 1500 feet a day, and flour-mill with 32 inch stones, cut from a granite boulder near by, which make good flour, a dwelling house 20 x 30, school house, weaving and spinning room store house and barn. Through the blessing of a kind Providence some 250 bushels of wheat, corn and peas are gathered and

stored, so that but little more than a garden will be needed for the next year or two. We have beef, pork and fowls, milk and butter more than we deserve. All grains and roots grow well in this country and especially in the valley come to maturity early by reason of the great heat. We are obliged to resort to irrigation more or less. Herds of all kinds increase rapidly and feed out for the winter without care except sheep which need care to protect them from wolves. Sheep produce regularly twice a year, which I believe is an exception to all other countries.[13]

There were fifteen to twenty acres under cultivation in 1834 and forty-four in 1847 (with 161 fruit trees). Mostly corn, wheat, peas, and potatoes were grown (Table 27). In 1845 from 450 to 500 bushels of wheat were produced.[14]

Table 27

CROP OUTPUT AT LAPWAI, 1838-43

Year	Corn	Wheat	Peas	Potatoes
1838[a]	100 bus.	90 bus.	58 bus.	1,500-2,000 bus.[b]
1839	200	60	?	500-700[c]
1840	120-135[d]	110-116	?	?
1841	?	300	60	?
1842		250		?
1843	500[e]	51[f]	?	?

For sources see *Sources for Tables*.
[a]Plus 21 bushels of barley, 20 of oats, 20 of buckwheat, and 50 of vegetables (Spalding to Allen, 22 September 1838, Spalding, "Papers.")
[b]Three hundred bushels per acre.
[c]Five and one-halffold the seed.
[d]Thirty bushels per acre.
[e]One hundred bushels per acre.
[f]Four and one-half bushels per acre.

Indians were hired for harvesting. They did the raking, binding, threshing, and winnowing. Nevertheless, wheat harvesting was "the severest labor of the whole year" for Spalding. No Indians were able to help him cradle; the wheat ripened so quickly in the intense heat that not all of it could be reaped before becoming overripe; and the heat was so great that he could spend only two-thirds of the day in the fields.[15]

Lapwai also had more success with livestock than Waiilatpu (Table 28). By the autumn of 1847 there were ninety-four cattle, thirty-nine horses, and thirty-one pigs at the mission.[16] The sheep were especially fertile; apparently the ewes lambed twice each year and usually had twins.

Table 28
NUMBER OF LIVESTOCK AT LAPWAI, 1840-43

Year	Cattle	Horses	Pigs	Sheep
summer, 1840	25[a]	42	6	26[b]
summer, 1841	21	35	?	41
1842	?	?	?	63
spring, 1843	24	36	?	67

For sources see *Sources for Tables*.
[a]Including nine cows and heifers.
[b]From five ewes and three rams in 1838.

At Tshimakain farming sputtered. Spalding visited the establishment in the summer of 1842 and reported:

> Messrs Walker & Eells have done almost nothing in the way of building or farming (Mr W being a very feeble man) live in very inconvenient cabins and pack their provisions from the station of Messrs Grey & Whitman a distance of 160 miles. They very much need an assistant to aid them in building & raising their provisions that their own time may be devoted more exclusively to teaching & preaching & preparing books.[17]

Spalding returned in the spring of 1843 and found "belonging to the station 34 head of cattle, 11 of horses some 40 hogs — One dwelling house of dobies well finished, a Blacksmith shop — Flour mill lately destroyed by fire — some 40 acres of land cultivated." Elkanah Walker and Cushing Eells themselves believed that "much of the land in this region is adapted to the cultivation of Wheat & Potatoes." Output was meagre, however, and every summer the missionaries harvested Seakwakin or Camas Ground, a camas prairie near the station.[18] Livestock multiplied slowly: from thirteen head of cattle and twenty-one to twenty-three horses and mules in the summer of 1840 to thirty cattle and about twenty horses and mules in the autumn of 1846.[19]

Waiilatpu, Lapwai, and Tshimakain dominated the agricultural efforts of the American Board. Altogether they harvested 450 bushels of coarse grains in 1842 and possessed 50 acres of improved land, 110 cattle, 100 sheep, 70 horses, and 50 pigs in the spring of 1843.[20] Kamiah added very little. It did have the advantage of the presence of Indians in three of the four seasons (they were absent hunting bison in summer, when Smith was free to farm), but sufficient summer rainfall was so infrequent ("only *one* out of *four* since the commencement of this mission") that irrigation was essential "& but few spots indeed can be found favorable for this."[21] In 1840 twenty-five to thirty-five acres were fenced but only six to eight acres were cultivated and twelve to fifteen were pastured. In the fall of the same year there were only nine cattle, five horses, three pigs, and one mule.[22]

Because the Catholic missionaries were more concerned with baptizing than farming and because the physical conditions of their missions were generally less propitious, they were far less successful at farming than the Protestants. The sole exception seems to have been St. Paul on the Willamette, but its records are fragmentary. It produced 400 bushels of wheat and 280 of coarse grains in 1842 and owned 150 acres of improved land, 24 horses, 20 cattle, 20 pigs, and 4 sheep in the spring of 1843.[23] The Cowlitz mission of St. Francis in 1842 owned up to 300 acres and farmed 30, which produced up to 650 bushels of grain that year, including up to 450 of wheat. The mission employed thirteen persons but the farming was done by one salaried man.[24] At the Flathead mission of St. Mary's a farm was begun in the winter of 1841-42. By the summer of 1846 it comprised twelve log houses, a flour mill and a sawmill, about forty head of cattle plus pigs and chickens, and "abundant crops" of wheat, oats, and potatoes. At St. Ignatius farming did not commence until the spring of 1845; by July the Indians had erected fourteen log houses and a barn, amassed thirty head of cattle plus some pigs and chickens, and planted more than three hundred fenced acres in grain.[25] None of the Catholic missions seems to have prospered. This failure, indeed, the failure of most mission farms, Protestant and Catholic alike, was attributable to a variety of human and physical obstacles.

11

DIVINE TESTING OR DEMONIC TEMPTING: THE OBSTACLES TO MISSION FARMING

You can not evangelize a people always on the wing.
REVEREND HENRY SPALDING, n.d.

There is too much show and parade in the Religion now practiced in the Columbia, not only at Vancouver but elsewhere. It is too puritanical, and as far as I can learn little good has hitherto resulted from it, at least there is little or no perceptible improvement in the Morals of the people either whites or blacks.
CHIEF TRADER JOHN WORK TO EDWARD ERMATINGER, 1839

The problems of mission farming were closely related to the problems of mission life in general. By 1850 each of the three missionary ventures, Wesleyan Methodist, Congregational-Presbyterian, and Roman Catholic, had been abandoned. By then Euroamerican settlement had usurped most of the lower Columbia and alienated most of the Indian inhabitants. Besides, there were not very many natives left to save, and most of them were no longer receptive to salvation. But even before this juncture mission farming had encountered several obstacles. Perhaps the most formidable was the fact that the missionaries themselves, especially the Catholics, simply were unable to spend enough time on agricultural operations. Their tasks were many and varied, and spiritual activities supposedly took precedence over secular pursuits. In particular, the semi-nomadism of the Indians conflicted with the sedentary proclivities of the missionaries. Narcissa Whitman pinpointed the problem: "the roving habits of the Indians make it necessary for us either to do so [itinerate], or else spend the greater part of our time alone during their absence from the station." She bemoaned "their wandering habits so little of the time at the station under the influence of truth — & their scattered situation in their wandering."[1] The reaction of the American Board missionaries was equivocal; some favoured itineration, others did not (indeed, their Oregon Mission was plagued by dissension, particularly in 1840). Henry Spalding detailed the opposing viewpoints:

there is conscientiously, a vast difference in our views as to the most efficient mode of prosecuting our labours. A few of us & a very few I believe, feel it to be an unyielding duty to do all we can with our limited means, & with a proper regard to our more appropriate labor as teachers & preachers, to call in as it were, the people from their wandering mode of living & settle them upon their lands. And our reasons are the following — On arriving in this country we found this people in a most pitiable situation as it respects the means of subsistance, depending entirely on roots fish & game. To obtain their living from these precarious sources, requires them to be constantly upon the move & moreover, requires an intensity of labor which but few white people could endure. In these circumstances, the people are cut off from permanent religious instructions, hear a religious discourse occasionally as they pass from one resort to another. True, during the winter, when the country is shut up by snows, they gather around our stations in considerable numbers. But does not all history & experience show us that a people in such circumstances, can not be essentially benefitted by religious instruction. We suffer most in this manner of life in our schools. Today a hundred interesting children may be gathered into schools, tomorrow half of them will be obliged to leave with their friends in search of food & the day after perhaps ten only remain who are soon to be displaced by strangers, & so on. I am often affected by tears on looking out in the morning & seeing many of my dear children whom I hoped to meet in school, climbing the hills in search of roots & weeds for food. — another reason is, the fact that their game & fish are fast passing away. Immediately on arriving at this place, I exerted myself to the extent of my limited means to settle the people around me....

But a great majority of this Mission conscientiously believe that this course will ruin the mission. They maintain that the means of tilling the ground should be kept from them & that they should not be encouraged to settle but continue the chase until they acquire moral principle enough to settle down on their lands & not become worldly minded. They would have their teachers travel with the people from one resort to another. One of our Methodist brethren has just written one of our number that the greatest calamity that has befallen missionary efforts in this country is the building of dwelling houses. He thinks he shall itinerate & give himself no concern about anything *except preaching*. The objections to this course are very serious [to] my mind. Admit these tribes to be in the most favorable situation for such a course; i.e., move as a tribe together in one band — the missionary who adopts this course must convert himself & family if he has any, into heathen, as it respects their manner of living. A faithful missionary of the Board, among the Indian tribes, who adopted this course before he was married (but has since abandoned it) informed one of our party when on our way to this country, that he had frequently given his shirt, which was swarming with living beings, to an Indian woman, to better its condition. No one who has been long in a heathen country would inquire what she done with the lice. You may be shocked at this statement but rest assured that no white man can adopt this course & take up his abode in a Indian lodge without being soon covered with fleas & lice. Even with a house, & a thorough cleansing of clothing & house frequently, we find it difficult to keep ourselves comfortable. It is a sore trial for a mother to bring up children in the midst of heathen & be in the most favorable circumstances. But what must be the trial of bringing up children in a wandering filthy Indian camp; washing could be done but seldom & then under the most unfavorable circumstances. And from whence could daily food be procured, suitable for children, to say nothing of many things which are considered necessaries by adults. Besides it would be out of the question to give them the first rudiments of an education. They would often be in great suffering by reason of the cold & storms, for the poor Indians are obliged to travel at all seasons for roots, fish or game. But all this & much more would be most cheerfully endured, did the salvation of this people depend upon such a course. In my opinion, everything would be lost. The Indians instead of moving in large bands, are scattered over the country in very small parties, almost constantly on the move. A few weeks in the spring, climbing the hills for roots, which are soon exchanged for the salmon fisheries & in a few weeks these give place to roots found in the plains where they make but a short tarry before the Buffalo season arrives & while half the tribe are absent four months in the Buffalo country, scattered in small parties, the remainder seek the upper waters for salmon or the mountains for Elk or Sheep, till the cold compels them to seek their respective countries, where they scarcely take time to exchange Buffalo robes for dry salmon, while those who remain are compelled by hunger to resort to the mountains, amid the snows of winter, for game . . . .

The middle system is still more objectionable to my mind; i.e., part civilized & part wandering, so far as it concerns the natives. While some in this country would have the missionary with his family upon the wing with the natives, others would have the missionaries themselves settled but the people continue to wander for their subsistance. I have but one objection to this system, which has greater weight on my mind, than all, urged against the other system. The missionary would settle down in some favorable place for building & cultivating (as we must raise our supplies by reason of the great expense of purchasing of the H.B. Co., as also their request that we should not depend on them) build his houses, cultivate his farm, increase his herds, eat & drink of the fat of the land & flock, but be sure & withhold from the poor starving natives who might occasionally pass by his establishment in their wanderings for subsistance, or gather around him in small numbers to assist in his secular labors, the means & encouragements for cultivating their lands. In this way he might escape the sin of introducing "worldly-mindedness," but in my opinion he would introduce a worse evil. This people who are not strangers to the fact that they are fast passing away for want of suitable means for subsistance, would see that their teachers, either did not understand their wants or cared not for them & consequently were not worthy of their confidence. From disregard, the steps would be few to distrust & envy. And the great houses full of all manner of goods things (as I am convinced that the missionary who can neglect the wants of the starving around him will not neglect his own) & the store house full of grains & the flocks ranging upon the hills & plains, would become a prey to a band of starving robbers. It is due to the brethren who advocate this system of missionary operation, to say that they would have the missionary make visits of a longer or shorter duration to the several resorts either with or without his family, as circumstances would admit. But if a missionary by wandering all the time with his people can do but little good, how much less by wandering occasionally.[2]

Increasingly the American Board missionaries rejected itineration, if only because of the hardships and dangers, so that they were able to pay more attention to farming. But the number of converts, even if only nominal, declined accordingly. Enrolment in the Lapwai mission school, for instance, fell from 234 pupils in the winter of 1842-43 to none in the winter of 1846-47, and the congregation from 500-800 to 108. The Indian agricultural settlement at the mission was "broken up" at the beginning of 1847. (Spalding even owned that "it is quite evident that my labors in future must be much in the way of itinerating".)[3] Fort Colvile's bourgeois, Archibald McDonald, reported in 1844 that "of the 17 Methodist preachers we at one time had in Oregon, there are not present in the field above, I believe, three or four — all back to Jonathan Land again." "Those sent out by the A.B.C.F.M. [American Board]," he added, "still hold on, but literally do nothing."[4] At Tshimakain Cushing Eells admitted in 1846 that "we have been here about seven years & I don't know that we have reason to think one soul had been converted to God."[5] The missionaries were disappointed and frustrated by the inconstancy of their Indian neophytes, whose piety seemed to be directly proportional to their material gain from conversion. Secretary David Greene of the American Board advised the missionaries to turn to the growing Euroamerican population. Marcus Whitman concurred and added his own rationalization:

> Your remarks and admonitions not to be discouraged & to look to a future sphere of usefulness among the white settlers are most satisfactory &

encouraging. It has been distinctly my feeling that we are not to measure the sphere of our action & hope of usefulness by the few natives of the country but by all that we can see in prospect both as it relates to a white population & Catholic influence.[6]

But most white newcomers were already Christians, if not altogether punctilious, and those who were not were as recalcitrant as the Indians.

Whitman also believed that by attempting agriculture at all of the American Board stations, the missionaries were overtaxing and overreaching themselves and perforce paying too much attention to temporal and too little to spiritual matters. It would be more efficient, he argued, for only one mission — his — to farm and to provision the rest: "I have felt ready, if I could have had a minister with me to be farmer not only to supply this station but any that could depend upon it for supplies. But why all must grow grain for themselves is more than I can say.... For every one to be his own farmer, house builder etc, is too much."[7] Whitman's criticism of what he termed the "extending system" (mission self-sufficiency) was partly aimed at his rival, Spalding:

> It will not be long before it will be apparent to all how foolish the extending system has been. And I believe all will soon see that this was the only station [Waiilatpu] that ought to have been called to cultivate for the use of the Mission & that only *here* — & not at Lapwai (that is Mr Spalding station) ought to have been all of the Mills Blacksmith shop Printing Office & Mechanicks. Here ought to have been as far as could have been the all & in all of the Nez Percés Mission.[8]

These remarks reveal some of the acrimonious factionalism that weakened the Oregon Mission.

In fact, Whitman was probably right, for not all of the mission locales were physically amenable to farming, and very few of the missionaries were experienced farmers. Indeed, Elkanah Walker was the only one of the seven American Board brethren in 1840 who came from a farm, and he was posted to Tshimakain, physically the most unsuitable station for agriculture. Walker himself acknowledged the problem:

> I should consider our station as greatly strengthened by the addition of a good judicious farmer, one who was willing to be a farmer when he had an appointment by the Board, as I should be if a minister were added to it, & even more so for ministers make poor farmers after having gone through a course of study.[9]

Asa Smith, at least, was downright averse to farming, which was apparently beneath him. He felt that his linguistic talent and his wife's health were being wasted at Kamiah:

> *We may not remain here long in this field.* Sarah's health is such & our destitute situation such that we cannot stand it long. Should we find it necessary to leave here for want of domestic assistance, we may go to the [Sandwich] Islands where our situation would be utterly different & we could devote all our time to the great work. I have spent too much time to be a mere *farmer*.[10]

Not long thereafter Kamiah was abandoned and the Smiths went to Hawaii. Yet even with these difficulties, the American Board missionaries seem to have been more experienced farmers than the Methodist brothers. None of Jason Lee's 1834 party of five was a farmer.[11] The first Methodist missionary farmer, Henry Brewer, did not arrive until the first reinforcement of the spring of 1837.

And farming was not the only task that was beneath some of the missionaries. Even the lowly Indians, the object of the religious activity, were patronized by the condescending evangelists. Narcissa Whitman, for example, referred to the natives as "benighted minds" who behaved "like children."[12] Such prejudice rationalized the unhappy fate of the Indians at the hands of Euroamerican settlers. As Marcus Whitman told his parents in 1844:

> Although the Indians have made and are making rapid advance in religious knowledge and civilization, yet it cannot be hoped that time will be allowed to mature either the work of Christianization or civilization before the white settlers will demand the soil and seek the removal of both the Indians and the Mission. What Americans desire of this kind they always effect, and it is equally useless to oppose or desire it otherwise. To guide, as far as can be done, and direct these tendencies for the best, is evidently the part of wisdom. Indeed, I am fully convinced that when a people refuse or neglect to fill the designs of Providence, they ought not to complain at the results; and so it is equally useless for Christians to be anxious on their account. The Indians have in no case obeyed the command to multiply and replenish the earth, and they cannot stand in the way of others in doing so.[13]

Earlier John McLoughlin had warned the American Board that "on no account ought the Whites to take Indians from their Lands."[14] The chief factor's jurisdiction did not extend to the missionaries, however, and his worst fears were realized nine years later in the Whitman Massacre.

Another problem with mission farming was the sharing of output. Farming at Waiilatpu and Wascopam was disadvantaged by their location on the homestretch of the Oregon Trail. Destitute overlanders consumed so much of the two missions' agricultural output that sometimes they did not have enough left for themselves. As Narcissa Whitman told her sister in 1840:

> We are emphatically situated on the highway between the States and Columbia River & are a resting place for the weary travelers consequently a greater burden rests upon us than any of [our] associates, to be always ready — & doubtless of those who are coming to this mission — their resting place will be with us untill they seek & find homes of their own among the soletary wilds of Oregon.[15]

In early October four years later, at the height of the "Oregon fever," Narcissa wrote to her brother that:

> The season has arrived when the Imigrants are beginning to pass us on their way to the Willamette. Last season there were such a multitude of starving people passed us that quite drained us of all our provisions except potatoes. Husband has been endeavouring this summer to cultivate so as to be able to impart without so much distressing ourselves.[16]

The following spring Marcus complained that "I have had much to do with supplying immigrants for the last two years."[17]

These human obstacles were aggravated by physical barriers, chiefly winter cold and summer drought. Late and early frosts were an annual hazard at Kamiah, where "the first year [1839] of this mission, crops were destroyed or injured from this source."[18] At Tshimakain frost could strike in any month of the year. As Walker reported to Secretary Greene in 1841:

> It is a fact that as a general thing the soil is not so productive as in the states or even in New England & crops of every kind are far more precarious than in the States not only from drought but on account of the late frost in the spring & early ones in the latter part of the summer months. Every year we have been here our corn has been killed or injured by frost in August. There is not a month in the year but we have frost at this station.[19]

Observations in the diaries and letters of the Walkers and Eellses speak for themselves: "We had a frost on friday night [4 June 1841] that killed all the corn potatoes etc."; "Sat. June 5. Was not disappointed in looking for frost. Everything in the garden froze stiff to the ground. The squash & melon vines escaped except a few I did not cover [with bark]"; "Every thing killed in our garden last night [30 August 1841]"; "There has been more or less frost at this place during every month this year [1841]"; "Sun. 31 [August 1845]. June, July & Aug. have [been] without frost. The longest period I think without frost since we came to this country"; "There was a hard frost last night [21 June 1846] which killed the corn & potatoes & every thing else that was susceptible of being injured by frost"; "We had a very hard frost last night [22 July 1846] which killed the vines & corn. This is the fifth we have had this month & the hardest"; "A cold night [9 September 1846] which killed everything down dead."[20] Little wonder that Smith concluded as early as 1840 that "it may ... be questionable whether agriculture can become general on account of the frost."[21] The severe winter of 1846-47 struck all of the mission farms east of the Cascades. At Tshimakain, reported Myra Eells, "it was only by unwearied labor and the greatest economy in feeding that enough of our cattle and horses were saved for present use. Only one horse has died, but we have lost twelve cattle."[22]

In addition to frost, the upper Columbia stations were also more susceptible to drought. The years 1837, 1839, and 1843 were especially dry in the interior. In 1839, 90 per cent of Lapwai's potatoes were no larger than a thumb and 50 per cent of Waiilatpu's no bigger than walnuts. The "severe drought" of the spring of 1843 reduced Lapwai's wheat crop.[23] Everyday living as well as farming suffered. In Spalding's words:

> the intense heat of this valley which owing to the position of the Bluffs and hills around is doubtless the hottest place in the world, much of the time for weeks the mercury ranging from 103 to 110 [°F.] in the shade and in the sun up to 150. No rain. Milk sours before noon unless kept in the River which during most of the hot season flows rappidly and cool from the snowy mts in full banks 480 feet wide. The house floor and every article of furniture becomes disagreeably hot and in a few days the house is more like an oven heated than a place for dwelling. At night we are obliegied to leave the house and sleep out in a tent.[24]

And at treeless Waiilatpu, Lapwai, and Kamiah fencing was difficult. There was little or no enclosure of Waiilatpu's livestock until 1845, when Whitman built a sawmill twenty miles away in the Blue Mountains to make lumber for housing and fencing. At Kamiah Smith reported that "it is usually very difficult to fence land here for want of timber, so that what is cultivated by the natives is liable to be destroyed by horses & cattle."[25] Until the spring of 1843 Tshimakain's fields were unfenced; consequently, the cattle and horses grazed the crops and strayed, and much time was spent finding them. Strays were easier prey for wolves and Indian dogs, which also killed mission chickens (the Indians themselves stole livestock and vegetables). Walker reported in 1839 that the wolves "are so thick & desperate that we dare not leave our horses out during the night. Quite a number of horses have been bitten & some actually killed." He added: "I should fear as much from the people [natives] as I should from the wolves. Some of them do not hesitate to eat horse flesh when they are hungry & they will not stop long to inquire to whom it belongs."[26] At Waiilatpu and Lapwai, too, wolves and Indian dogs were a menace. In the fall of 1839 Spalding asked the American Board to send enough strychnine to kill one thousand wolves, which threatened the mission's sheep in particular.[27] The following spring he requested enough poison to kill twenty thousand wolves, declaring that "they are becoming very destructive to my domestic animals." From 1838 through 1841 Lapwai lost thirty-nine sheep, mostly lambs, to wolf attacks and milk shortages.[28] Indian dogs were easier to combat, and Spalding's vigourous campaign against these marauders illustrates not only his determination to eradicate them but also his uncompromising attitude towards the natives, an attitude that contributed to Indian alienation, as Smith reported:

> Another way Mr. Sp. got into difficulty with the Indians was in waging a war upon their dogs. He & Doct. W. were in such haste to introduce all the arts of civilization among the Indians at the very onset, they encumbered themselves with sheep; but the camp was so full of dogs that the poor harmless sheep could have no peace but were in danger of being destroyed at once. The Indians were fond of their dogs & unwilling to give them up & there was great excitement throughout the camp. Mr. Sp. offered a reward for every dog they would kill & some undertook the work of destruction, tho' at the risk of getting the hate of their own people. All this was known to the gentlemen of the H.H.B. Co. so that when Mr. [Frank] Ermatinger met us at the Rendezvous, he remarked to one of our number that "Mr. Sp. had got so that he did nothing but whip Indians & kill dogs."[29]

The Methodist missions in the Willamette Valley were more fortunate. They were shielded by the high Cascade Mountains from frigid continental air masses; at the same time, the low Coast Range could not block the mild, moist Westerlies, and they rarely experienced drought. Moreover, the denser Euroamerican settlement of the Willamette lessened the number of predators, bestial and Indian alike.

12

FROM NOBLE SAVAGE TO STURDY YEOMAN: INDIAN FARMING

> *let those who wish to do good to Indians — teach them to get their food in a different way than at present — in short teach them Agriculture While they are instructing them in Religion —.*
>
> CHIEF FACTOR JOHN McLOUGHLIN TO
> EDWARD ERMATINGER, 1836

The entire Indian way of life, not just native beliefs, was the target of evangelical activity, although even this goal was not unanimously supported by the factious American Board missionaries. Marcus Whitman and especially Henry Spalding were convinced that the mobile Indians had to be settled, and this meant farming to replace hunting, fishing, and gathering. Asa Smith, however, opposed this policy, arguing that acculturation would be disastrous. He believed that it was foolish to sedentarize their charges because the American Board could not afford the attempt and the natives would not survive it. "Rev. Mr. Smith preaches against all efforts to settle the poor Indians, thinks they should be kept upon the chase to prevent their becoming wordly minded," noted Spalding in 1839.[1] In the same year Smith argued that "there is a very great danger of introducing the habits of civilized life faster than the natives are capable of appreciating them."[2] Smith elaborated his view to Secretary David Greene the following year:

> At present a school can be sustained here [Kamiah] only about six months in a year & that liable to numerous interruptions. The prospect that the people will change their habits & become a settled people & thus be [in] a situation to be instructed is not so favorable as may be supposed. The encouragement to do so is not so great as you may apprehend....
>
> It is usually very difficult to fence land here for want of timber, so that what is cultivated by the natives is liable to be destroyed by horses & cattle. Cultivation too, in order to be successful, requires a great amount of labor, so that we have good reasons to fear that an idle wandering people will soon be discouraged in their attempts to cultivate. As yet their cultivation has

scarcely diminished their wandering at all. Seed time & harvest confines them where their fields are & between these periods they are usually wandering not looking after their fields & when they return find little or nothing. . . .

That the people will change their habits & become a settled people enjoying the blessings of civilization before the gospel savingly affects their hearts, is something which the history of missions does not warrant us to expect. But how can we bring the gospel to bear on them while they are living a wandering life? Some may say we ought to wander with them. But here we meet difficulties. They wander usually in small clans. But few can be found together & they are so occupied as to furnish but little opportunity for giving instruction. Again *our* situation is such as to render it impossible. We can only stay where we are & instruct what we can find.[3]

Smith found that missionizing was hamstrung by the "wandering habits" of the Indians, their "scattered condition," their "small number," their difficult language, their self-righteousness, their "lack" of a religion for comparison with Christianity, and their "lack" of formal government and law.[4] Partly because of his convictions and partly because of the physical constraints on agriculture at his mission, Smith's Kamiah farm was a subsistence operation for his family, whereas those at Waiilatpu and Lapwai were demonstration farms for the Indians.

Whitman and Spalding prevailed, however. To them the problem was simple: it was impractical to roam with savages and evangelize en route. The Indians were often on the move in search of food, shifting every few months from bison pastures to salmon waters to camas meadows. So migratory were they that Smith asserted that even at Kamiah, where the Indians were less nomadic, "one half of the year we are almost entirely alone."[5] And Spalding complained that Lapwai's "congregation on the sabbath varies at different seasons of the year, & must continue to, till the people find a substitute in the fruits of the earth & herbs, for their roots, game & fish, which necessarily requires much wandering."[6] Even the Catholic fathers, who itinerated more than the Protestants, advocated sedentarization. Father Pierre De Smet wrote:

> It would be impossible to do any solid and permanent good among these poor people, if they continue to roam about from place to place, to seek their daily subsistence. They must be assembled in villages — must be taught the art of agriculture, consequently must be supplied with implements, with cattle, with seed.[7]

Father François Blanchet agreed:

> As long as these natives are not at all gathered into one great mission, to learn agriculture and to secure the things necessary for life under the direction of some missionaries, they will learn nothing or will lose very quickly from sight the holy truths which they have learned.[8]

Not only would sedentary Indian farmers be easier to instruct; farming would also improve their miserable lot. Spalding, for example, felt that "this suffering must continue till they become settled & raise their subsistance by cultivating the soil."[9] More importantly, farming was seen as necessary to their very survival. Whitman and Spalding elaborated this point to Secretary Greene:

> We find the natives of this country, poor in the extreme, depending entirely upon roots, fish & game for subsistance. We see them fast decreasing in numbers, not from any prevailing disease, but actually from the want of the necessaries of life — & we see causes now in existance, which were not in former days, which must to all human appearance, result in their entire annihilation, unless a redeeming power be interposed — We believe that redeeming power to be the means of civilization, & a permanent subsistance; & therefore while we point them with one hand to the Lamb of God which taketh away the sins of the world, we believe it to be equally our duty to point with the other to the hoe, as the means of saving their famishing bodies from an untimely grave & furnishing the means of subsistance to future generations.[10]

To some extent the two missionaries may have been right. Game, at least, particularly bison, if not deer as well, were decreasing in response to both the encroachment on pasture and browse and the hunting of herds by Euroamerican settlers, trappers, guides, and explorers. So any food supplement from agriculture was helpful and perhaps even vital.

At Waiilatpu in the first and second years of its existence hoes were given to the Cayuses; for three years they were given seed, and subsequently ploughs were lent or sold to them.[11] Narcissa Whitman wrote to her brother in 1838 that:

> Peas, corn, & potatoes, are given them for as much land as they will cultivate, To some we give garden vegetables such as carriots turnips beats etc of which they are very fond. The Lord did not smile upon us for naught last summer, in causing the earth to bring forth in abundance, for he is disposing the hearts of the heathen to till the land for sustanance.[12]

The agricultural efforts of the Indians changed their seasonal round, as Marcus Whitman reported in 1843:

> The months of February & March bring a return of the Indians from their winter dispersion in order to commence the agricultural year & also to avail themselves of such provisions as we stored the previous fall; of which potatoes, corn & wheat form an important part. The Months February, March & April are mostly occupied with the labour of preparing & ploughing the ground.[13]

In 1842 sixty to seventy Cayuses cultivated fifty to sixty acres. The following year about fifty Indian families tilled one-quarter to four acres each and together owned fifty to seventy head of cattle.[14] Their adoption of cattle may have been facilitated by their experience in horse rearing. Smith noted upon his arrival in 1838 that "thousands of horses are raised here in these plains. Some individuals of the Kayuses, I have been told, have a thousand horses apiece." "As we came to this place [Waiilatpu]," he added, "we saw large bands of horses here & there grazing on the plains. Horses constitute the principal wealth of the people." Father De Smet observed in 1845 that the Cayuses and Nez Percés were the wealthiest Indian tribes in the Oregon Country, thanks to their livestock, and some families owned 1,500 horses.[15] Cayuse farming increased steadily. In the summer of 1842 Narcissa Whitman reported that:

> The success of the Kayuses in farming is pleasing beyond description — There is scarcely an individual of them but what has his little farm some where — & every year extending it farther & farther — A large number of the Walla Walla tribe are doing the same .... The Nez Percés also are advancing as rapidly as possible with their means — & their is a universal desire for cattle.

A month later Narcissa wrote that:

> The Kayuses almost to a man have their little farms now in every direction in this valley & are adding to it as their means & experience increases.... A few of the Lahaisies have done considerable at cultivation both here & on some parts of the Snake River where tillable land can be found. The Nez Percés are extending themselves in every direction throughout their own country & are cultivating in great numbers. As we have had but little & sometimes no rain during the summers both our own plantations & many of the Indians are watered by irrigation.[16]

By 1845, reported the two Whitmans, "the Indians are doing more this year at farming than before & fencing much better — a thing much needed — for most of them are now getting more or less cows — & other cattle." They added that "there are but few who have not cattle — a number have sheep & nearly all have plantations more or less — These things will be likely to deter them from acts of violence to the White lest they become the greatest sufferers in case of war."[17] This belief was belied, of course, by the violent death of both Whitmans two years later at the hands of their charges.

At Lapwai more Indians cultivated more land. Spalding sold hoes and calves to the Nez Percés, and in 1838 sixty to eighty Indian families tilled wheat, barley, corn, peas, and potatoes with hoes. That year they raised 251 bushels of grain, 58 of peas, 800 of potatoes, and 50 of vegetables and reared 40 pigs and 12 cattle. In 1839 some 100 Nez Percé families hoed potatoes around Lapwai.[18] By 1842, 110 to 200 families worked one-quarter to eight acres each, and up to ninety acres altogether, and eight of them had ploughs; each of sixty families grew 100 bushels of grain and one chief 100 bushels of corn, 176 of peas, and 300 to 400 of potatoes, and twenty-two individuals owned fifty cattle each and seven individuals eighteen sheep each. Spalding also recorded that seventeen Indians owned forty pigs, thirty-two cattle, and ten sheep.[19] He reported:

> Probably half of the cultivation is in this valley i.e. upon a small stream putting into the river at this place. Many are cultivating as they find good land on their respective streams, while many are producing considerable quantities of grain on the Snake or Lewis river as they are able to find small pieces of from a half to four acres between the river and the bluffs under some small rivulet gushing out of the rocks, perhaps 1000 feet above.[20]

From 1842 to 1843 the number of Nez Percés cultivators increased by one-third, and Spalding estimated that the Indians were enlarging their cropland 25 per cent annually. They were keeping their best furs to trade for light ploughs as they became available. By 1845 Nez Percés were travelling to the Willamette to swap horses for cattle. They owned 400 to 500 head of cattle in 1846 and grew "thousands of bushels of different grains."[21]

Elsewhere, Indian farming at Tshimakain was described by Smith in 1840:

> Efforts at agriculture have been carefully encouraged. Hoes have been sold for less than the usual price, & in one instance exchanged for other goods which had been previously given. Hoes have occasionally been lent, tho' it has not been practiced very generally. The principal chief has been presented with apparatus which together with a few cords of his own made a convenient harness for two horses, & then the only plough used at the station loaned him with which to prepare his ground. When the members of the mission located at Tshimakain found it necessary to send to Clear

Water for potatoes to plant, one third of all that were obtained were furnished to the natives, tho' it was stated expressly that the whole were designed for the members of the mission. Many of the smaller seeds, particularly such as were thought to be the more useful have been freely distributed to such as desire to plant them. Care has been taken to assist in the selection of the most suitable land for tillage, & instruction as to the manner of cultivation & securing the crops etc has been faithfully given.[22]

These efforts were in vain, however. Walker reported in 1841 that "in regard to the settling the Indian & inducing them to cultivate we have not as yet been able to do much & it is very evident that they do not manifest much disposition to do it." In 1842 none of Tshimakain's Indians cultivated and only one owned any cattle.[23]

The Methodist missionaries were even less successful at converting the natives into farmers than at transforming them into Christians. Jason Lee reported to Indian Agent Elijah White in 1843 that "an attempt was made in the Willamette valley to induce the Indians to cultivate their lands but did not succeed, and we have made but little effort of the kind in our other Stations."[24] At nearby Fort Vancouver in 1837 two hundred Klikitats cultivated corn, peas, and potatoes on a "large plain" fourteen miles from the post with seed furnished by John McLoughlin, but the outcome is unknown.[25]

The earliest and most successful adoption of agriculture by the Indians of the Oregon Country was the acceptance of the potato by the coastal natives. It was undoubtedly introduced by a fur trading vessel, English or more likely American, perhaps before the end of the eighteenth century. The tuber was consistent with the root gathering activity of the Indians and compatible with the light, sandy soils and the cool, moist summers of the Northwest Coast. The Haidas of the Queen Charlotte Islands and the Tlingits of the Alexander Archipelago and adjacent mainland became especially proficient at growing potatoes. McLoughlin rated the Haidas the best potato growers on the coast. By the late 1820's the Russian-American Company was buying potatoes at Sitka from the Tlingits for food and seed, and by the middle 1830's the Hudson's Bay Company was doing likewise at Forts Simpson and McLoughlin. The potato diffused rapidly down the coast. It was being cultivated by Indians at Fort Langley by 1835, at Fort Nisqually by 1839, and on Whidbey Island by 1840. In 1841 Lieutenant Charles Wilkes found that the natives on the coast and islands of Puget Sound raised "extremely fine" potatoes in "great abundance" and that potatoes formed a "large portion" of their diet.[26] By the early 1840's even Fort Kamloops was procuring potatoes from local Indians.

While the coastal Indians retained potato growing, the few interior Indians of the upper Columbia who had adopted cultivation began to abandon it in the mid-1840's. Here the pressure of Euroamerican settlement on Indian land and life was greatest, and the Indians became increasingly dissillusioned and disenchanted, discarding both their newfound religion and their newfound livelihood. The American pioneer Peter Burnett believed that prior to the influx of American settlers in the early 1840's, Indian-Euroamerican relations were amicable for several reasons: McLoughlin was an honest and just administrator; Hudson's Bay Company trade goods (blankets, guns, and so forth) were uniformly low in price and high in quality; the company occupied very little Indian land and then mainly for trading posts of mutual benefit; some company employees were related by marriage to the Indians; and it was to the company's commercial advantage to preserve peace with the Indians since, as suppliers of furs and buyers of goods, they formed the mainstay of its business.[27] Burnett added that Indian hostility increased with the arrival of American immigrants because they came in considerable numbers and because they came not to establish mutually beneficial trade but to take and farm lands exclusively for themselves, so that:

> Every succeeding fall [when the annual American immigration arrived] they found the white population about doubled, and our settlements continually extending, and rapidly encroaching more and more upon their pasture and camas grounds. They saw that we fenced in the best lands, excluding their horses from the grass, and our hogs ate up their camas.[28]

As Narcissa Whitman observed in 1844, "the Indians are roused a good deal at seeing so many Emigrants."[29] Three years later that arousal erupted in the Whitman Massacre and an Indian war.

Farming was, moreover, antithetical to the traditional native lifestyle. As Elkanah Walker, perhaps the most realistic American Board missionary on the scene, explained to Secretary Greene:

> They are placed in a country abounding with natural food & one which does not offer them much encouragement to labor for the fruits of civilized life & so they will be slow to engage in the cultivation of the soil. They would rather lounge in their lodges half feed and half clothed a part of the time to being in the field gathering their food by the sweat of the brow. They have no fears of starving to death so long as the pine tree stands from which they can gather moss to make skalapkan.[30]

He added:

> I am pretty well convinced in my own mind that there are not ten Indians in this country that are desirous of settling down & remaining fixed in their manners of life & it is perfectly natural that they should love their wandering

habits. It is as hard & unnatural for him to lead a settled life as it would be for a New England farmer to change & lead a wandering life.³¹

Farming was also at odds with the Indians' spiritual doctrine, for they believed in the inviolable oneness of humanity and nature and in the supreme chieftaincy of the earth, their mother. Some Euroamerican pursuits, including cultivation, violated this belief, as Smohalla, a Wanapum shaman and prophet, so poignantly protested:

> You ask me to plow the ground! Shall I take a knife and tear my mother's bosom? Then when I die she will not take me to her bosom to rest.
> You ask me to dig for stone! Shall I dig under her skin for her bones? Then when I die I cannot enter her body to be born again.
> You ask me to cut grass and make hay and sell it, and be rich like white men! But how dare I cut off my mother's hair?³²

Similarly, Chief Seattle was to make an equally poignant plea to American settlers to husband, not exploit, the land.

Thus, the Indians, who had halfheartedly adopted agriculture anyway, soon abandoned it as part of their larger rejection of Euroamerican values. And this abandonment was to be confirmed by the Euroamericans themselves when they subsequently established Indian reserves and reservations which were environmentally more suited to the traditional native pursuits of hunting, fishing, and gathering than to farming.

Epilogue
DIVIDING THE OREGON COUNTRY

> There is every appearance to my mind, without too fancifully indulging in the spirit of prophecy, that these regions are ere long destined to become great in the annals of the world; and if so, the foundation of this their greatness will have been laid by the Honble Company, who will not, I hope, leave the superstructure entirely to the exertions of others, who are yearly gaining a firmer footing on this side of the Continent to the evident detriment of the long established Traders. I fear the Americans will soon make a grand effort to oust us out of this place altogether, as they claim to be proprietors of the soil to the forty-ninth parallel of latitude. In this case, they will act the part of the ungrateful snake in the fable. They could never have existed here a day without our assistance. It is bad policy to encourage them.
>
> REVEREND HERBERT BEAVER, 1837

Reverend Beaver exaggerated both the future of the Oregon Country and the helplessness of its American pioneers, and he did not foresee that the Hudson's Bay Company's efforts to win the northern bank of the Columbia River for the British flag would in the end be undermined by faltering official resolve in London. When, after nearly three decades of so-called "joint occupation," the Columbia Department was finally divided between Great Britain and the United States by the Oregon Treaty or Treaty of Washington on 15 June 1846, farming as well as fishing and lumbering and even some manufacturing had long since been successfully established, thanks mainly to the industry of the Hudson's Bay Company, and its subsidiary, the Puget's Sound Agricultural Company, at the widely scattered "King's posts."[1] In the autumn of 1845 the Honourable Company had 628 servants at forts and on ships in the Columbia Department, plus 3,005 acres under cultivation and 16,900 head of livestock (Table 29).[2] Under the resolute direction of George Simpson, the "emperor of the plains," and John McLoughlin, the "emperor of the west," the company succeeded in producing enough wheat, beef, and butter

Table 29

NUMBER OF EMPLOYEES, CULTIVATED ACREAGE, AND NUMBER OF LIVESTOCK AT THE POSTS OF THE HUDSON'S BAY COMPANY AND THE PUGET'S SOUND AGRICULTURAL COMPANY IN THE COLUMBIA DEPARTMENT, 1845

Post	Employees	Cultivated Acreage	Cattle	Horses	Sheep	Pigs
Fort Vancouver	200	1,200	1,377	702	1,991	1,581
Cowlitz	30	1,000	579	103	1,062	0
Fort Colvile	30	118	96	350	0	73
Fort Langley	20	240	195	15	0	180
Fort Victoria	20	120	23	7	0	1
Nisqually	20	100	1,857	198	5,795	0
Fort Simpson	20	8	0	0	0	0
Fort Hall	20	5	95	171	0	0
Thompson's River	15	6	0	0[a]	0	0
Fort Alexandria	10	46	0	0	0	0
Fort George (N.C.)	10	30	0	0	0	0
Stuart's Lake	10	15	94	39	0	14
Fort Nez Percés	10	12	23	68	0	12
Fort Umpqua	8	50	64	46	0	45
Fort Boise	8	2	27	17	0	0
Fort George (Col.)	6	4	0	0	0	0
Fraser's Lake	5	20	0	0	0	0
McLeod's Lake	5	10	0	0	0	0
Babine Lake	5	8	0	0	0	0
Connolly's Lake	5	4	0	0	0	0
Chilcotin Post	5	0	0	0	0	0
Flathead Post	5	0	0	0	0	0
Fort Okanagan	2	7	0	0[a]	0	0
coasting vessels	140	—	—	—	—	—
freemen	19	—	—	—	—	—
	628[b]	3,005	4,430	1,716	8,848	1,906

For sources see *Sources for Tables*.
[a]These figures are undoubtedly erroneous, as there must have been some horses kept for the Columbia brigade.
[b]The total of 643 in the original represents an error in addition.

to meet not only its own needs but also those of the rival Russian-American Company and some of those of the Sandwich Islands (where "Columbia Country Produce" fetched "very good prices"), as well as the emergency requirements of starving Indians (as at Fort Nez Percés), arriving migrants, overland travellers, company ships, and visiting vessels and expeditions. In the winter of 1833-34, for example, one hundred hogs were salted down at Fort Vancouver for the company's coastal vessels, and in 1841 the United States Exploring Expedition bought thirty-five hundred pounds of supplies at Fort Vancouver.[3] McLoughlin estimated in 1840 that thirteen thousand bushels of wheat were needed annually to satisfy the demands of both the Oregon Country and Russian America. A year later he expected wheat crops of four thousand bushels at Fort Vancouver and six to seven thousand bushels at Cowlitz Farm, plus six thousand bushels from the Willamette settlers.[4] In 1842 "this side the mountains" the company produced ten thousand bushels of wheat and ten thousand pounds of butter and owned seven thousand sheep, two thousand cattle, and two thousand horses. By 1845 the company had forty thousand bushels of wheat in store on the lower Columbia, enough to meet the 1840 requirements for more than two years.[5] Essential to this success was the ample output of the Willamette settlers, most of whom were Americans after 1843. The company needed them as much as they needed the company.

Indeed, throughout the period of joint occupancy, farming and fur trading were generally compatible.[6] There was some spatial conflict since crop growing and stock rearing entailed the clearing of woodland, the main habitat of fur bearers, but in the Oregon Country there was ample room for both activities, given the low population pressure on the soil and most game species. Besides, hunting and trapping occurred largely on land ill-suited to agriculture. For this reason, farming thrived at only a few of the Hudson's Bay Company's posts, and for the same reason, most of the Red River settlers abandoned the Cowlitz Portage for the Willamette Valley. In a wider sense this disjunction also helps to explain why Canada, to a large extent a creature of the fur trade, was less attractive to agricultural immigrants than the United States, where the fur trade did not loom as large.

It was not farming and fur trading *per se* that clashed; rather, it was the freehold farm and the company's monopoly. The firm's charter shielded it from competition in Rupert's Land but not in the unincorporated Oregon Country. Here the United States threatened the company's exclusive privileges. As James Douglas warned in 1838, "the interests of the [Willamette] Colony, and Fur Trade will never harmonize, the former can flourish, only, through the protection of equal laws, the influence of free trade, the accession of respectable inhabitants; in short by establishing a new order of things, while the fur Trade, must suffer by each innovation."[7] The Willamette Settlement's equality and competition (even though the latter tended to defeat the former) appealed to the company's servants. Douglas reported in 1838:

> The only perceptible effect yet produced on our affairs, by the existence of

the Settlement, is a restless desire, in the Companys servants, to escape from our service to the Colony, but in the present state of the country, when the Settlers are so entirely dependent on us, that every man must go down to Vancouver, to sharpen his share, his coulter, and his mattock, we have no reason to fear desertion, however, when the introduction of foreign capital terminates this dependence such events may be expected; and our general influence will decline as the wants of the Settlement find a provision in other sources. The feelings of the Colonists generally are now favourable to the Company, the Canadians being attached from habit and association, identify as far as possible their interest with ours; the vagrant Americans respect power and integrity.[8]

This situation changed with the influx of American migrants via the Oregon Trail. They soon provided the "foreign capital" that Douglas feared, as well as loyalty to another polity. The newcomers brought hostility towards Great Britain and particularly towards the monopolistic Hudson's Bay Company, and although this enmity waned as they received assistance from McLoughlin and became dependent upon the company as a market for their surplus wheat, their national loyalty remained strong and some resented their reliance upon the company.

By the time the American migrants arrived, the Bay men had already tested and proved the soil, water, and forest resources of *les pays d'en haut*, a vast and varied cordillera that was to become one Canadian province and three American states. Moreover, in doing so they ironically paved the way for the company's ouster by laying the groundwork for agrarian development. As Lieutenant Charles Wilkes said of the Puget's Sound Agricultural Company, "I cannot but view the Industry and labours of this Company but as tending to forward greatly the advantages to be derived from it by the future possession of the soil — enabling emigration to go on with much greater ease profit and rapidity."[9] Wilkes added that by securing the territory and succouring immigrants, the mother company had facilitated settlement:

> I must first bear testimony that the officers of the Hon. Compy. service have not only quieted the country but their operations have been so admirably conducted that they have opened the country to safe and secure emigration, and provided it with the means necessary to the success of emigrants, and rendered the task an easy one to its peaceful possession and at a very moderate expense.[10]

In other words, the company rendered an American takeover of the Willamette Valley, and eventually of the entire lower Columbia, safe, easy, and cheap.

The loss of the "garden of the Columbia," however, came as no surprise to the Hudson's Bay Company. From the mid-1820's, when Fort Vancouver was founded, the company had assumed that the Columbia River would become the boundary and the left or southern bank American territory. And with good reason, since the right bank had been explored, settled, and exploited primarily by British subjects. An early American settler noted in 1832 that "the National boundary had not then been settled beyond the [Rocky] mountains, and these [Hudson's Bay Company] traders claimed that the river would be the boundary, and called the south side American."[11] The American settlers themselves believed likewise, according to McLoughlin. He wrote that the arrivals of 1843 "firmly believe the Columbia will be the Boundary." In 1845, just a year before the boundary treaty, McLoughlin confidently told some American settlers that "they might depend the British Government will not give up its right to the north bank of the Columbia."[12]

McLoughlin obviously miscalculated, for Great Britain did cede the northern bank of the lower half of the river. To explain why requires an examination of the "Oregon question." The diplomatic aspect has been thoroughly aired by Frederick Merk, at least from the American standpoint.[13] In essence, the expansionist and confident United States was probably willing to go to war over the "Oregon triangle" between the Columbia and the forty-ninth parallel because of its conviction that the land rightfully belonged to the American flag. The grounds, however tenuous, were discovery (of the Columbia River by Captain Robert Gray in 1792), exploration (by Lewis and Clark in 1805-6), contiguity, Manifest Destiny, and access to Puget Sound's safe harbours.[14] At the same time, a somewhat overextended imperial Great Britain was not willing to wage war over the disputed territory because of its lack of vested national interest and its receipt of unflattering official reports on the region, as well as its reluctance to disturb commercial (and even consanguinal) links with its former thirteen colonies. British resolve weakened from 1841, when the Conservative government of Sir Robert Peel, a peace prime minister with a peace cabinet, came to power and the conciliatory Lord Aberdeen replaced the truculent Lord Palmerston as foreign secretary. Meanwhile, American determination stiffened under President James Polk, who was elected in 1844 on an "All Oregon" platform.[15] British fears of war were heightened by the more passionate and even incendiary tenor of American opinion, including Senator Benton's threat of 1843 to place from thirty to forty thousand rifles in Oregon. And British opinion was perhaps mollified by President Polk's moderate tariff policy, embodied in the Walker Tariff of 1846.

Washington's primary concern was access by sea to the Oregon Country for naval and mercantile purposes. The Columbia with its notorious bar and fearsome gorge was unsatisfactory, and the coast from Cape Mendocino to Cape Flattery offered only exposed, shallow roadsteads. That left Puget Sound and Georgia Strait with their sheltered inlets and bays. William Sturgis, a prominent

Boston merchant and a participant in the maritime fur trade of the Northwest Coast, warned in a well-known public lecture in January 1845 that the Columbia was an unsatisfactory waterway whereas Puget Sound was "easy of access, safe, and navigable at all seasons and in any weather."[16] The U.S. Navy's desire for Puget Sound was undoubtedly reinforced by the losses at the Columbia's mouth of the *Peacock* under Captain William Hudson in 1841 and the *Shark* under Lieutenant Neil Howison in 1846. The United States Exploring Expedition made a point of the desirability of gaining the harbours of Puget Sound. In his report to Secretary of the Navy Abel Upshur in June 1842, Lieutenant Wilkes noted that "the entrance to the Columbia is inpracticable two-thirds of the year, and the difficulty of leaving is equally great."[17] He added:

> This, however, is not the case with the harbors found within the strait of Juan de Fuca, of which there are many, and no part of the world affords finer inland sounds or a greater number of harbors than can be found here capable of receiving the largest class of ships, and without a danger to them that is not visible. From the rise and fall of the tide (18 feet) all facilities are afforded for the erection of works for a great maritime nation.[18]

Wilkes effused that "nothing can exceed the beauty of these waters, and their safety" and that "there is no country in the world that possesses waters equal to these." By contrast, Wilkes said, the coast north of Puget Sound was uninviting. He reported: "The coast of the main land North of the parallel of 49° is broken up by numerous inlets, called canals, having perpendicular sides and very deep water in them, affording no harbors and but few commercial inducements to frequent." "There is no point on the coast," he added, "where a settlement could be formed between Fraser River or 49° north and the northern boundary of 54° 40′ North, that would be able to supply its own wants."[19] In other words, the strategic and economic plum of the Oregon Country's seaboard, Captain Vancouver's "New Georgia," was to become American and Great Britain was to be left with the dross, Vancouver's "New Cornwall," of the coast north of the forty-ninth parallel. And that is precisely what occurred.

But why was Great Britain willing to cede the right bank of the Columbia? One reason was that its own official reconnaissance by land and sea in 1845 yielded a negative assessment of the territory. Lieutenants Henry Warre and Mervyn Vavasour were dispatched overland from the Canadas

> to acquire a knowledge of the character & resources of the country situated between the Sault de Ste. Marie and the shores of the Pacific, and of the practicability of forming military stations therein and conveying troops thither, with a view, should it hereafter become necessary, to the occupation thereof for military operations.[20]

The two officers painted a gloomy picture of the Columbia Department. Its posts were "calculated only to resist the sudden attacks of the Indians"; the upper country was "desolute in the extreme"; the lower country (with the exception of the "fertile" Willamette Valley, where American citizens outnumbered British subjects four to one) was "rugged and impracticable"; passage on the Columbia was obstructed by the bar and the gorge; and soldiers could reach the territory overland more easily from the United States than from the Canadas.[21] The naval inspectors were just as pessimistic. Captain John Gordon and Lieutenant William Peel felt that the Oregon Triangle was not worth a fight, and McLoughlin allegedly expressed the same opinion. Captain Gordon asserted that "the Country up to the South side of the Straits of Fuca, being almost all covered with Fir Woods, is of little value," and Fort Nisqually was "a very poor place without any kinds of defences" and with "very poor land."[22] Of Vancouver Island Gordon is alleged to have told Roderick Finlayson of the Hudson's Bay Company that "I would not give the most barren hills in the Highlands of Scotland for all I see around me"; Fort Victoria, he reported, was "almost impossible for a Stranger to find." It was inferred at the time that the naval officers' opinion of the region was "not very encouraging."

Meanwhile, the Hudson's Bay Company's view of the Oregon Country was more positive. Simpson recognized the value of Puget Sound's harbours:

> The country situated between the northern bank of the Columbia river . . . and the southern bank of Frazer's river . . . is remarkable for the salubrity of its climate and excellence of its soil, and possesses, within the Straits of De Fuca, some of the finest harbours in the world, being protected from the weight of the Pacific by Vancouver's and other islands. To the southward of the Straits of De Fuca . . . there is no good harbour nearer than the bay of St. Francisco . . . as the broad shifting bar off the mouth of the Columbia, and the tortuous channel through it, renders the entrance of that river a very dangerous navigation even to vessels of small draught of water.[24]

But that evaluation was written in 1837. Within three years the fur returns of the Columbia Department were decreasing and with them the territory's usefulness to the company. Long before then, however, even during the boom years of the fur trade, the department was not held in high estimation by most company servants. Few of them liked the Oregon Country, with the exception of the Willamette Valley, and many of them roundly detested the cordillera.[25] John Tod referred to it as "this vile Country." John Work was equally contemptuous: "Would to God my means admitted of my quitting this wretched country of which I have so long tired." Work often called the department a "cursed country," as well as a "barbarous country."[26] But to those men, like Work and Tod, with Indian wives and half breed children, the Canadas or the British Isles were *verboten*, and they had to retire in the Indian Country itself, usually the Willamette Valley or Vancouver Island.

New Caledonia was particularly disreputable. Following the coalition of 1821, Bay men, who had never been stationed in New Caledonia, "lived in fear of being transferred thither" because the "Northwesters gave such a poor account of the country."[27] Company service there was marked by disease, murder, and desertion. The district was infamous for its "misery and privation" and "Starvation & Solitude," as well as "extra duty" — "difficulties, to which the business of no other part of the country was exposed."[28] Chief Factor William Connolly, who was then in charge of New Caledonia, informed the Governor and Council of Rupert's Land in 1827 that

> the labor performed by the Columbia Men and those in this district [New Caledonia] will bear no comparison: the duties the latter have now to perform altho' not exactly of the same description as heretofore are nevertheless almost equally severe, and in regard to living dry salmon as usual is their only fare.[29]

He added that it was difficult for the company to hire men for the Columbia Department and to re-engage them there because there was "nothing but hard labor and bad fare before their eyes." All of the New Caledonia servants whose terms of service expired in the spring of 1827 intended to leave the district and the department.[30] Simpson acknowledged that in New Caledonia

> nearly the whole year round was occupied in transport, and the unceasing labourious duties of the people added to the privations to which they were exposed from the poverty of the Country in the means of living, rendered the Service the most painful and harassing in North America and operated as a check to enterprise and exertion.[31]

"There is not a District in the country," he added, "where the Servants have such harassing duties or where they undergo so many privations; to compensate for which, they are allowed a small addition to the Wages of other Districts [£2 or from 9 to 12 per cent more per annum]."[32] Such were the "hard duties and still harder fare" that, according to Tod, "New Caledonia was then looked on in the light of another Botany Bay, Australia, the men were in dread of being sent there," and engagés in New Caledonia were sometimes referred to as "exiles." The damp and dreary coast was just as uninviting; recruits had a "dread" of being posted there.[34]

The Columbia Department's disrepute among Bay men stemmed from its isolation and privation and, especially, its "precious Bill of Mortality." Twenty-six men were killed in the department in 1830 alone.[35] Archibald McDonald noted that the territory was "fertile in disasters"; "man's life now in the Columbia is become mere lottery," he added in 1832, thanks to the bar at the Columbia's mouth, the rapids at the Dalles, "intermittent fever," salmon failures, and Indian attacks.[36] Simpson acknowledged the department's "considerable sacrifice of life among the Company's officers and servants, owing to the fierce, treacherous, and blood-thirsty character of its population, and the dangers of the navigation." For instance, Chief Factor Samuel Black of Fort Kamloops was shot and killed by an Indian in 1841, and John McIntosh, the clerk in charge of McLeod's Lake, was murdered by two Indians in 1844. Until 1834 the Indian danger was such that it was considered unsafe to travel up or down the Columbia with fewer than sixty armed men.[37]

The river itself took a heavier toll, however. As Reverend Henry Spalding of Lapwai noted, "the Columbia is the most frightful river I ever saw navigated by any craft." "Many lives are lost in this river," he added. "None but Canadians and Indians would ever think of navigating this terrible Columbia."[38] The deceptive bar, the turbulent Dalles or Narrows, where the Columbia Gorge narrowed to seventy-five feet, and other rapids both downriver and upriver took many lives. In 1830, for example, sixteen servants were drowned in the Columbia (nine at the Dalles in July and seven at the Okanagan Dalles in October). Narcissa Whitman, whose own daughter was drowned at Waiilatpu, contended at the end of 1836 that more than one hundred lives had been lost at the Dalles alone.[39] And in the spring of 1838 she wrote: "O the dangers of that river! scarcely a year passes without the loss of several lives, we have just been told that the Company have lost upwards of three hundred men in the Columbia." Mary Walker of Tshimakain was equally appalled. "There is probably more danger in going down the River Col. to Van. than in the whole journey across the Mts. [on the Oregon Trail]," she asserted.[40] In October 1838 twelve persons, including the botanists Peter Banks and Robert Wallace, were drowned while trying to run the Dalles des Morts on the upper Columbia. And during the five years from mid-1838 through mid-1843, thirty people were drowned in the lower Columbia and Willamette. The heavy loss of lives and goods by brigades while shooting rapids, particularly on the Columbia, prompted the Council of the Northern Department in 1840 to pass a resolution forbidding guides and steersmen from running rapids or falls that entailed the slightest danger.[41]

The Columbia Department's high mortality was also attributable to disease, which took many Indian lives: syphilis ("Chinook lover fever"), tuberculosis, smallpox, and especially "intermittent" or "trembling" fever or "cold sick." This last "dreadful visitation" was either malaria or influenza, perhaps aggravated by typhus. It struck the lower Columbia twice each year in the first half of the 1830's and again in the first half of the 1840's. The naturalist John Townsend described the disease in 1834: "The symptoms are a general coldness, soreness and stiffness of the limbs and body, with violent tertian ague. Its fatal termination is attributable to its tendency to attack the liver, which is generally affected in a few days after the first symptoms are developed."[42] It first appeared in 1829 or 1830 (the Indians blamed Captain John Dominis of the American brig *Owhyhee* for introducing it), and every summer from 1830 through 1835 Fort Vancouver and vicinity were afflicted, with various operations being hampered. In late September of 1830, McLoughlin reported, "half of our people are laid up with it," including Peter Ogden, one of McLoughlin's most able lieutenants, who nearly died.[43] In 1831 all but seven at the post contracted the disease, and some had three or even four bouts; it raged so fiercely that "for a time [it] put an entire stop to all our Business."[44] In 1832 everybody at Fort Vancouver save eight were affected, and in 1833 more than three-quarters of Work's Southern Expedition were struck, including Work himself, who was "reduced to a perfect skeleton and could scarcely walk."[45] Few Euroamericans actually died, however, whereas the Indian fatality rate probably averaged 75 per cent and may have reached 90 per cent.[46] Father Bolduc, a Catholic missionary, reported that "the savages die very frequently because they cannot resist the temptation of drinking cold water and when the fever overcomes them they at once run and dive into the river which causes instant death." Entire villages were annihilated. In October 1830 McLoughlin reported that the disease had "carried off three fourths of the Indn. population in our vicinity."[47] A year later Duncan Finlayson observed that "there has been & still continues a great mortality among the natives ... consequently, the trade is not so brisk as at former periods." In 1834 McLoughlin acknowledged that "the Mortality among the Natives has been Immense."[48] From 1834 through 1837 in particular the Indians of the Willamette Valley were decimated by fever. The American agent William Slacum put the Indian toll at five to six thousand by 1837. Dr. William Bailey, an English surgeon, told Lieutenant Wilkes that one-quarter of the Indians of the Willamette Valley died every year from "fever and ague," and Wilkes's own surgeon reported that "nearly all the natives on the Columbia or its branches have been destroyed by it."[49] High morbidity was not confined to the lower Columbia, however. The dank Northwest Coast was particularly conducive to respiratory ailments. At Fort McLoughlin one-quarter to one-third of the men were on the sick list in winter. Worst of all was the 1835-38 smallpox epidemic, which killed one-third of the Tlingits, Haidas, and Tsimshians.[50]

Disease meant fewer Indian sellers of furs and buyers of goods for the company, as well as fewer obstacles to Euroamerican land seekers. The hardships of service in the Columbia Department, including fatal accidents on the Columbia, meant a shortage of employees for the company. The "generally esteemed" Chief Trader John Harriott reported in early 1831 that "the Columbia has again been prolific in misfortunes, in fact more so than ever," with twenty-four engagés being killed in the last two months of 1830 and the first two months of 1831.[51] "I am afraid," he wrote, "it will prevent people from volunteering for this side of the mountains. I am not however aware that many were over anxious on that score but on the reverse rather desirous of keeping from it as long as they could." He added:

> a number of recruits will be wanted but really I do not see where they will come from, there are now very few who have not tasted the sweets of the Columbia and New Caledonia and it would certainly be considered hard to send those back who have already passed three or four years & have had the satisfaction of getting safe out of it.[52]

Simpson deplored "that system of violence which has so often been noticed as prevailing on the West side of the mountains" and "the tendency of habitual severity to render the service unpopular."[53] In the middle and late 1830's and early 1840's there was considerable discontent among Bay men in the Oregon Country, thanks to the privations, dangers, and falling profits. "Porkeaters" (new employees) were increasingly difficult to find, and posts were shorthanded. From the late 1830's company operations were impeded by the fact that fewer and poorer men were entering the service. McLoughlin informed Simpson in 1843 that "from year to year our men are falling off so much that out of those [who have] come [here] these last four years we have been able to get only one to make a Boute [bowsman or steersman] and yet he is a poor one." In 1845 the Columbia Department required sixty recruits but obtained only twenty-five.[54]

In the meantime the United States was steadfastly insisting upon at least the forty-ninth parallel as the boundary, a convenient extension of the line that already marked the border across the Great Plains and, in the words of Henry Clay, American Secretary of State, would "consent to no other line more favourable to Great Britain."[55] At the same time, British steadfastness, under Peel, was becoming appeasement of American expansionism, at the expense of not only the Hudson's Bay Company but also the future Canadian dominion. Even when the convention on joint occupancy was renewed in 1827, the British plenipotentiaries were wavering. The American minister to London, Albert Gallatin, reported immediately after the signing that although "national pride" and "public opinion" did not allow the British to accept a division of the Oregon Country that the United States "have a right to claim," they did not have "any wish to colonise it," viewed it "rather with indifference," and seemed willing "to let the Country gradually and silently slide into the hands of the United States," being "anxious that it should not, in any case, become the cause of a rupture between the two Powers." He added that American acquisition probably would not occur while the Hudson's Bay Company had "still sufficient influence" and until American citizens had gained a "respectable footing" in the territory.[56]

By the mid-1840's these two conditions had both been met. On the one hand, American settlers had won a "respectable footing" by overwhelming the

Canadian settlement in the Willamette Valley. Whitehall, for whatever reasons, did not encourage or organize British migration to the Oregon Country from either Britain or Canada, even though McLoughlin was certain in 1836 that Canadian settlers would leave Upper and Lower Canada for the Willamette "if it was not so difficult to get here — and that the Country was better Known." "You have not so fine a Country in Canada as the Willamette," he added to Edward Ermatinger, a former clerk at Fort Vancouver who had retired to St. Thomas, Upper Canada.[57] A year later Simpson reported from the Columbia Department to the Governor and Committee that:

> The possession of that country to Great Britain may become an object of very great importance, and we are strengthening that claim to it (independent of the claims of prior discovery and occupation for the purpose of Indian trade) by forming the nucleus of a colony through the establishment of farms, and the settlement of some of our retiring officers and servants as agriculturalists.[58]

Early in 1842 Simpson told McLoughlin that there was almost a "mania" among the company's Canadian personnel to retire as farmers on the Pacific slope, and he instructed McLoughlin to discourage legitimate retirees from settling on the "South Side of the River" and to encourage them to homestead at Cowlitz, Nisqually, and the southern end of Vancouver Island.[59] Although the British Government did not augment this company effort at colonization, it should be noted that American settlement on the right bank of the Columbia was so negligible as to be virtually non-existent. The first American settler to cross the Columbia was Dr. John Richmond, a Methodist missionary who with his wife and four children resided at Nisqually in 1840 but moved to the Willamette in 1841. The first permanent white settler north of the Columbia was likely John Jackson, an Englishman, who located several miles north of Cowlitz's Catholic mission in 1844. He was followed a year later by Michael Simmons, an American, whose party of nine included a mulatto barred from the Willamette Settlement by its provisional laws. They went at the encouragement of the Hudson's Bay Company, which had employed them at Fort Vancouver as shingle makers.[60] By 1846 this group comprised five American families and "some single men," twenty-eight persons altogether, on the openings between the Cowlitz River and Puget Sound, mainly Simmon's Prairie and Bush's Prairie.[61] They were outnumbered by the Hudson's Bay Company's and Puget's Sound Agricultural Company's employees and the remaining Red River freeholders. On the whole, however, American citizens outnumbered British subjects two to one by 1844, three to one by 1845, and as much as five to one by 1846 (see Table 24), although their numerical superiority was restricted to the southern side of the Columbia.

On the other hand, by the mid-1840's, too, the Hudson's Bay Company's influence had waned in proportion to its economic downturn. Governor Simpson had sensed early that the old order of the Indian Country, centred on the fur trade, was changing and that new sources of profit like farming, fishing, lumbering, mining, and whaling had to be developed as alternatives. And his response — diversification — succeeded. In 1830, for instance, he reported that:

> The great exertions made to gain a share of the Northwest coasting trade have been attended with success, though the outlay of money in the face of powerful opposition has been large. Other branches of trade such as timber and salmon, together with the returns of the furs have defrayed all expenses, so that what has usually been regarded as the regular trade of the Columbia Department has not been affected.[62]

American competition on the Northwest Coast and in the Flathead Country, however, drove fur prices up and company profits down in the mid-1830's.[63] In 1841 a member of the United States Exploring Expedition reported that "it is stated that the income of the H.B.C. from their fur trade is gradually diminishing; but that the deficit is more than supplied by their exports of flour, beef and agricultural products generally, with the lumber, of which they export a considerable quantity."[64] By the mid-1840's, however, the company's farms were being outproduced by those of the Willamette Settlement, and the firm had even become dependent upon at least the Canadian settlers' output of wheat to meet the requirements of the Russian contract.

Meanwhile, what Douglas termed "the decline and wreck of fur trade affairs" was continuing.[65] That trend had begun at least as early as the mid-1830's in the Columbia Department. As William Tolmie noted at the beginning of 1839:

> The profits [of the Hudson's Bay Company] have been steadily decreased since that period [mid-1830's] owing in part, to the depreciation in the value of Beaver, occasioned by improvements in the manufacture of Silk Hats; but chiefly, to diminished Returns; caused by the exterminating system of hunting pursued, which if not checked, will speedily eventuate in the destruction of the more valuable Fur-bearing animals.[66]

Capricious fashion was changing; silk hats were replacing felt, or beaver, hats, and the value of beaver was falling accordingly. As Tod lamented, "beaver were valuable before silk hats came into use."[67] Frank Ermatinger informed his brother, Edward, in 1844 that "the Columbia trade is getting from bad to worse, and upon the other side [of the mountains] it is no much better"; "besides," he added, "the price of Beaver is falling, owing to the use of silk." The popularity of silk hats was so great that in 1837 Work complained of "the almost worthlessness of beaver."[68]

As Tolmie had noted, however, the main culprit was overhunting (indeed, the decreasing supply of beaver pelts may have prompted the introduction of silk hats). That, in turn, stemmed from the company's view of the Columbia Department. It was regarded as a monopolistic game preserve — a "fur nursery" — whose returns would offset the depletion of Rupert's Land. And when it became clear to Simpson in the mid-1820's that the left bank of the Columbia would eventually become American territory, he hastened to scour it by means of the annual Snake and Southern Expeditions. There would be no point in letting the beaver restock themselves if their habitat were to become the property of American competitors. "How much more ought we to endeavour to get all we can out of it [the Snake Country] when we are certain we will be deprived of it," declared McLoughlin in 1828.[69] Four years later he reported that "it is certain the Snake country is getting nearly exhausted, and the new Country between this and S. Francisco does not afford employment for a party sufficiently strong to protect themselves."[70] Ogden's 1830 trapping expedition, which had probed as far as the Gulf of California, had "found Beaver very scarce," and Work's 1831 foray had suffered from the "exhausted state" of the Snake Country. ("The country to the southward," Work added, "is ruined so much that little or nothing is to be done in it," and only the Flathead Country offered " a likelihood of making anything.")[71] "Furs are already becoming scarce & the present supply is obtained by an almost exterminating system of hunting," remarked Tolmie in 1833.[72] Three years later McLoughlin acknowledged that "I consider that the Tenure which the Company have of the South side of the Columbia is of that precarious nature, that we are certain of being deprived of it when a Boundary line is run between the British and American Government, and that it is the interest of the Company to make all they can out of it while it is in our power."[73]

It was not long before returns were diminished by this *Raubwirtschaft* (Table 30). Overall, the Columbia Department's fur take decreased from a high in 1839; the Columbia District's (lower Columbia's) catch also peaked in 1839, but its share of total returns fell fairly steadily from 1826, demonstrating an increasing reliance upon New Caledonia. At the same time the Columbia District's share of beaver returns remained rather stable between 39 and 47 per cent, reflecting Simpson's special attention. Meanwhile, the average annual beaver catch of New Caledonia, which was being held in reserve, slumped from about 6,500 in the last half of the 1820's to about 5,100 in the last half of the 1830's and about 3,300 in the last half of the 1840's.[74] Lieutenant Wilkes was told at Fort Vancouver in 1841 that the fur trade had declined 50 per cent, and indeed it had since 1838 and 1839, according to Table 30. Simpson admitted to the Governor and Committee in 1841 that "the Returns in many of the districts [of the Columbia Department] . . . are rapidly declining, owing to the very close manner in which the country has been hunted."[75] He added:

this is the natural result of the exertions that have been made to prosecute the trade with vigour, with the double object of benefitting by immediate results, & of rendering the country less inviting to the numerous United States trapping parties, who formerly threatened to overrun the whole of the accessible country on the West Side the Rocky Mountains.[76]

Table 30

HUDSON'S BAY COMPANY FUR RETURNS OF THE COLUMBIA DEPARTMENT AND THE COLUMBIA DISTRICT, 1826-46

Year	All Furs[a]			Beaver		
	Col. Dept.[b]	Col. Dist.[c]	Share	Col. Dept.[b]	Col. Dist.[c]	Share
1826	34,748	19,438	56%	17,483	11,242	64%
1827	37,297	15,784	42	16,538	7,708	47
1828	39,339	19,226	49	18,435	8,406	46
1829	37,417	17,355	46	16,887	7,825	46
1830	36,520	16,359	45	19,037	9,121	48
1831	55,428	22,289	40	21,746	8,801	40
1832	58,751	22,293	38	21,717	8,296	38
1833	70,616	28,546	40	21,290	8,933	42
1834	67,954	31,488	46	21,431	9,051	42
1835	62,364	24,586	39	19,964	7,844	39
1836	70,342	23,048	33	21,253	8,559	40
1837	73,217	23,495	32	21,780	9,656	44
1838	86,704	32,909	38	20,157	9,244	46
1839	87,731	36,303	41	20,970	9,887	47
1840	55,576	19,502	35	19,481	8,540	44
1841	59,171	21,763	37	18,599	7,323	39
1842	58,012	22,487	39	15,779	6,789	43
1843	69,329	25,878	37	17,725	6,517	37
1844	57,349	20,165	35	12,096	4,265	35
1845	71,292	20,148	28	14,437	5,604	39
1846	80,163	20,040	25	12,958	5,400	42

For sources see *Sources for Tables*.
[a]Excluding beaver coating, castoreum, marten robes, sea otter tails, swan skins, ox hides, deer hides, sheep hides, whale oil, whalebone, salmon, isinglass, tortoise shell, pearl shell, and gold dust.
[b]Including coastal vessels and, from 1841, the Hawaii and California establishments.
[c]The Columbia District or Lower Columbia includes Fort Vancouver, Colvile, Nez Percés, and Nisqually, the Snake and Southern Parties, and the Hawaii and California establishments.

Beaver House wrote to McLoughlin in 1843 that the "continually decreasing price, when considered in connection with a constantly decreasing supply, holds out no very cheering prospect for the future, unless the tide of fashion change."[77] But fashion did not change back to felt, and even if it had, not enough beaver remained anyway. As the irascible John Tod concluded gloomily in 1844, "there appears to be no longer any prospect of either profit or pleasure here. A general scarcity of furs prevails at present all over the country, so that even a ready Market in China would in the present exhausted State of the Country, be of little or no advantage to us."[78]

This decline was reflected, of course, in company profits and dividends. Following the "destructive contest" with the North West Company of the years 1800-21, when the Hudson's Bay Company's dividends were reduced to 4 per cent (1800-7), nil (1808-13), and 4 per cent again (1814-21), the merger of 1821 and Simpson's management brought renewed, if shortlived, prosperity: half-year dividends of 5 per cent from 1824 through 1838, with an annual bonus of 10 per cent in 1828-32, 6 per cent in 1833-36, and nil in 1837.[79] The decline of the bonus was symptomatic of the turnabout in returns of the mid-1830's. The income of a chief trader (like a chief factor not a salaried employee but a shareholder in the company with a vested interest in its performance) amounted to some £500 per year at the end of the 1820's but fell to about £300 at the end of the 1830's.[80] The shareholders in the field were disappointed, if not disgruntled. Archibald McDonald complained in 1837 that "the general profits are annually decreasing & will continue to decrease." In 1841 Frank Ermatinger told his retired brother that "the trade is getting worse every year at all the [Oregon Country] posts . . . and the profits of course are lower."[81] And in 1844 John Work, one of the stalwart servants of the Columbia Department, wrote: "The universal depression of business [of 1841-43, probably the century's worst] has at length reached even the Indian Country, furs have been selling badly and what is worse greatly decreased in number, this is an evil that there is little prospects of being remedied. The dividends are fallen off greatly."[82] The Columbia accounts showed losses of £4,003 in 1842 and £3,156 in 1843, compared with a profit of £1,474 in 1841 (and £20,000 in 1832).[83] The fur trade was rapidly fading from a mainstay to a sideline of the economy of not only the Oregon Country but all of British North America as well.

As a result of all of these pressures, the Hudson's Bay Company developed a contingency plan a decade before the boundary settlement. This plan assumed the loss of both banks of the Columbia to the United States and the company's withdrawal to New Caledonia and the northern coast. Simpson knew from his 1828 trip down the Fraser River that it was an impracticable route to the interior, and at the same time he was determined to dominate the coast trade, so an island base was sought to replace the headquarters at Fort Vancouver. In the spring of 1837 Captain William McNeill reconnoitered the southeastern end of Vancouver Island "to ascertain its capabilities for an Establishment on an extensive scale with all the suitable requisites for farming, rearing cattle, a good harbour, and wood fit for building." McNeill found Camosun, an "excellent harbour and a fine open country along the sea shore apparently well adapted for both tillage and pasturage."[84] In 1842 Simpson ordered the building of a new depot here and the closing of Forts Taku and McLoughlin.[85] Construction of Fort Victoria began in 1843, the year that saw the Canadian settlers of the Willamette Valley swamped by American migrants. By then Simpson had warned that the company should be "prepared for the worst," that is, a settlement of the boundary dispute in favour of the United States.[86] On 16 June 1846 the governor instructed the Board of Management at Fort Vancouver:

> we feel that, from the number & desperate character of the American population by whom we are surrounded, our interests in the Columbia are in a very hazardous position, & with that feeling, we are very anxious that, the Company's business should be as much concentrated & put upon as small a scale as possible, to which end, we have a desire that, no more goods be kept in depôt at Fort Vancouver than may be absolutely necessary to meet the immediate & pressing demands of the Fur trade, & ... instead of increasing our flocks & herds at Vancouver, the Cowelitz & Nisqually, we think it desirable to direct our attention to the rearing of stock on an extended scale on Vancouver's Island, near Ft. Victoria.[87]

Unknown to Simpson, the United States Senate had ratified the Oregon Treaty the day before. It confirmed the governor's fear of a surrender. The forty-ninth parallel was extended across the Rockies to the Pacific, thereby placing Forts Vancouver, Colvile, Okanagan, Nez Percés, George, Nisqually, Boise, Hall, and Umpqua and Cowlitz and Nisqually Farms in American territory. The Bay men were shocked. Douglas and Ogden, both veteran and esteemed servants, understated that the settlement was "more favourable to the United States than we had occasion to anticipate."[88] Douglas himself assailed "this monstrous treaty" as being "unfavourable to the character and interests of the British Government."[89] He realized that the treaty was not explicit enough to protect the property of British subjects in American territory, and he correctly predicted that it would be a pretext for "every species of oppression" (such as import duties) against the company's trade via the Columbia. His reaction echoed the sentiments of the botanist David Douglas, of "Oregon pine" fame, who reportedly told an American pioneer that "it makes me mad to see how Jonathon has always got the advantage of John Bull in all negotiations relating to territory."[90] The Maritimers of the St. John River Valley and the Assiniboians of the Red River Valley would have understood, as would British Columbians a quarter of a century later during the San Juan Islands boundary farce.

At least some, if not most, American settlers had by 1846 fully expected a settlement at 49° or even farther north. Their attitude was expressed by Peter Burnett:

The final settlement of the conflicting claims of the two governments in this manner [at 49°] did not surprise any sensible man in Oregon, so far as I remember. It was what we had every reason to expect. We knew, to a moral certainty, that the moment we brought our families, cattle, teams, and loaded wagons to the banks of the Columbia River in 1843, the question was practically decided in our favor. Oregon was not only accessible by land from our contiguous territory, but we have any desirable number of brave, hardy people who were fond of adventure, and perfectly at home in the settlement of new countries. We could bring into the country ten immigrants for every colonist Great Britain could induce to settle there.[91]

On this point Burnett was right. Apart from the few Red River families, no British settlers had flocked to the Oregon Country from the Canadas or the British Isles; in fact, in 1845 only from one-fifth to one-quarter of the population of the Willamette Valley was non-American (see Table 24). This was partly because the Canadas themselves were still sparsely settled, partly because they were not yet an independent nation with a strong sense of identity and role to match that of the successful and confident American republic, and partly because the Hudson's Bay Company's primary purpose was the fur trade, not agricultural colonization. Some farming was necessary for subsistence and even exportation, but more would alienate, if not decrease, Indian middlemen and would diminish fur beavers. At such a widespread level the fur trade and agriculture were incompatible. Burnett added, however, that "neither claim could be properly called a plain indisputable right, because much could be and was said on both sides of the question. But, while our *title* might be disputed, there was no possible doubt as to the main fact, that *we had settled the country*."[92] On this score he was wrong. Americans had settled the Willamette Valley (and even there they were not the first Euroamerican settlers and not in a majority until late 1843) but not the right bank of the Columbia, or even the left bank above the Snake or either bank of the Snake itself. The boundary dispute had been reduced to either Great Britain yielding the Columbia routeway or the United States ceding the harbours of Puget Sound. As McLoughlin phrased the issue in 1845, "if Britain gave up the Columbia, it gave up all water communication with the interior of the country, and if the American Government gave up Puget Sound, they would have no good harbour in the Pacific."[93] In the end, of course, the United States stood firm and Great Britain gave way, yielding not only all of the lower Columbia but also all of Puget Sound. None of the latter's harbours were kept for British North America, which was fortunate to retain Vancouver Island. (By the end of the decade the United States had also acquired San Francisco Bay, the best harbour on the entire coast.) The Oregon Treaty was *not* a fair compromise; there was no division of the "Oregon triangle," all of which went to the United States. Even the division of territory between

42° N. (the Adams-Onís line of 1819 marking the northern limit of Spanish territory) and 54° 40′ N. (the southern limit of Russian America and the extreme limit demanded by American warhawks) was not exactly equal (that would have been 48 °20′).[94] London was willing to cede because "the British people knew nothing and cared less about Oregon" and "British statesmen took little interest in what one [Lord Aberdeen, British Foreign Secretary] described as 'a few miles of pine swamp' "; Aberdeen "saw nothing worth fighting for in the comparatively small area [Oregon Triangle] that was really in dispute."[95] And the future of the rebellious Canadian colonies was not high on the list of priorities of the Peel government (1842-46), which was preoccupied with the repeal of the Poor and Corn Laws and the maintenance of Anglo-American trade.[96]

Thus, present-day Canadians have valid reasons for regretting and even resenting the Oregon settlement, since the British claim to the territory north of the Columbia-Snake-Clearwater river system was at least as good as, if not better than, that of the United States on the grounds of discovery, exploration, and settlement, and since the future Canadian Dominion was deprived of any harbour on Puget Sound (as a concession to the "reasonableness" of the unwavering American demand for a Pacific outlet there). They should remember, however, that in view of the absence of a Canadian vested interest in the territory, and the prevailing moods of appeasement in Great Britain and expansionism in the United States, they were perhaps fortunate to inherit as much as they did. Indeed, if President Polk had not been as intent on the "re-annexation" of Texas and the seizure of California from Mexico as he was on the "re-occupation" of Oregon, even less of the Cordillera may have remained British. Canadians may find solace, too, in the fact that despite the strident slogan of American extremists, "Fifty-four forty or fight," they neither gained that parallel nor fought when they were denied it. But Canadians should not forget that they were dispossessed of part of their rightful Columbia heritage, a heritage whose economic potential in general and agricultural possibilities in particular were initially and successfully demonstrated by the Hudson's Bay Company. They should also remember that whenever it is tritely declared that Canada and the United States share the longest undefended border in the world, it is so mainly because the stronger American republic won its northern boundary disputes at the expense of its weaker neighbour, just as its southern boundary was gained at the expense of a weaker Mexico. It may come as a surprise to Americans, and even to Canadians, that unlike the United States, Canada has an irredentist legacy, to which the Oregon "compromise" was the chief contributor.

Abbreviations

HBCA = Hudson's Bay Company Archives (Winnipeg). (N.B. The author's folio numbers refer to handwritten numbers whenever handwritten numbers only have been used in the manuscripts but to stamped numbers whenever both handwritten and stamped numbers have been used.)
Mc. = Microcopy
NARS = National Archives and Records Service (Washington, DC)
OHS = Oregon Historical Society (Portland)
PABC = Provincial Archives of British Columbia (Victoria)
PAC = Public Archives of Canada (Ottawa)
PRO = Public Record Office (London)

Sources for Tables

TABLE 1

Fleming, *Minutes of Council*, pp. 50, 80, 110, 149, 182, 208, 239, 257, 277.

TABLE 2

HBCA, B.239/g/61-62, B.239/1/1a-1b: 2-16.

TABLE 3

Anonymous, "Reminiscences of Fort Vancouver," p. 75; [De Mofras], *Duflot De Mofras' Travels*, vol. 2, p. 98; Elliott, "British Values in Oregon, " pp. 30, 30-32; HBCA, D.4/59: 90, D.4/100: 25v.; "Papers relative to the Expedition of Lieuts. Warre and Vavasour," 72v.; Peel, "Lieutenant Peel's Report," p. [2]; Rich, *Letters of John McLoughlin*, vol. 1, pp. 205, 283; United States, Congress, Senate, "Message from the President of the United States," p. 10; Wyeth, *Correspondence and Journals*, p. 177.

TABLE 4

Anonymous, "Documents," (1907), p. 40; Anonymous, "Documents," (1908), p. 168; Barker, *Letters of Dr. John McLoughlin*, pp. 94, 186; British and American Joint Commission, [*Proceedings*], vol. 2, p. 52; Dease, "Diary," p. 7; McLoughlin to E. Ermatinger, 3 March 1837, "Ermatinger Papers"; Ermatinger, "Papers" vol. 1, p. 188; Drury, *First White Women*, vol. 1, p. 102; Elliott, "Letters of Dr. John McLoughlin," p. 371; HBCA, A.11/70: 185v., 229v., D.4/59: 90, D.4/67: 85v., D.4/100: 6v., D.4/110: 25v.-26, D.4/120: 54, D.4/121: 37v., D.4/123: 96v., D.4/125: 78v.; McLeod, "John McLeod Papers," vol. 1, pp. 296, 338; [McLoughlin], "Letter of John McLoughlin," p. 206; Merk, *Fur Trade and Empire*, pp. 270, 291, 324; Parker, *Journal of an Exploring Tour*, p. 172; Rich, *Letters of John McLoughlin*, vol. 1, pp. 16, 31, 51, 61, 67, 105, 113, 143, 206, 228, vol. 3, pp. 36, 148; Rich, *Part of Dispatch from George Simpson*, p. 69; Spalding, "Letter," 1837 and "Letters," p. 374, in Spalding, "Papers."

TABLE 5

Anonymous, "Documents," (1908), 168; Barker, *Letters of Dr. John McLoughlin*, p. 37; British and American Joint Commission, [*Proceedings*], vol. 2, p. 52; HBCA, B.223/d/2b: 17, B./223/d/5: 56-56v., B.223/d/13: 14v.-15, B.223/d/22: 11, B.223/d/42: 7v., B.223/d/49: 22v., 38v., B.223/d/59: 27, 35v., B.223/d/71: 30, 35v., B.223/d/80: 38, 45, B./223/d/93: 52, B.223/d/105b: 50v., B.223/d/115: 70-70v., B.223/d/126: 63, B. 223/d/136: 68v., B.223/d/144: 92, B.223/d/148: 68, B.223/d/155: 87v.-88; McLeod, "John McLeod Papers," vol. 1, p. 338; McLoughlin, "Copy of a Document," pp. 46, 51; McLoughlin, "Private papers," pt. 2, p. 1; Merk, *Fur Trade and Empire*, pp. 270, 301; Powers, Hopkins, and Ball, *John Ball*, p. 94; Rich, *Letters of John McLoughlin*, vol. 1, pp. 79, 207, 285; [Scouler], "Dr. John Scouler's Journal, " p. 174; United States, Congress, Senate, "Messsge from the President," p. 22.

210   Farming the Frontier

TABLE 6

Anonymous, "Documents," (1908), 255; Drury, *First White Women*, vol. 2, p. 151; Drury, *Nine Years with the Spokane Indians*, p. 75; Farnham, *Travels in the Great Western Prairies*, vol. 2, p. 182; HBCA, B.223/d/49: 49v.; McLeod, "John McLeod Papers," vol. 1, pp. 199, 225, 363; Spalding to Greene, 4 September 1837, Walker to Green, 15 October 1838, "Papers of the American Board of Commissioners," vol. 1; McDonald to McKinlay, 13 October 1841, [Fort Colville], "Correspondence of Archibald McDonald"; [John Work], "Work Spalding, Correspondence, p. 56; Spalding, "Letter," 1838, "Papers"; Spalding, "Diary 1838," p. 9, "Diaries."

TABLE 7

Anonymous, "Documents," (1908), 255; Brackenridge, *Brackenridge Journal*, p. 31; Drury, *First White Women*, vol. 1, p. 103; Eells, "Letters to A.B.C.F.M., " p. 75; Farnham, *Travels in the Great Western Prairies*, vol. 2, p. 182; HBCA, A.11/17: 1v., B.223/d/5: 70, B.223/d/13: 33-33v., B.223/d/22: 27v., B.223/d/49: 49v., B.223/d/59: 61, B.223/d/71: 39v., B./223/d/80: 58, B.223/d/93: 82v., B.223/d/105b: 72, B.223/d/115: 102v., B.223/d/126: 78, B.223/d/136: 88v., B.223/d/144: 127v., B.223/d/155: 136, B.223/d/160: 107v., B.223/z/4: 196v.; McLeod, "John McLeod Papers," vol. 1, p. 363; Spalding to Greene, 4 September 1837, "Papers of the American Board of Commissioners," vol. 1; "Papers relative to the Expedition of Lieuts. Warre and Vavasour," pp. 67v., 86.

TABLE 8

Dease, "Diary," p. 5;[Fort Langley], "Correspondence," p.[1];[Fort Langley], "Fort Langley Journal," pp. 84-86, 89-90, 148; HBCA, B.113/a/2: 19v., B.113/a/3: 20, 22v., B.223/d/49: 35v., D.4/123: 69, 72, D.4/125: 63; Reid, "Fort Langley Correspondence," 187; Rich, *Part of Dispatch from George Simpson*, p. 43; [Work], *Journal of John Work*, p. 85.

TABLE 9

[Fort Langley], "Correspondence," p. [22]; HBCA, B.223/d/49: 35v., B.223/d/59: 45, B.223/d/71: 34v., B.223/d/80: 41v., B.223/d/93: 63v., B.223/d/105b: 58v., B.223/d/115: 88, B.223/d/126: 70v., B.223/d/136: 74, B.223/d/144: 105, B.223/d/148: 81, B.223/d/155: 92, B.223/d/165: 41v.-42.

TABLE 10

HBCA, B.223/d/2b: 25v., B.223/d/5: 63, B.223/d/13: 23v., B.223/d/22: 18v., B.223/d/71: 36v., B.223/d/80: 48, B.223/d/93: 72v., B.223/d.105b: 62, B.223/d/115: 97v., B.223/d/126: 75v., B.223/d/136: 80, B.223/d/144: 101-101v., B.223/d/148: 91, B.223/d/155: 126v., B.223/d/160: 102, B.223/d/165: 46v.-47.

TABLE 11

HBCA, B.223/d/2b: 29, B.223/d/5: 67v., B.223/d/13: 26v., B.223/d/22: 31, B.223/d/59: 51, B.223/d/71: 45v., B.223/d/80: 50v., B.223/d/93: 79-79v., B.223/d/105b: 63, B.223/d/115: 105, B.223/d/126: 80v., B.223/d/136: 82.

TABLE 12

HBCA, B.223/d/13: 23v., 32, B.223/d/41: 25, B.223/d/49: 43v., B.223/d/59: 53v., B.223/d/71: 42v., B.223/d/80: 54v., B.223/d/93: 77v., B.223/d/105b: 67v., B.223/d/126: 83v., B.223/d/136:

85, B.223/d/144: 122v., B.223/d/155: 132.

TABLE 13

HBCA, B.223/d/71: 32, B.223/d/80: 43v., B.223/d/93: 66-66v., B.223/d/105b: 56v., B.223/d/115: 83-83v., B.223/d/126: 66, B.223/d/136: 71v., B.223/d/144: 115-115v., B.223/d/148: 85.

TABLE 14

[Clark], "Puget Sound Agricultural Company," 59; HBCA, D.4/59: 85, D.4/110: 23, F.12/2: 8Bv., 39v.-40, 74, 145v.-146; Schafer, "Letters of Sir George Simpson," 78; Wolfenden, "John Tod," 211.

TABLE 15

HBCA, F.8/1: 28, 63, F.12/1: 468, 560, F.12/2: 33.

TABLE 16

HBCA, B.223/d/133: 38d, B.223/d/150: 35d, B.223/d/158: 76d, B.223/d/161: 39, B.223/d/166: 9d, B.223/d/179: 8d, 9, B.223/d/183a: 11d, B.223/d/190: 15.

TABLE 17

HBCA, B.223/d/115: 91, B.223/d/126: 65, D.4/59: 85, D.4/110: 22v., F.8/1: 33, F.12/1: 564.

TABLE 18

HBCA, B.223/d/126: 67v., F.8/1: 33, F.12/1: 564, F.23/1: 15v.

TABLE 19

[Clark], "Puget Sound Agricultural Company," 59; HBCA F.8/1: 28, 33, 39, 46, 51v., 57, F.12/1: 466v.-467v., 564, F.15/6: 68-69, F.15/7: 36v.-38v., F.15/8: 50-53, F.15/9: 58-60, F.23/1: 15v., F.26/1: 101.

TABLE 20

[Clark], "Puget Sound Agricultural Company," 58; HBCA, F.15/32: 3, 5v.-6v., 7v.-8v., 9v., F.26/1: 101.

TABLE 21

[Clark], "Puget Sound Agricultural Company," 59; HBCA, F.12/2: 9A-9Av., 34v., 39-39v., 75.

TABLE 22

[Slacum], *Report*, p. 210; Willamette Settlers, "Letters," pp. 143, 166.

TABLE 23

Allen, *Ten Years in Oregon*, p. 240; Anonymous, "Documentary," 190-91; Applegate, "A Day with the Cow Column," p. 58; Bolduc, *Mission of the Columbia*, p. 129; Burnett, *Recollections and Opinions*, pp. 141-42; Carey, "Diary of Reverend George H. Gray," 293; [Crawford], *Journal of Medorem Crawford*, p.7; Crawford, "Occasional Address," p. 10;

212   Farming the Frontier

Deady, "Annual Address," p. 35; Douglas to Ross, 12 March 1844, [James Douglas], "Correspondence outward, 1844-1857," pt. 1; Douglas, "Private Papers," series 2, pp. 11-13, series 3, pp. 6-7; Drury, *First White Women*, vol. 1, p. 263; Emmons, "Journal," vol. 3, n.p.; Ermatinger, "Edward Ermatinger Papers," vol. 1, p. 218; McLoughlin to Edward Ermatinger, 4 March 1845, "Ermatinger Papers," pt. 1; Evans, "Annual Address," p. 26; McNeill to Governor and Committee, 17 February 1845, [Fort Nisqually], "Correspondence outward, May 23, 1841 - Sept. 17, 1842"; Glazebrook, *Hargrave Correspondence*, p. 385; HBCA, A.11/70: 101, 183, B.223/e/4: 3v.-4v., 7, 526v.-527, 530, B.223/z/4: 202, D.4/67: 85v., D.5/8: 159v.; "Diary of Mrs. William Henry Gray," William Henry Gray Papers, p. 22; Holmes, "Journal," vol. 3, p. 10; Landerholm, *Notices & Voyages*, pp. 202, 236; Lee, "Articles on the Oregon Mission," pt. 5, p. 3; Rogers to Leslie, 15 October 1842, Leslie and Leslie Papers; McClane, "First Wagon Train," p. 11; McLoughlin, "Copy of a Document," p. 52; McLoughlin to Simpson, 20 March 1843, McLoughlin to Tolmie, 15 October 1843, McLoughlin, "Correspondence"; Minto, "Occasional Address," p. 42; White to the Commissioner of Indian Affairs, 15 October 1842, [Office of Indian Affairs], "Letters Received"; Blanchet to Demers, 5 November 1840, "Catholic Church in Oregon"; *Oregon Spectator*, 21 March, 9 July, 10 December 1846; "Papers relative to the Expedition of Lieuts. Warre and Vavasour," p. 76; Rich, *Letters of John McLoughlin*, vol. 2, pp. 139-40, vol. 3, pp. 143, 178, 288-89; Rowland to his sister, 9 January 1844, Rowland, "Miscellaneous Material"; Letter from Shaw, n.d., Shaw, "Diary & Letters"; Spalding to the Presbyterian Church of Kinsman, Ohio, 15 September 1845, Spalding, "Papers"; White, "Government of and the Emigration to Oregon," pp. 18-19; Narcissa Whitman to Jane Prentiss, 1 October 1841, Narcissa Whitman to Stephen Prentiss, 6 October 1841, Marcus Whitman to Porter, 1843, Whitman and Whitman, "Correspondence."

TABLE 24

Allen, *Ten Years in Oregon*, p. 255; De Smet, *Letters and Sketches*, p. 386; Edwards, *Sketch of the Oregon Territory*, p. 18; Ermatinger, "Edward Ermatinger Papers," vol. 1, p. 128; McLoughlin to Edward Ermatinger, 1 February, 4 March 1843, "Ermatinger Papers"; Hafen and Hafen, *To the Rockies and Oregon*, pp. 116, 298; HBCA, B.223/z/4: 205v., 231, D.4/59: 93-94, D.4/67: 86, D.4/110: 28; Jennings, *Mission of the Columbia*, vol. 1, p. 106; Landerholm, *Notices & Voyages*, pp. 94, 145, 173; National Intelligencer, "From Oregon," p. 1; Blanchet to White, 15 March 1843, [Office of Indian Affairs], "Letters Received"; "Papers relative to the Expedition of Lieuts. Warre and Vavasour," pp. 47v., 78v.-79; Pierce and Winslow, *H.M.S. Sulphur*, p. 66; Rich, *Letters of John McLoughlin*, vol. 1, p. 240, vol. 3, p. 47; Schafer, "Letters of Sir George Simpson," 80, 82; Letter from Shaw, n.d., Shaw, "Diary & Letters"; Wiggins, "Reminiscences," p. 1; Wilkes, "Oregon Territory," p. 20.

TABLE 25

Carey, "Mission Record Book," 240; Lee, "Articles on the Oregon Mission," pt. 2, p. 8v.; Leslie, "Oregon Mission," n.p.; Shepard to the Missionary Society, 28 September 1835, Shepard to his brother and sister, 26 May 1838, Shepard, "Letters to and from"; Shepard, "Oregon Mission,; n.p.; [Slacum], *Report*, p. 194; Williams, *Narrative of a Tour*, pp. 62-63.

TABLE 26

Drury, *Diaries and Letters*, p. 88; Drury, *First White Women*, vol. 3, p. 158; Hafen and Hafen, *To the Rockies and Oregon*, p. 251; Munger and Munger, "Diary," 404; A. Smith to Greene, 15 September 1838, M. Whitman to Greene, 10 May 1839, 15 October 1840, "Papers of the American Board of Commissioners," vol. 1; M. Walker to Richardson, 15 September 1838, Walker and Walker, "Diaries"; Williams, *Narrative of a Tour*, p. 61.

TABLE 27

Drury, *Diaries and Letters*, pp. 248, 260n, 261, 276-77, 297, 299; Drury, *Nine Years with the Spokane Indians*, p. 92; Spalding to Greene, 10 December 1838, 2 October 1839, 22 September 1840, "Papers of the American Board of Commissioners," vol. 1; Diary, November 1838 - November 1839, p. 29, Diary, December 1839 - January 1841, pp. 49, 52, Diary, February 1841 - February 1843, p. 72, Spalding to Allen, 22 September 1838, 29 April, 29 July 1843, Spalding to Hinsdale, 17 August 1842, Spalding to Greene, 26 August 1843, Spalding, "Papers"; Walker and Walker, "Diary 1838," p. 27, "Diaries."

TABLE 28

Drury, *Diaries and Letters*, p. 186; Spalding to White, n.d., [Office of Indian Affairs], "Letters Received"; A. Smith to Greene, 28 September 1840, Spalding to Greene, 12 July 1841, "Papers of the American Board of Commissioners," vol. 1; Spalding to Allen, 18 February 1842, Spalding, "Papers."

TABLE 29

HBCA, B.223/z/4: 211; "Papers relative to the Expedition of Lieuts. Warre and Vavasour," p. 86.

TABLE 30

[Fort Vancouver], "Fort Vancouver - Fur Trade Returns"; HBCA, B.223/d/18: 27, 33, 58, 67, 75, B.239/h/4: 6.

Bibliography

BIBLIOGRAPHIES

Lowther, Barbara J. *A Bibliography of British Columbia.* Victoria: University of Victoria, 1968.

Olsen, Michael L., comp. *A Preliminary List of References for the History of Agriculture in the Pacific Northwest and Alaska.* Davis: Agricultural History Center of the University of California at Davis, 1968.

Oregon Historical Society. *A Bibliography of Pacific Northwest History.* Portland: Oregon Historical Society, 1958.

Smith, Charles W. *Pacific Northwest Americana.* 3d ed. Portland: Oregon Historical Society, 1950.

Strathern, Gloria M. *Navigations, Traffiques & Discoveries 1774-1848.* Victoria: University of Victoria, 1970.

PRIMARY SOURCES: UNPUBLISHED

Abernathy, Mrs. Anne. "The Mission Family and Governor Abernethy the Mission Steward." 1878. Bancroft Library, Ms. P-A 1.

Anderson, A.C. "A.C. Anderson's Memo Relating to the Cowlitz Farm, etc., 1841." OHS, Ms. 1502.

——. "The Origin of the Puget Sound Agricultural Company." 1865. OHS, Ms. 1502.

——. "History of the Northwest Coast." 1878. Bancroft Library, Ms. P-C 2.

Aulick, John H. "Journal." OHS, Ms. 101.

Baillie, Thomas. "Enclosure No. 1 to Letter No. 18, 4 March 1845." PRO, Ms. A.D.M. 1/5550.

Ball, John. "Papers." OHS, Ms. 195.

Black, Samuel. "List of the Mineral Samples Sent to Geo. Simpson by Sam Black." British Museum of Natural History, Mineral Deposits Library, Geological Society Collection, Richardson Papers.

[Briskoe, William]. "Journal of William Briskoe, Armorer Aboard the *Relief* and the *Vincennes* August 18, 1838 – March 23, 1842." NARS, File Microcopy 75 ("Records of the United States Exploring Expedition Under the Command of Lieutenant Charles Wilkes, 1838-1842").

"Catholic Church in Oregon." OHS, Ms. 1580.

Cooke, Amos S. "Papers." OHS, Ms. 1223.

Cooper, James. "Maritime Matters on the Northwest Coast . . . ." 1878. Bancroft Library, Ms. P-C 6.

216 Farming the Frontier

Deady, M.P. "History & Progress of Oregon after 1845." 1878. Bancroft Library, Ms. P-A 24.
Dease, John Warren. "Diary of John Warren Dease." OHS, Ms. 560.
Douglas, James. "Correspondence outward, 1844-1857." PABC, Ms. B40, pt. 1.
―――――. "Journal of Sir James Douglas 1840-1841." Bancroft Library, Ms. P-C 11.
―――――. "Private Papers of Sir James Douglas, First Series." PABC, Ms. B20 1858.
Eells, Cushing. "Letters to A.B.C.F.M. 1843-1859." OHS, Ms. 1218.
Eld, Henry, Jr. "Journal Statistics etc. in Oregon & California." Yale University Library, Western Americana Ms. 161.
Emmons, George Foster. "Journal kept while attached to the Exploring Expedition . . . ." 3 vols. Yale University Library, Western Americana Ms. 166.
Ermatinger, Edward. "Edward Ermatinger Papers 1820-1874." PAC, Ms. 19, series A2(2), vol. 1.
Ermatinger, Francis. "Letters of Francis Ermatinger 1823-1853." Huntington Library.
"Ermatinger Papers." PABC, Ms. AB40 Er. 62.3.
Finlayson, Roderick. "The History of Vancouver Island and the Northwest Coast." 1878. Bancroft Library, Ms. P-C 15.
[Fort Colvile]. "Correspondence of Archibald McDonald relating to Fort Colvile." PABC, Ms. AB20 C72M1.
[Fort Kamloops]. "Journal, Aug. 3, 1841-Dec. 18, 1843, kept by John Tod." PABC, Ms. AB20 K12A 1841-1843.
[Fort Langley]. "Correspondence relating to Fort Langley, from the Hudson's Bay Company's Archives, 1830-1859." PABC, Ms. AB20 L3A.
―――――. "Fort Langley Journal June 27, 1827 — July 30, 1830." PABC, Ms. AB20 L2A2M.
[Fort Nisqually]. "Correspondence outward, May 23, 1841-Sept. 17, 1842." PABC, Ms. AB20 Ni2.
[Fort Vancouver]. "Fort Vancouver — Fur Trade Returns — Columbia District & New Caledonia, 1825-1857." PABC, Ms. AB20 V3.
Foster, Philip. "Papers." OHS, Ms. 996.
Gervais, Joseph (et al.). "Letters." OHS, Ms. 83.
Gordon, John. "Captain J. Gordon's Report to the Secretary of the Admiralty, October 19, 1845." PRO, Ms. A.D.M. 1/5564, no. 8.
Gray, William Henry. "Papers." OHS, Ms. 1202.
Hancock, Samuel. "Thirteen Years' Residence on the north-west coast . . . ." Bancroft Library, Ms. P-B 29.
Harriott, John Edward. "Memoirs of Life and Adventure in the Hudson's Bay Company's Territories, 1819-1825." Yale University Library, Western Americana Ms. 245.
Harvey, Mrs. David. "Life of John McLoughlin . . . ." 1878. Bancroft Library, Ms. P-B 12.
Holmes, Silas. "Journal kept by Assist. Surgeon Silas Holmes . . . ." 3 vols. Yale University Library, Western Americana Ms. 260.
Hudson's Bay Company Archives. Ms. A.6/22, A.6/23, A.6/24, A.6/25, A.6/26, A.8/2, A.11/17, A.11/69, A.11/70, A.11/72, A.12/2, A.12/7, B.5/a/2, B.5/a/4, B.5/a/5, B.5/a/6, B.5/a/7, B.45/e/3, B.76/e/1, B.76/z/1, B.97/e/1, B.113/a/1, B.113/a/2, B.113/a/3, B.146/e/1, B.146/e/2, B.188/a/17, B.201/a/3, B.223/a/3, B.223/b/12, B.223/b/24, B.146/e/2, B.188/a/17, B.201/a/3, B.223/a/3, B.223/b/12, B.223/b/24, B.223/b/26,

B.223/b/26, B.223/b/27, B.223/b/28, B.223/b/29, B.223/b/41, B.223/c/1, B.223/d/2b, B.223/d/5, B.223/d/13, B.223/d/18, B.223/d/22, B.223/d/41, B.223/d/42, B.223/d/49, B.223/d/59, B.223/d/71, B.223/d/80, B.223/d/93, B.223/d/105b, B.223/d/115, B.223/d/126, B.223/d/133, B.223/d/136, B.223/d/144, B.223/d/148, B.223/d/150, B.223/d/155, B.223/d/158, B.223/d/160, B.223/d/161, B.223/d/165, B.223/d/166, B.223/d/179, B.223/d/183a, B.223/d/190, B.223/e/1, B.223/e/4, B.223/z/4, B.223/z/5, B.235/b/2, B.239/1/1a-1b, B.239/g/61-62, B.239/h/4, C.4/1, D.4/5, D.4/10, D.4/21, D.4/25, D.4/58, D.4/59, D.4/60, D.4/66, D.4/67, D.4/68, D.4/88, D.4/90, D.4/100, D.4/106, D.4/110, D.4/120, D.4/121, D.4/123, D.4/125, D.4/126, D.4/127, D.5/6, D.5/8, D.5/9, E.8/3, E.12/2, F.8/1, F.11/1, F.12/1, F.12/2, F.15/6, F.15/7, F.15/8, F.15/9, F.15/29, F.15/32, F.23/1, F.26/1, F.29/2.

"Information Concerning Fort Colville, Washington." (n.d.) Washington State University Library, Ms. 979: 725 H869i.

Lee, Daniel. "Articles on the Oregon Mission." 8 pts. OHS, Ms. 1211.

Lee, Jason. "Journal" and "Letters on Oregon Mission, 1834-35." 5 pts. OHS, Ms. 1212.

Leslie, David, and Adelia Judson Leslie. "Papers." OHS, Ms. 1216.

McBride, [John]. "Oregon in 1846." OHS, Ms. 458.

McClane, John Burch. "The First Wagon Train to Oregon." 1878. Bancroft Library, Ms. P-A 46.

[McDonald, Archibald]. "McDonald Correspondence." PABC, Ms. AB20 M142A.

———. "Correspondence, etc. re Puget Sound Agricultural Company." PABC, Ms. AB25 M14.

McKay, Joseph William. "Recollections of a Chief Trader in the Hudson's Bay Company." 1878. Bancroft Library, Ms. P-C 24.

McLeod, John. "John McLeod Papers 1811-1837." PAC, Ms. 19, series A23, vol. 1.

McLoughlin, John. "Correspondence." University of Washington Library, Ms. 271 A-G.

———. "Private papers, 1825-1856." Pt. 2. Bancroft Library, Ms. P-A 155.

Meek, Joe. "Census of Oregon 1845." OHS, Ms. 1226.

[Office of Indian Affairs]. "Letters Received by the Office of Indian Affairs 1824-81." NARS, Mc. 234, roll 607.

"Papers of the American Board of Commissioners for Foreign Missions." Houghton Library, Ms. ABC: 18.5.3, vol. 1.

"Papers relative to the Expedition of Lieuts. Warre and Vavasour to the Oregon Territory." PRO, Ms. F.O. 5/457.

Peel, William. "Lieutenant Peel's Report to Captain J. Gordon, September 27, 1845." PRO, Ms. A.D.M. 1/5564, no. 8.

Pettygrove, F.W. "Oregon in 1843." 1878. Bancroft Library, Ms. P-A 60.

Roberts, George B. "Recollections of George B. Roberts." (n.d.) Bancroft Library, Ms. P-A 83.

Rowland, Richard. "Miscellaneous Material." OHS, Ms. 1180.

[Russian-American Company]. "Records of the Russian-American Company 1802-67: Correspondence of Governors General." Mc. 11, rolls 7, 8, 12, 18, 19, 34, 42, 43, 46, 51, 52, 55, 136.

Scouler, John. "Dr. John Scouler's Journal of a Voyage to N.W. America . . . ." OHS, Ms. 949B.

Shaw, A.C.R. "Diary & letters, 1843-1848." OHS, Ms. 941.

Shepard, Cyrus. "Letters To and From." OHS, Ms. 1219.

Spalding, Henry Harmon. "Papers." OHS, Ms. 1201.
Stuart, John. "Letter Books and Journals . . . ." OHS, Ms. 1502.
Sturgis, Josiah. "Extracts from his Journal, 1818." OHS, Ms. 153.
Sylvester, Edmund. "Founding of Olympia." 1878. Bancroft Library, Ms. P-B 22.
Talbot, Theodore. "Journal, 1843-44." vol. 1. OHS, Ms. 773.
Tod, John. "History of New Caledonia and the Northwest Coast." PABC.
Tolmie, William Fraser. "Correspondence." University of Washington Library, Ms. AIA 8/3.
―――――. "History of Puget Sound and the Northwest Coast." 1878. Bancroft Library, Ms. P-B 25.
Walker, Elkanah and Mary. "Diaries." OHS, Ms. 1204.
Walker, Joel P. "Narrative of Adventures . . . ." 1878. Bancroft Library, Ms. C-D 170.
Waller, Alvan F. "Papers." OHS, Ms. 1210.
White, Elijah. "Government of and the Emigration to Oregon." 1879. Bancroft Library, Ms. P-A 76.
―――――. "Papers." OHS, Ms. 1217.
Whitman, Marcus and Narcissa. "Correspondence." OHS, Ms. 1203.
Wiggins, William. "Reminiscences . . . ." 1877. Bancroft Library, Ms. C-D 175.
Wilkes, Charles. "Oregon Territory." OHS, Ms. 56.
Williamette Settlers. "Letters to the Bishop of Juliopolis." Bancroft Library, Ms. P-A 305.
Work, John. "Journals" and "Letters." OHS, Ms. 319.
―――――. "Work Correspondence." PABC, Ms. AB40.

PRIMARY SOURCES: PUBLISHED

Allen, A.J., comp. *Ten Years in Oregon: Travels and Adventures of Doctor E. White and Lady West of the Rocky Mountains.* Ithaca: Andrus, Gauntlett, & Co., 1850.
Anonymous. "Documentary." *Oregon Historical Quarterly* 2 (1901): 187-203.
Anonymous. "Documents." *Washington Historical Quarterly* 1 (1907): 40-43, 256-66.
Anonymous. "Documents." *Washington Historical Quarterly* 2 (1908): 161-68, 254-64.
Anonymous. "Reminiscences of Fort Vancouver . . . ." *Transactions of the Oregon Pioneer Association 1881,* pp. 75-80.
Applegate, Jesse. "A Day with the Cow Column in 1843." *Transactions of the Oregon Pioneer Association 1876,* pp. 57-65.
Bagley, Clarence B., ed. *Early Catholic Missions in Old Oregon.* 2 vols. Seattle: Lowman & Hanford, 1932.
―――――. "Journal of Occurrences at Nisqually House, 1833 [1833-35]." *Washington Historical Quarterly* 6 (1915): 179-97, 264-78 and 7 (1916): 59-75, 144-67.
Ball, John. "Oregon Expedition." *Zion's Herald,* 15 January 1834.
Barker, Burt Brown, ed. *Letters of Dr. John McLoughlin Written at Fort Vancouver 1829-1832.* Portland: Oregon Historical Society, 1948.
[Bishop, Charles]. *The Journals and Letters of Captain Charles Bishop . . . .* Edited by Michael Roe. London: Hakluyt Society, 1967.
[Black, Samuel]. *A Journal of A Voyage From Rocky Mountain Portage in Peace River To the Sources of Finlays Branch And North West Ward In Summer 1824.* Edited by E.E. Rich. London: Hudson's Bay Record Society, 1955.

Bolduc, Jean Baptiste Zacharie. *Mission of the Columbia*. Edited and translated by Edward J. Kowrach. Fairfield, WA: Ye Galleon Press, 1979.

Bowsfield, Hartwell, ed. *Fort Victoria Letters 1846-1851*. Winnipeg: Hudson's Bay Record Society, 1979.

Brackenridge, W.D. *The Brackenridge Journal for the Oregon Country*. Edited by O.B. Sperlin. Seattle: University of Washington Press, 1931.

Brewer, Henry Bridgman. "Log of the Lausanne — V." *Oregon Historical Quarterly* 30 (1929): 111-24.

British and American Joint Commission for the Final Settlement of the Claims of the Hudson's Bay and Puget's Sound Agricultural Companies. [*Proceedings.*] Vols. 2-4. Montreal: J. Lovell, 1868.

Burnett, Peter H. *Recollections and Opinions of an Old Pioneer*. New York: D. Appleton and Company, 1880.

Carey, Charles Henry, ed. "Diary of Reverend George H. Gray." Pts. 2-3. *Oregon Historical Quarterly* 24 (1923): 153-85, 269-333.

————. "Methodist Annual Reports Relating to the Willamette Mission (1834-1848)." *Oregon Historical Quarterly* 23 (1922): 303-64.

————. "The Mission Record Book of the Methodist Episcopal Church, Willamette Station, Oregon Territory, North America, Commenced 1834." *Oregon Historical Quarterly* 23 (1922): 230-66.

Chernykh, Ye. "O zemledelii v verkhney Kalifornii' ["Concerning Agriculture in Upper California"]. *Zhurnal selskavo khozyaistva i ovtsevodstva* (1841): 234-65.

[Clark, R.C., ed.] "Puget Sound Agricultural Company." *Washington Historical Quarterly* 18 (1927): 57-59.

Corney, Peter. *Early Voyages in the North Pacific 1813-1818*. Fairfield, WA: Ye Galleon Press, 1965.

Coues, Elliott, ed. *The Manuscript Journals of Alexander Henry . . . and of David Thompson . . . .* 2 vols. in 1. Minneapolis: Ross and Haines, 1965.

Cowie, Isaac. *The Minutes of the Council of the Northern Department of Rupert's Land, 1830 to 1843 . . . .* State Historical Society of North Dakota, n.d.

Cox, Ross. *The Columbia River . . . .* Edited by Edgar I. Stewart and Jane R. Stewart. Norman: University of Oklahoma Press, 1957.

[Crawford, Medorem]. *Journal of Medorem Crawford*. Edited by F.G. Young. Sources of the History of Oregon, vol. 1, no. 1. Eugene: Star Job Office, 1897.

————. "Occasional Address." *Transactions of the Oregon Pioneer Association 1881*, pp. 9-19.

[De Mofras, Duflot]. *Duflot de Mofras' Travels on the Pacific Coast*. Translated and edited by Marguerite Eyer Wilbur. Vol. 2. Santa Ana, CA: The Fine Arts Press, 1937.

De Smet, Pierre-Jean. *Letters and Sketches: With a Narrative of a Year's Residence Among the Indian Tribes of the Rocky Mountains*. Philadelphia: M. Fithian, 1843.

————. *Origin, Progress, and Prospects of the Catholic Mission to the Rocky Mountains*. Fairfield, WA: Ye Galleon Press, 1972.

————. *Origin, Progress, and Prospects of the Catholic Mission to the Rocky Mountains*. Fairfield, CA: Ye Galleon Press, 1972.

Deady, Matthew P. "The Annual Address." *Transactions of the Oregon Pioneer Association 1875*, pp. 17-41.

Drury, Clifford Merrill, ed. *The Diaries and Letters of Henry H. Spalding and Asa Bowen Smith relating to the Nez Perce Mission 1838-1842*. Glendale, CA: Arthur H. Clark, 1958.

_____. *First White Women Over the Rockies.* Glendale, CA: Arthur H. Clark, 1963-66. 3 vols.

_____. *Nine Years with the Spokane Indians: The Diary 1838-1848, of Elkanah Walker.* Glendale, CA: Arthur H. Clark, 1976.

Dunn, John. *History of the Oregon Territory....* 2d ed. London: Edwards and Hughes, 1846.

Edwards, P.L. *Sketch of the Oregon Territory: or, Emigrants' Guide.* Liberty, MO: The "Herald" Office, 1842.

Elliott, T.C., ed. "Journal of John Work [1825-26]..." *Washington Historical Quarterly* 5 (1914): 83-115, 163-91, 258-87, and 6 (1915): 26-49.

_____. "Letters of Dr. John McLoughlin." *Oregon Historical Quarterly* 23 (1922): 365-71.

Evans, Elwood. "The Annual Address." *Transactions of the Oregon Pioneer Association 1877*, pp. 13-37.

Farnham, Thomas J. *Travels in the Great Western Prairies, the Anahuac and Rocky Mountains, and in the Oregon Territory.* 2 vols. London: Richard Bentley, 1843.

Fleming, R. Harvey, ed. *Minutes of Council Northern Department of Rupert Land, 1821-31.* Toronto: Champlain Society, 1940.

Flett, John. "A Sketch of the Emigration from Selkirk's Settlement to Puget Sound in 1841." *Tacoma Daily Ledger*, 18 February 1885.

Franchère, Gabriel. *Journal of a Voyage on the North West Coast....* Translated by Wessie Tipping Lamb. Toronto: Champlain Society, 1969.

[Frémont, John Charles]. *The Expeditions of John Charles Frémont.* Edited by Donald Jackson and Mary Lee Spence. Vol. 1. Urbana: University of Illinois Press, 1970.

Gairdner, Meredith. "Observations during a Voyage from England to Fort Vancouver, on the North-West Coast of America." *Edinburgh New Philosophical Journal* 16 (1834): 290-302.

Gay, Theressa. *Life and Letters of Mrs. Jason Lee.* Portland: Metropolitan Press, 1936.

Glazebrook, G.P. de T., ed. *The Hargrave Correspondence 1821-1843.* Toronto: Champlain Society, 1938.

Glover, Richard, ed. *David Thompson's Narrative 1784-1812.* Toronto: Champlain Society, 1962.

Great Britain, Parliament, House of Commons, Select Committee on the Hudson's Bay Company. *Report from the Select Committee on the Hudson's Bay Company....* London, 1857.

Hafen, LeRoy R., and Hafen, Ann W., eds. *To the Rockies and Oregon 1839-1842.* Glendale, CA: Arthur H. Clark, 1955.

[Harmon, Daniel Williams]. *Sixteen Years in the Indian Country: The Journal of Daniel Williams Harmon 1800-1816.* Edited by W. Kaye Lamb. Toronto: Macmillan, 1957.

Heath, Joseph. *Memoirs of Nisqually.* Fairfield, WA: Ye Galleon Press, 1979.

Himes, Geo. H., ed. "Letters Written by Mrs. Whitman from Oregon to her Relations in New York." *Transactions of the Oregon Pioneer Association 1891*, pp. 79-179.

_____. "Mrs. Whitman's Letters." *Transactions of the Oregon Pioneer Association 1893*, pp. 53-219.

Holmes, Kenneth L., ed. and comp. *Covered Wagon Women: Diaries & Letters from the Western Trails 1840-1890.* Vol. 1. Glendale, CA: Arthur H. Clark, 1983.

Howison, Neil M. "Report of Lieutenant Neil M. Howison on Oregon, 1846." *Oregon Historical Quarterly* 14 (1913): 1-60.

Jennings, Tess E., trans. *Mission of the Columbia ....* 2 vols. Seattle: Works Progress Administration, 1937.

Jessett, Thomas E., ed. *Reports and Letters of Herbert Beaver 1836-1838*. Portland: Champoeg Press, 1959.

Kane, Paul. *Wanderings of an Artist* . . . . Edmonton: M.G. Hurtig, 1968.

[Klotz, Otto]. *Certain Correspondence of the Foreign Office and of the Hudson's Bay Company* . . . . Pt. 2. Ottawa: Government Printing Bureau, 1899.

Landerholm, Carl, trans. *Notices & Voyages of the Famed Quebec Mission to the Pacific Northwest*. Portland: Oregon Historical Society, 1956.

Lee, Jason. "Diary." *Oregon Historical Quarterly* 17 (1916): 116-46, 240-76, 397-430.

Lee, D., and Frost, J.H. *Ten Years in Oregon*. New York: The Authors, 1844.

Leslie, David. "Oregon Mission." *Christian Advocate and Journal*, 30 September 1840.

Mackenzie, Alexander. *Voyages from Montreal* . . . . Edmonton: M.G. Hurtig, 1971.

[McDonald, Archibald]. *Peace River. A Canoe Voyage from Hudson's Bay to Pacific* . . . . Edited by Malcolm McLeod. Edmonton: M.G. Hurtig, 1971.

McDonald, Lois Halliday, ed. *Fur Trade Letters of Francis Ermatinger* . . . . Glendale, CA: Arthur H. Clark, 1980.

[McLean, John]. *John McLean's Notes of a Twenty-Five Year's Service in the Hudson's Bay Territory*. Edited by W.S. Wallace. Toronto: Champlain Society, 1932.

McLoughlin, John. "Copy of a Document Found among the Private Papers of the late Dr. John McLoughlin." *Transactions of the Oregon Pioneer Association 1880*, pp. 46-55.

———. "Letter of John McLoughlin." *Oregon Historical Quarterly* 15 (1914): 206-7.

Mengarini, Gregory. *Recollections of the Flathead Mission*. Edited and translated by Gloria Ricci Lothrop. Glendale, CA: Arthur H. Clark, 1977.

Merk, Frederick, ed. *Fur Trade and Empire: George Simpson's Journal . . . 1824-25*. Rev. ed. Cambridge, MA: Belknap Press, 1968.

Minto, John. "The Occasional Address." *Transactions of the Oregon Pioneer Association 1876*, pp. 35-50.

Morgan, William James, Tyler, David B., Leonhart, Joye L., and Loughlin, Mary F., eds. *Autobiography of Rear Admiral Charles Wilkes, U.S. Navy 1798-1877*. Washington, DC: Department of the Navy, 1978.

Munger, Asahel, and Munger, Eliza. "Diary of Asahel Munger and Wife." *Oregon Historical Quarterly* 8 (1907): 387-405.

Munnick, Harriet Duncan, ed. *Catholic Church Records of the Pacific Northwest: Vancouver Volumes I and II and Stellamaris Mission*. Translated by Mikell De Lores Wormell Warner. St. Paul, OR: French Prairie Press, 1972.

National Intelligencer. "From Oregon." *New York Herald Tribune*, 18 January 1842.

Oregon Spectator, 19 February 1846, 21 March 1846, 9 July 1846, 6 August 1846, 10 December 1846, 7 January 1847.

Palmer, Joel. *Journal of Travels over the Rocky Mountains* . . . . Cincinnati: J.A. and U.P. James, 1847.

Parker, Samuel. *Journal of an Exploring Tour beyond the Rocky Mountains . . . 1835*, '36, and '37. Minneapolis: Ross and Haines, 1967.

Pierce, Richard A., and Winslow, John H., eds. *H.M.S. Sulphur on the Northwest and California Coasts, 1837 and 1839*. Kingston, Ont.: Limestone Press, 1979.

Pipes, Nellie B., ed. "Journal of John H. Frost, 1840-43." *Oregon Historical Quarterly* 35 (1934): 348-75.

[Point, Nicolas]. *Wilderness Kingdom: Indian Life in the Rocky Mountains: 1840-1847: The Journals & Paintings of Nicolas Point, S.J.* Translated by Joseph P. Donnelly. New York: Holt, Rinehart and Winston, 1967.

Powers, Kate Ball, Hopkins, Flora Ball, and Ball, Lucy, eds. *John Ball . . . Autobiography*. Glendale, CA: Arthur H. Clark, 1925.

Quaife, M.M. ed. "Letters of John Ball, 1832-1833." *Mississippi Valley Historical Review* 5 (1919): 450-68.

Reid, Robie L., ed. "Fort Langley Correspondence." *British Columbia Historical Quarterly* 1 (1937): 187-94.

Rich, E.E., ed. *The Letters of John McLoughlin from Fort Vancouver to the Governor and Committee*. 3 vols. London: Hudson's Bay Record Society, 1941-44.

――――. *Part of Dispatch from George Simpson Esqr Governor of Ruperts Land to the Governor & Committee of the Hudson's Bay Company London March 1, 1829*. Toronto: Champlain Society, 1947.

Ross, Alexander. *Adventures of the First Settlers on the Oregon or Columbia River*. London: Smith, Elder and Co., 1849.

――――. *The Fur Hunters of the Far West*. Edited by Kenneth A. Spaulding. Norman: University of Oklahoma Press, 1956.

――――. *The Red River Settlement* . . . . Minneapolis: Ross and Haines, 1957.

Sage, W.N., ed. "Peter Skene Ogden's Notes on Western Caledonia." *British Columbia Historical Quarterly* 1 (1937): 45-56.

Schafer, Joseph, ed. "Letters of Sir George Simpson, 1841-1843." *American Historical Review* 14 (1908): 70-94.

[Scouler, John]. "Dr. John Scouler's Journal of a Voyage to N.W. America." *Oregon Historical Quarterly* 6 (1905): 54-75, 159-205, 276-87.

Shepard, Cyrus. "Oregon Mission." *Christian Advocate and Journal*, 19 August 1836.

Shur, L.A., ed. *K beregam Novovo Sveta* . . . . [*To the Shores of the New World* . . . .]. Moscow: "Nauka," 1971.

Simpson, Sir George. *Narrative of a Journey Round the World*. 2 vols. London: Henry Colburn, 1847.

Slacum, William A. "Memorial of William A. Slacum . . . ." *Senate Document 24*, 25th Congress, 2d Session.

[Slacum, William A.] "Mr. Slacum's Report." *Reports Committees*, Vol. 1 (1838-39), 25th Congress, 3d Session.

Spalding, Eliza, and Spalding, Henry. "Letters of Reverend H.H. Spalding and Mrs. Spalding . . . ." *Oregon Historical Quarterly* 13 (1912): 371-79.

Spalding, H.H. "Letter." *Missionary Herald*, October 1837.

――――. "Letter." *Missionary Herald*, October 1838.

[Strange, James]. *James Strange's Journal and Narrative* . . . . Madras: Government Press, 1928.

Thwaites, Reuben Gold, ed. *Original Journals of the Lewis and Clark Expedition 1804-1806*. Vols. 3-4. New York: Arno Press, 1969.

[Tolmie, William Fraser]. *The Journals of William Fraser Tolmie: Physician and Fur Trader*. Vancouver: Mitchell Press Ltd., 1963.

Townsend, John Kirk. *Narrative of a Journey Across the Rocky Mountains to the Columbia River*. Fairfield, WA: Ye Galleon Press, 1970.

United States, Congress, Senate. "Message from the President of the United States . . . ." *Senate Document No. 39*, 21st Congress, 2d Session.

Wagner, Henry R. *Spanish Explorations in the Strait of Juan de Fuca*. New York: AMS Press, 1971.

Wallace, W. Stewart, ed. *Documents Relating to the North West Company.* Toronto: Champlain Society, 1934.

Warre, H.J. *Overland to Oregon in 1845: Impressions of a Journey across North America.* Edited by Madeleine Major-Frégeau. Ottawa: Public Archives of Canada, 1976.

———. *Sketches in North America and the Oregon Territory.* Barre, MA: Imprint Society, 1970.

West, John. *The Substance of a Journal during a Residence at the Red River Colony.* New York: Johnson Reprint Corp., 1966.

[Wilkes, Charles]. *Diary of Wilkes in the Northwest.* Edited by Edmond S. Meany. Seattle: University of Washington Press, 1926.

Wilkes, Charles. *Narrative of the United States Exploring Expedition During the Years 1838, 1839, 1840, 1841, 1842.* Vol. 4. Philadelphia: Lea & Blanchard, 1845.

Willamette Settlers, "Letters to the Bishop of Juliopolis, March 22, 1836 and March 8, 1837." *Les cloches de Saint-Boniface* 31 (1932): 143-44, 165-66.

Williams, Glyndwr, ed. *Hudson's Bay Company Miscellany 1670-1870.* Winnipeg: Hudson's Bay Record Society, 1975.

———. *London Correspondence Inward from Sir George Simpson 1841-42.* London: Hudson's Bay Record Society, 1973.

Williams, Joseph. *Narrative of a Tour from the State of Indiana to the Oregon Territory in the Years 1841-2.* New York: Edward Eberstadt, 1921.

Wolfenden, Madge, ed. "John Tod: 'Career of a Scotch Boy.'" *British Columbia Historical Quarterly* 28 (1954): 132-238.

[Work, John]. *The Journal of John Work January to October, 1835.* Edited by Henry Drummond Dee. Archives of British Columbia Memoir No. X. Victoria: Charles F. Banfield, 1945.

[Wyeth, Nathaniel J.] *The Correspondence and Journals of Captain Nathaniel J. Wyeth 1831-6.* Edited by F.G. Young. Sources of the History of Oregon, vol. 1, nos. 3-6. Eugene: University Press, 1899.

SECONDARY SOURCES: BOOKS, BOOKLETS, DISSERTATIONS

Binns, Archie. *Peter Skene Ogden: Fur Trader.* Portland: Binfords & Mort, 1967.

Bowen, William A. *The Willamette Valley: Migration and Settlement on the Oregon Frontier.* Seattle: University of Washington Press, 1978.

Brown, Jennifer S.H. *Strangers in Blood: Fur Trade Company Families in Indian Territory.* Vancouver: University of British Columbia Press, 1980.

Cline, Gloria Griffen. *Peter Skene Ogden and the Hudson's Bay Company.* Norman: University of Oklahoma Press, 1974.

Cole, Jean Murray. *Exile in the Wilderness: The Biography of Chief Factor Archibald McDonald 1790-1853.* Don Mills, Ont.: Burns & MacEachern, 1979.

Cook, Warren L. *Flood Tide of Empire: Spain and the Pacific Northwest, 1543-1819.* New Haven and London: Yale University Press, 1973.

Cullen, Mary K. *The History of Fort Langley, 1827-96.* Canadian Historic Sites: Occasional Papers in Archaeology and History No. 20. Ottawa: Parks Canada, 1979.

Dillon, Richard. *Siskiyou Trail: The Hudson's Bay Company Route to California.* New York: McGraw-Hill, 1975.

Galbraith, John S. *The Hudson's Bay Company as an Imperial Factor, 1821-1869.* Berkeley: University of California Press, 1957.

———. *The Little Emperor: Governor Simpson of the Hudson's Bay Company*. Toronto: Macmillan, 1976.

Gibson, James R. *Imperial Russia in Frontier America: The Changing Geography of Supply of Russian America, 1784-1867*. New York: Oxford University Press, 1976.

Gough, Barry M. *The Royal Navy and the Northwest Coast of North America, 1810-1914: A Study of British Maritime Ascendancy*. Vancouver: University of British Columbia Press, 1971.

Howay, F.W., Sage, W.N., and Angus, H.F. *British Columbia and the United States: The North Pacific Slope from Fur Trade to Aviation*. Edited by H.F. Angus. New York: Russell & Russell, 1970.

Hussey, John A. *Champoeg: Place of Transition*. Portland: Oregon Historical Society, 1967.

———. *The History of Fort Vancouver and Its Physical Structure*. Portland: United States National Park Service and Washington State Historical Society, 1957.

Josephy, Alvin M., Jr. *The Nez Perce Indians and the Opening of the Northwest*. Abridged ed. New Haven: Yale University Press, 1971.

Kaye, Barry. "The Historical Geography of Agriculture and Agricultural Settlement in the Canadian Northwest, 1774-ca. 1830." Ph.D. diss., University of London, 1976.

Lent, D. Geneva. *West of the Mountains: James Sinclair and the Hudson's Bay Company*. Seattle: University of Washington Press, 1963.

Loewenberg, Robert J. *Equality on the Oregon Frontier: Jason Lee and the Methodist Mission 1834-43*. Seattle: University of Washington Press, 1976.

Merk, Frederick. *The Oregon Question: Essays in Anglo-American Diplomacy and Politics*. Cambridge, MA: Belknap Press, 1967.

Morton, Arthur S. *A History of the Canadian West to 1870-71*. Edited by Lewis G. Thomas. 2d ed. Toronto and Buffalo: University of Toronto Press, 1973.

———. *Sir George Simpson: Overseas Governor of the Hudson's Bay Company* . . . . Toronto and Vancouver: J.M. Dent and Sons (Canada), 1944.

Olsen, Michael Leon. "The Beginnings of Agriculture in Western Oregon and Western Washington." Ph.D. diss., University of Washington, 1970.

Parkman, Francis. *The California and Oregon Trail: Being Sketches of Prairie and Rocky Mountain Life* . . . . 1st ed. New York and London: G.P. Putnam, 1849.

Phillips, Clifton Jackson. *Protestant America and the Pagan World: The First Half Century of the American Board of Commissioners for Foreign Missions, 1810-1860*. Harvard East Asian Monographs No. 32. Cambridge, MA: Harvard University East Asian Research Center, 1969.

Rich, E.E. *The Fur Trade and the Northwest to 1857*. The Canadian Centenary Series. Toronto: McClelland and Stewart, 1967.

Sturgis, William. *The Oregon Question: Substance of a Lecture Before the Mercantile Library Association, Delivered January 22, 1845*. Boston: Jordan, Swift & Wiley, 1845.

Tikhmenev, P.A. *A History of the Russian-American Company*. Translated and edited by Richard A. Pierce and Alton S. Donnelly. Seattle: University of Washington Press, 1978.

Unruh, John D., Jr. *The Plains Across: The Overland Emigrants and the Trans-Missisippi West, 1840-60*. Urbana: University of Illinois Press, 1979.

Van Kirk, Sylvia. *"Many Tender Ties": Women in Fur-Trade Society in Western Canada, 1670-1870*. Winnipeg: Watson and Dwyer, 1980.

SECONDARY SOURCES: ARTICLES

Archer, Christon I. "Spanish Exploration and Settlement of the Northwest Coast in the

18th Century." *Sound Heritage* 7 (1978): 32-53.

———. "The Transient Presence: A Re-Appraisal of Spanish Attitudes toward the Northwest Coast in the Eighteenth Century." *BC Studies* (Summer 1973): 11-19.

Barry, J. Neilson. "Agriculture in the Oregon Country in 1795-1844." *Oregon Historical Quarterly* 30 (1929): 161-68.

———. "Early Oregon Country Forts: A Chronological List." *Oregon Historical Quarterly* 46 (1945): 101-11.

Beidleman, Richard G. "Nathaniel Wyeth's Fort Hall." *Oregon Historical Quarterly* 58 (1957): 197-250.

Betts, William J. "From Red River to the Columbia." *The Beaver* (Spring 1971): 50-55.

Boyd, Robert T. "Another Look at the 'Fever and Ague' of Western Oregon." *Ethnohistory* 22 (1975): 135-54.

Cannon, Miles. "Fort Hall on the Saptin River." *Washington Historical Quarterly* 7 (1916): 217-32.

Carlos, Ann. "The Birth and Death of Predatory Competition in the North American Fur Trade: 1810-1821." *Explorations in Economic History* 19 (1982): 156-83.

Commager, Henry. "England and [the] Oregon Treaty of 1846." *Oregon Historical Quarterly* 28 (1927): 18-38.

Cook, S.F. "The Epidemic of 1830-1833 in California and Oregon." *University of California Publications in American Archaeology and Ethnology* 43 (1955): 303-25.

Crean, J.F. "Hats and the Fur Trade." *Canadian Journal of Economics and Political Science* 28 (1962): 373-86.

Creech, E.P. "Brigade Trails of B.C." *The Beaver* (Spring 1953): 10-15.

Drury, Clifford M. "The Spokane Indian Mission at Tshimakain, 1838-1848." *Pacific Northwest Quarterly* 67 (1976): 1-9.

Elliott, T.C., ed. "British Values in Oregon, 1847." *Oregon Historical Quarterly* 32 (1931): 27-45.

Ermatinger, C.O. "The Columbia River Under Hudson's Bay Company Rule." *Washington Historical Quarterly* 5 (1914): 192-206.

Galbraith, John S. "The Early History of the Puget's Sound Agricultural Company, 1838-43." *Oregon Historical Quarterly* 55 (1954): 234-59.

Gibson, James R. "Food for the Fur Traders: The First Farmers in the Pacific Northwest, 1805-1846." *Journal of the West* 7 (1968): 18-30.

———. "Russian Expansion in Siberia and America." *Geographical Review* 70 (1980): 127-36.

———. "Smallpox on the Northwest Coast, 1835-1838." *BC Studies* (Winter 1982-83): 61-81.

Gough, Barry M. "Corporate Farming on Vancouver Island: The Puget's Sound Agricultural Company, 1846-1857." *Canadian Papers in Rural History* 4 (1984): 72-82.

———. "The Royal Navy and the Oregon Crisis, 1844-1846." *BC Studies* (Spring 1971): 15-37.

Grant, L.S. "Fort Hall under the Hudson's Bay Company, 1837-1856." *Oregon Historical Quarterly* 41 (1940): 34-39.

Haines, Francis D. "The Western Limits of the Buffalo Range." *Pacific Northwest Quarterly* 31 (1940): 389-98.

Hodgson, Edward. "The Epidemic on the Lower Columbia." *The Pacific Northwesterner* 1 (1957): 1-8.

Idaho Historical Society. "Fort Boise: From Imperial Outpost to Historic Site." *Idaho Yesterdays* 6 (1962): 15-16, 33-39.

Ireland, W.E. "Early Flour Mills in British Columbia." Pt. 1. *British Columbia Historical Quarterly* 5 (1941): 89-109.

Johannessen, Carl L., Davenport, William A., Millet, Artimus, and McWilliams, Steven. "The Vegetation of the Willamette Valley." *Annals of the Association of American Geographers* 61 (1971): 286-302.

Johnson, F. Henry. "Fur-trading Days at Kamloops." *British Columbia Historical Quarterly* 1 (1937): 171-85.

Lawson, M. "The Beaver Hat and the Fur Trade." In *People and Pelts*, edited by M. Bolus. Winnipeg: Peguis Publishers, 1972, pp. 27-37.

Meany, Edmond S. "First American Settlement on Puget Sound." *Washington Historical Quarterly* 7 (1916): 136-43.

Merriam, Willis B. "The Role of Pemmican in the Canadian Northwest." *Yearbook of the Association of Pacific Coast Geographers* 17 (1955): 34-38.

Moodie, D.W. "Agriculture and the Fur Trade." In *Old Trails and New Directions: Papers of the Third North American Fur Trade Conference*, edited by Carol M. Judd and Arthur J. Ray. Toronto: University of Toronto Press, 1980, pp. 272-90.

Morton, Arthur S. "The Place of the Red River Settlement in the Plans of the Hudson's Bay Company, 1812-1825." *Annual Report of the Canadian Historical Association*, 1929, pp. 103-9.

Morton, W.L. "Agriculture in the Red River Colony." *Canadian Historical Review* 30 (1949): 305-22.

Olsen, Michael L. "Corporate Farming in the Northwest: The Puget's Sound Agricultural Company." *Idaho Yesterdays* 14 (1970): 18-23.

Ormsby, Margaret A. "Agricultural Development in British Columbia." *Agricultural History* 19 (1945): 11-20.

_____. "The History of Agriculture in British Columbia." *Canadian Journal of Agricultural Science* 20 (1939): 61-72.

Reid, Robie L. "Early Days at Old Fort Langley." *British Columbia Historical Quarterly* 1 (1937): 71-85.

Roe, F.G. "The Red River Hunt." *Transactions of the Royal Society of Canada* 29 (1935): 171-218.

Ross, Frank E. "The Retreat of the Hudson's Bay Company in the Pacific North-West." *Canadian Historical Review* 18 (1937): 262-80.

Sage, Walter N. "The Oregon Treaty of 1846." *Canadian Historical Review* 27 (1946): 349-67.

_____. "The Place of Fort Vancouver in the History of the Northwest." *Pacific Northwest Quarterly* 39 (1948): 83-102.

Schafer, Joseph. "The British Attitude toward the Oregon Question, 1815-1846." *American Historical Review* 16 (1911): 272-99.

Taylor, Herbert, and Hoaglin, Lester. "The Intermittent Fever Epidemic of the 1830's on the Lower Columbia River." *Ethnohistory* 9 (1962): 160-78.

Walker, Peter. "The Origins, Organization and Role of the Bison Hunt in the Red River Valley." *Manitoba Archaeological Quarterly* 6 (1982): 62-68.

Wrinch, Leonard A. "The Formation of the Puget's Sound Agricultural Company." *Washington Historical Quarterly* 24 (1933): 3-8.

Notes

NOTES TO THE PROLOGUE

1. For Spanish interest in, and activity on, the Northwest Coast, see Warren L. Cook, *Flood Tide of Empire: Spain and the Pacific Northwest, 1543-1819* (New Haven and London: Yale University Press, 1973). See also Christon I. Archer, "The Transient Presence: A Re-Appraisal of Spanish Attitudes toward the Northwest Coast in the Eighteenth Century," *B.C. Studies* (Summer 1973): 11-19, and "Spanish Exploration and Settlement of the Northwest Coast in the 18th Century," *Sound Heritage* 7 (1978): 32-53.
2. In 1775 Spanish Captain Bruno de Hezeta became the first Euroamerican to sight (not enter or ascend) the Columbia River.
3. In 1827 the company's Columbia and New Caledonia Districts were united to form the Columbia Department, which joined the Northern, Southern, and Montreal Departments as one of the four vast hunting and trading preserves of the company in North America (in 1836 they contained 136 posts and 1,404 employees) (see Frederick Merk, ed., *Fur Trade and Empire: George Simpson's Journal . . . 1824-25*, [Cambridge, MA: Belknap Press, 1968], pp. 337-38). The company described the territory variously as the Columbia or Western or Northwest Department. By 1841 the Columbia District included Forts Colvile, George, Hall, Langley, McLoughlin, Nez Percés, Nisqually, Okanagan, Simpson, Stikine, Taku, Thompson's River, Umpqua, and Vancouver; the New Caledonia District comprised Alexandria, Babine, Chilcotin, Connolly's Lake, Fort George, Fraser's Lake, McLeod's Lake, and Stuart's Lake. In the same year members of the United States Exploring Expedition estimated that Old Oregon held 20,000 Indians and 700 to 800 whites and halfbreeds (including 350 settlers and 100 missionaries) (see George Foster Emmons, "Journal kept while attached to the Exploring Expedition . . . ," [Yale University Library, Western Americana Ms. 166], vol. 3, n.p.; and Charles Wilkes, "Oregon Territory," OHS, Ms. 56, p. 29). By the end of the period of joint occupation, there were more than 600 company servants in the Columbia Department, including 200 at Fort Vancouver, the frontier's capital (HBCA, B.223/z/4: 53).
4. The few studies include: J. Neilson Barry, "Agriculture in the Oregon Country in 1795-1844," *Oregon Historical Quarterly* 30 (1929): 161-68; John S. Galbraith, "The Early History of the Puget's Sound Agricultural Company, 1838-43," *Oregon Historical Quarterly* 55 (1954): 234-59; James R. Gibson, "Food for the Fur Traders: The First Farmers in the Pacific Northwest, 1805-1846," *Journal of the West* 7 (1968): 18-30; Michael Leon Olsen, "The Beginnings of Agriculture in Western Oregon and Western Washington" (Ph.D. diss., University of Washington, 1970); and Leonard A. Wrinch, "The Formation of the Puget's Sound Agricultural Company," *Washington Historical Quarterly* 24 (1933): 3-8. The Barry and Wrinch articles are very skimpy; for his article Galbraith failed to use the extensive records of the Puget's Sound Agricultural Company in the archives of the Hudson's Bay Company, and he wrongly concludes that the former firm was both a political and an economic failure; Gibson's article is too compendious and marred by typographical errors; and Olsen's dissertation ignores the archives of the Hudson's Bay Company and relies too much on the unreliable evidence of claims presented to the British and American Joint Commission and as a result gives undue weight to American settlers (see his Ch. 1). For a more general

study of the development of farming in the northern half of the Oregon Country, see Margaret A. Ormsby, "Agricultural Development in British Columbia," *Agricultural History* 19 (1945): 11-20, and "The History of Agriculture in British Columbia," *Canadian Journal of Agricultural Science* 20 (1939): 61-72.
5. Samuel Black, "List of the Mineral Samples Sent to Geo. Simpson by Sam Black," British Museum of Natural History, Mineral Deposits Library, Geological Society Collection, Richardson Papers, p. [16].

NOTES TO CHAPTER 1

1. In the Oregon Country bison did not range below American Falls on the upper Snake River.
2. Arthur S. Morton, "The Place of the Red River Settlement in the Plans of the Hudson's Bay Company, 1812-1825," *Annual Report of the Canadian Historical Association, 1929*, p. 103.
3. Alexander Ross, *The Fur Hunters of the Far West*, ed. Kenneth A. Spaulding (Norman: University of Oklahoma Press, 1956), p. 56.
4. HBCA, E.8/3: 12.
5. Alexander Ross, *The Red River Settlement* .... (Minneapolis: Ross and Haines, 1957), pp. 17-18.
6. Arthur S. Morton, *A History of the Canadian West to 1870-71*, ed. Lewis G. Thomas (Toronto and Buffalo; University of Toronto Press, 1973), pp. 55-59.
7. Ross Cox, *The Columbia River* .... ed. Edgar I. Stewart and Jane R. Stewart (Norman: University of Oklahoma Press, 1957), pp. 301-2.
8. On the bison hunt see F.G. Roe, "The Red River Hunt," *Transactions of the Royal Society of Canada* 29 (1935): 171-218, and Peter Walker, "The Origins, Organization and Role of the Bison Hunt in the Red River Valley," *Manitoba Archaeological Quarterly* 6 (1928): 62-68.
9. John Tod, "History of New Caledonia and the Northwest Coast," PABC, pp. 3-4.
10. On the rivalry between the two companies during the first and second decades of the nineteenth century see Ann Carlos, "The Birth and Death of Predatory Competition in the North American Fur Trade: 1810-1821," *Explorations in Economic History* 19 (1982): 156-83.
11. John West, *The Substance of a Journal during a Residence at the Red River Colony* (New York: Johnson Reprint Corp., 1966), pp. 62-63.
12. R. Harvey Fleming, ed., *Minutes of Council Northern Department of Rupert Land, 1821-31* (Toronto: Champlain Society, 1940), pp. 393-94.
13. Frederick Merk, ed., *Fur Trade and Empire: George Simpson's Journal . . . 1824-25* (Cambridge, MA: Belknap Press, 1968), p. 179.
14. Wheat and barley returned from twenty to twenty-five bushels per acre and potatoes yielded thirtyfold the seed. (West, *The Substance of a Journal*, p. 108.)
15. Merk, *Fur Trade and Empire*, p. 250.
16. Cox, *Columbia River*, pp. 385-86.
17. United States, Congress, Senate, "Message from the President of the United States ...." *Senate Document No. 39*, 21st Congress, 2d Session, p. 12
18. James Douglas, "Private Papers of Sir James Douglas," 1st series, Bancroft Library, Ms. P-C 2, pp. 79-80.
19. Merk, *Fur Trade and Empire*, p. 339; "Papers relative to the Expedition of Lieuts. Warre and Vavasour to the Oregon Territory," PRO, Ms. F.O. 5/457, pp. 60v., 136.
20. "Papers relative to the Expedition of Lieuts. Warre and Vavasour," p. 7.
21. G.P. de T. Glazebrook, ed., *The Hargrave Correspondence 1821-1843* (Toronto: Champlain Society, 1938), pp. 386-87. On farming at the Red River Settlement see Barry Kaye, "The Historical Geography of Agriculture and Agricultural Settlement in the Canadian Northwest, 1774-ca. 1830" (Ph.D. diss., University of London, 1976) and W.L. Morton, "Agriculture in the Red River Colony," *Canadian Historical Review* 30 (1949): 305-22.
22. [Daniel Williams Harmon], *Sixteen Years in the Indian Country: The Journal of Daniel Williams Harmon 1800-1816*, ed. W. Kaye Lamb (Toronto: Macmillan, 1957), pp. 210, 219, 228.
23. [James Strange], *James Strange's Journal and Narrative* .... (Madras: Government Press, 1928), p. 21.
24. Henry R. Wagner, *Spanish Explorations in the Strait of Juan de Fuca* (New York: AMS Press, 1971), p. 162.
25. [Charles Bishop], *The Journals and Letters of Captain Charles Bishop* .... ed. Michael Roe (London: Hakluyt Society, 1967), pp. 58, 124, 129.
26. Gabriel Franchère, *Journal of a Voyage on the North West Coast* .... trans. Wessie Tipping

Lamb (Toronto: Champlain Society, 1969), pp. 75, 89; Alexander Ross, *Adventures of the First Settlers on the Oregon or Columbia River* (London: Smith, Elder and Co., 1849), p. 80.
27. Franchère, *Journal*, pp. 96, 117; Cox, *Columbia River*, p. 213.
28. Cox, *Columbia River*, 254; Elliott Coues, ed., *The Manuscript Journals of Alexander Henry ... and of David Thompson* .... (Minneapolis: Ross and Haines, 1965), vol. 2, pp. 889, 903, 909.
29. By 1826 Simpson had mastered his job, and in that year the Governor and Committee wrote to him that "the Fur Trade is very much indebted for its prosperous state to your talents for distinct businesslike arrangement, and to your indefatigable zeal and perseverance in making yourself master of all the minute details, as well as of the general arrangement of the business, indeed we consider that you have acquired a more perfect knowledge of the Indian Trade than perhaps was ever possessed by any one Individual or even by any body of Men who have been engaged in it" (Merk, *Fur Trade and Empire*, pp. 285-86).
30. [Archibald McDonald], *Peace River. A Canoe Voyage from Hudson's Bay to Pacific ....* ed. Malcolm McLeod (Edmonton: M.G. Hurtig, 1971), p. 75. In following Simpson's career through his correspondence one is struck by his wide practical knowledge, his great influence with the Governor and Committee, his staunch loyalty to company and crown, his enormous energy and diligence, his ruthlessness and exactingness, his thousands of miles of tireless travel, and his reams of detailed letters, reports, and journals. He expected every employee to live up to the company's motto, *pro pelle cutem* — "risk one's skin for a skin." On Simpson see John S. Galbraith, *The Little Emperor: Governor Simpson of the Hudson's Bay Company* (Toronto: Macmillan, 1976) and Arthur S. Morton, *Sir George Simpson: Overseas Governor of the Hudson's Bay Company* (Toronto and Vancouver: J.M. Dent and Sons [Canada], 1944.).
31. [Samuel Black]. *A Journal of a Voyage from Rocky Mountain Portage in Peace River To the Sources of Finlays Branch And North West In Summer 1824*, ed. E.E. Rich (London: Hudson's Bay Record Society, 1955), pp. xxv-xxvi.
32. For example, the beaver returns at Thompson's River (Fort Kamloops) and Fort Okanagan increased annually from the former's founding in 1811 to a peak of 2,946 pelts in 1822, whereupon they decreased steadily to 1,051 pelts in 1826 (HBCA, B.97/e/1: 3). In 1827 Clerk Archibald McDonald reported from Thompson's River that "the beaver ... is but rare enough considering the extent of country — A person can walk for days together without seeing the smallest quadruped, the little brown squirrel excepted" (E.E. Rich, ed., *Part of Dispatch from George Simpson Esqr Governor of Ruperts Land to the Governor & Committee of the Hudson's Bay Company London March 1, 1829* [Toronto: Champlain Society, 1947], p. 226.) By comparison New Caledonia's beaver returns fluctuated only slightly around six thousand skins between 1822 and 1826 (HBCA, B.239/h/1: 3-4). Regarding Simpson's orders to evaluate the "Columbia quarter," see Fleming, *Minutes of Council*, p. 302 and Merk, *Fur Trade and Empire*, p. 175.
33. Merk, *Fur Trade and Empire*, pp. 243-44.
34. HBCA, D4/5: 18; D.4/88: 21v., 22, 31. See also Merk, *Fur Trade and Empire*, p. 72.
35. Merk, *Fur Trade and Empire*, p. 65.
36. Ibid., pp. 47-48.
37. HBCA, B.76/z/1: 17-18.
38. Merk, *Fur Trade and Empire*, pp. 128-29.
39. Ibid., p. 50.
40. Roderick Finlayson, "The History of Vancouver Island and the Northwest Coast," 1878, Bancroft Library, Ms. P-C 15, pp. 72-73.
41. Franchère, *Journal*, p. 190.
42. HBCA, D.4/125: 50. William Kittson, incidentally, was the brother of Norman Kittson, the millionaire financial associate of the American railroad magnate J.J. Hill.
43. Merk, *Fur Trade and Empire*, pp. 66, 264, 338.
44. Simpson also advised that Fort George be abandoned and that the "principal Depot" be situated at the mouth of the Fraser River because it "appears to be formed by nature as the grand communication with all our Establishments on this side the mountains" (Merk, *Fur Trade and Empire*, p. 76). However, Simpson subsequently discovered, during his trip down the Fraser in 1828, that it "can no longer be thought of as a practicable communication with the interior; ... I should consider the passage down, to be certain Death, in nine attempts out of Ten." Hell's Gate, or "Simpson's Falls," in the Fra-

ser Canyon was the main bottleneck in the "New Caledonia River." Simpson added: "if the Navigation of the Columbia is not free to the Hudsons Bay Company... they must abandon and curtail their Trade in some parts, and probably be constrained to relinquish it on the West side of the Rocky Mountains altogether" (Rich, *Part of Dispatch from George Simpson*, pp. 38-39, 174).

Until 1822 New Caledonia received supplies from, and dispatched furs to, Fort George via the Columbia on North West Company boats and pack horses (Peter Corney, *Early Voyages in the North Pacific 1813-1818* [Fairfield, WA: Ye Galleon Press, 1965], pp. 141, 176; [Harmon], *Sixteen Years in the Indian Country*, passim; HBCA, B.188/a/1, B.188/b/5; Alexander Ross, *Adventures of the First Settlers on the Oregon or Columbia River* [London: Smith, Elder and Co., 1849], passim). Thereafter the district was outfitted from York Factory via the Saskatchewan and Peace Rivers to McLeod's Lake. But in 1825 Simpson described this supply line as "the most tedious harrassing and expensive transport in the Indian Country" (HBCA, D.4/88: 8). So it was replaced in 1826 by the Columbia brigade, which conveyed supplies by boat from Fort Vancouver to Fort Okanagon, by horse from Fort Okanagon to Fort Alexandria via Thompson's River, and by boat from Fort Alexandria to McLeod's Lake, the depot of New Caledonia. Furs were sent southward by the same means. The brigade left McLeod's Lake in the middle of April, reached Fort Vancouver in mid-June, departed in mid-July, and returned to McLeod's Lake at the end of September. But New Caledonia continued to receive some items, such as leather, hide, and cord, from the Saskatchewan District. Leather, that is, dressed moose and deer skin, was more abundant on the eastern than on the western side of the Rockies, where it was much in demand for pack straps (colliers), "a most important instrument in that world of portages." It was received via Yellowhead Pass, which was commonly called Leather Pass by the Bay men ([McDonald], *Peace River. A Canoe Voyage*, pp. 44, 113). From 1848 New Caledonia was outfitted via the lower reaches of the Fraser to Thompson's River. Besides the Columbia brigade, a light and fast "annual express" linked Fort Vancouver and York Factory via Athabasca Pass (Rocky Mountain Portage). The spring or York express left Fort Vancouver in mid-March with dispatches and retirees and reached Hudson Bay in late June or early July. It left York Factory in late July as the fall or Vancouver express with recruits, orders, and news and reached the Columbia in late October or early November. In addition, from the early 1830's the company's coastal vessels transported outfits to and from the coastal posts, provisions and manufactures to New Archangel (Sitka), timber and fish to Hawaii, and tallow and hides from California.

45. E.E. Rich, ed., *The Letters of John McLoughlin from Fort Vancouver to the Governor and Committee* (London: Hudson's Bay Record Society, 1941-44), vol. 1, p. ci, Glyndwr Williams, ed., *Hudson's Bay Company Miscellany 1670-1870* (Winnipeg: Hudson's Bay Record Society, 1975), p. 176.
46. Finlayson, "History of Vancouver Island," p. 5.
47. Rich, *Part of Dispatch from George Simpson*, p. 68.
48. Merk, *Fur Trade and Empire*, p. 139. On Fort Langley see Mary K. Cullen, *The History of Fort Langley 1827-96* (Ottawa: Parks Canada, 1979).
49. The profits of the company's fur trade were divided into one hundred parts, of which sixty went to the stockholders and forty to the "wintering partners" (the chief factors and chief traders in the field, the former usually in charge of departments and stationed at departmental headquarters and the latter commonly "bourgeois" in command of important posts). The forty parts were themselves subdivided into eighty-five shares, of which two went to each chief factor and one to each chief trader. The lower ranks (clerks, postmasters, guides, interpreters, boatmen, and others) were salaried in money and kind. See Madge Wolfenden, ed., "John Tod: 'Career of a Scotch Boy'," *British Columbia Historical Quarterly* 28 (1954): 190.
50. Rich, *Letters of John McLoughlin*, vol. 1, p. 235.
51. Ibid., vol 3, p. 113.
52. HBCA, C.4/1: 25v.-27. From 1847 their destination was Fort Victoria.
53. Finlayson, "History of Vancouver Island," p. 37.
54. HBCA, A.6/25: 124v.; Wilkes, "Oregon Territory," p. 38. The mark-up on company

goods over prime cost (England prices) at Fort Vancouver was double that at York Factory. The goods were sold at Fort Vancouver for 80 per cent above the London wholesale price and at the other Columbia posts for 100 per cent in order to cover transport expenses. Still, these articles were no more expensive and perhaps even cheaper than in the United States in the early 1840's (Wilkes, "Oregon Territory," p. 39).

55. Rich, *Letters of John McLoughlin*, vol. 1, pp. 1, 25, 137.
56. James Cooper, "Maritime Matters on the Northwest Coast...," 1878, Bancroft Library, Ms. P-C 6, p. 1.
57. John Kirk Townsend, *Narrative of a Journey Across the Rocky Mountains to the Columbia River* (Fairfield, WA: Ye Galleon Press, 1970), pp. 313-14.
58. HBCA, D.4/67: 55v.
59. Samuel Parker, *Journal of an Exploring Tour beyond the Rocky Mountains ... 1835, '36, and '37* (Minneapolis: Ross and Haines, 1967), p. 147. The Governor and Committee felt as early as 1837 that the Columbia Bar was so hazardous that Fort Vancouver should be replaced as the entrepôt of the Columbia Department by a new base on the Strait of Juan de Fuca (HBCA, A.6/24: 62-62v.), but this new headquarters — Fort Victoria — was not to appear for another decade.
60. HBCA, D.4/59: 99-100; "Papers relative to the Expedition of Lieuts. Warre and Vavasar," p. 45v.
61. HBCA, B.223/a/3: 47. Father P.J. De Smet said that "*Dalle* is an old French word, meaning a trough, and the name is given by the Canadian voyageurs to all contracted running waters, hemmed in by walls of rocks." Pierre-Jean De Smet, *Oregon Missions and Travels over the Rocky Mountains, in 1845-46* (New York: Edward Dunigan, 1847), p. 214.
62. HBCA, B.76/e/1: 5.
63. Merk, *Fur Trade and Empire*, pp. 76-77.
64. Rich, *Part of Dispatch from George Simpson*, p. 68.
65. Cox, *Columbia River*, p. 343.
66. Fleming, *Minutes of Council*, p. 343.
67. HBCA, B. 223/e/1: 1v.
68. Wilkes, "Oregon Territory," p. 22.
69. De Smet, *Oregon Missions*, p. 109.
70. Anonymous, "Documents," *Washington Historical Quarterly* 1 (1907): 259; John Dunn, *History of the Oregon Territory* .... 2d ed.

(London: Edwards and Hughes, 1846), p. 216; Rich, *Letters of John McLoughlin*, vol. 1, p. 37.
71. Cox, *Columbia River*, p. 217.
72. McLean contended that when salmon were scarce, rabbits were plentiful, so that the Indians seldom starved. But starving natives were often observed by travellers. When salmon failed in New Caledonia, the Indians had no bait for their marten traps and tried to kill company horses as a substitute. See [John McLean], *John McLean's Notes of a Twenty-Five Year's Service in the Hudson's Bay Territory*, ed. W.S. Wallace (Toronto: Champlain Society, 1932), pp. 152-53.
73. Lois Halliday McDonald, ed., *Fur Trade Letters of Francis Ermatinger* .... (Glendale, CA: Arthur H. Clark, 1980), p. 64.
74. Carefully cured salmon would keep two or three years but pemmican almost indefinitely; however, only at Fort Hall and Flathead Post was much pemmican available. See W.N. Sage, ed., "Peter Skene Ogden's Notes on Western Caledonia," *British Columbia Historical Quarterly* 1 (1937): 52. See also [McDonald] *Peace River. A Canoe Voyage*, p. 31, and McDougall to McLeod, 8 March 1828, "John McLeod Papers 1811-1837," PAC, Ms. 19, series A23, vol. 1.
75. [McDonald], *Peace River. A Canoe Voyage*, p. 29.
76. [Harmon], *Sixteen Years in the Indian Country*, pp. 141, 146.
77. Rich, *Part of Dispatch from George Simpson*, p. 228.
78. HBCA, B.97/e/1: 2v. Fort Alexandria shipped salmon to Thompson's River and even to Fort Okanagan. And in the early 1840's Thompson's River procured between ten and twelve thousand dried salmon annually from the Indians of the Fraser Canyon above Spuzzum (Wolfenden, "John Tod," p. 225).
79. HBCA, B.146/e/1: 1-1v.
80. Sage, "Ogden's Notes," p. 51.
81. In addition to fish, various roots, particularly camas and wappatoo (arrowhead), were eaten. Camas was an Indian staple on the lower Columbia. Roots and berries were commonly mixed with salmon; two dried salmon and one quart of roots or berries was one man's daily ration at Fort Colvile (HBCA, B.45/e/3: 10-10v.).
82. Rich, *Part of Dispatch from George Simpson*, p. 19.
83. Ibid., pp. 17-18.

84. HBCA, D.4/100: 8v.-9.
85. Wolfenden, "John Tod," p. 227.
86. For example, by the last half of the 1830's at Forts George and Alexandria in New Caledonia, grain crops were offsetting salmon failures (Sage, "Ogden's Notes," p. 52).
87. [McDonald], *Peace River. A Canoe Voyage*, p. 108.
88. Dears to E. Ermatinger, 5 March 1831, "Ermatinger Papers," PABC, AB40 Er62.3.
89. [John Work], "Work Correspondence," PABC, AB40, p. 6.
90. Glazebrook, *Hargrave Correspondence*, p. 63.
91. Rich, *Part of Dispatch from George Simpson*, p. 25. Drunkards were regularly exiled to New Caledonia. For example, Simpson characterized one clerk as "a low Drunken Worthless fellow ... a blackguard ... he has therefore been removed to New Caledonia this Season [1832] where he cannot be otherwise than sober" (Williams, *Hudson's Bay Company Miscellany 1670-1870*, pp. 216-17).
92. Merk, *Fur Trade and Empire*, p. 78.
93. Ibid., p. 86.
94. In fact, by the late 1830's American traders had largely withdrawn from the fur trade on the coast and in the interior. But they were quickly replaced by agricultural migrants who proved more numerous and more permanent than the mountain men.
95. Merk, *Fur Trade and Empire*, p. 266.

NOTES TO CHAPTER 2

1. John Stuart, "Letter Books and Journals ...," OHS, Ms. 1502, reel 1, n.p.
2. Alexander Caulfield Anderson, "History of the Northwest Coast," 1878, Bancroft Library, Ms. P-C 2, p. 88.
3. The new post was named Fort Vancouver and was placed on the right bank in order to underline the British claim to the territory north of the Columbia. For a detailed account of Fort Vancouver see John A. Hussey, *The History of Fort Vancouver and Its Physical Structure* (Portland: United States National Park Service and Washington State Historical Society, 1957).
4. Frederick Merk, ed., *Fur Trade and Empire: George Simpson's Journal ... 1824-25* (Cambridge, MA: Belknap Press, 1968), pp. 123-24.
5. HBCA, B.76/e/1: 7v.
6. Merk, *Fur Trade and Empire*, p. 87. On 30 March 1806, on their return east from Fort Clatsop, Lewis and Clark camped on the site of Fort Vancouver, "a beautiful prairie above a large pond," and Lewis asserted that this "Columbian valley" was "the only desirable situation for a settlement which I have seen on the West side of the Rocky mountains" (Reuben Gold Thwaites, ed., *Original Journals of the Lewis and Clark Expedition 1804-1806* [New York: Arno Press, 1969], vol. 4, p. 220).
7. HBCA, D.4/88: 33v.
8. Ibid., 29-29v.
9. John McLoughlin, "Copy of a Document Found among the Private Papers of the late Dr. John McLoughlin," *Transactions of the Oregon Pioneer Association 1880*, p. 46; John McLoughlin, "Private Papers, 1825-1856," Bancroft Library, Ms. P-A 155, pt. 2, p. 5.
10. Lois Halliday McDonald, ed., *Fur Trade Letters of Francis Ermatinger ...*, (Glendale, CA: Arthur H. Clark, 1980), p. 61.
11. [John Work], "Work Correspondence," PABC, AB40, p. 29.
12. John Kirk Townsend, *Narrative of a Journey Across the Rocky Mountains to the Columbia River* (Fairfield, WA: Ye Galleon Press, 1970), pp. 298-99.
13. P.L. Edwards, *Sketch of the Oregon Territory: or, Emigrants' Guide* (Liberty, MO: The "Herald" Office, 1842), pp. 18-19; William A. Slacum, "Memorial of William A. Slacum ...," *Senate Document 24*, 25th Congress, 2d Session, p. 6; [William A. Slacum], "Mr. Slacum's Report," *Reports Committees*, 25th Congress, 3d Session, p. 186.
14. Richard A. Pierce and John H. Winslow, eds., *H.M.S. Sulphur on the Northwest and California Coasts, 1837 and 1839* (Kingston, Ont.: Limestone Press, 1979), p. 66; Charles Wilkes, "Oregon Territory," OHS, Ms. 56, pp. 37-38.
15. HBCA, B.223/z/4: 199v.-200.
16. E.E. Rich, ed., *The Letters of John McLoughlin from Fort Vancouver to the Governor and Committee* (London: Hudson's Bay Record Society, 1941-44) vol. 1, p. 260.
17. Fort Vancouver caught most of its fish in the Willamette River at Salmon (or Multnomah or Willamette) Falls. See [William Fraser Tolmie], *The Journals of William Fraser Tolmie: Physician and Fur Trader* (Vancouver: Mitchell Press, 1963), p. 175. One thousand barrels were salted in 1842, the most abundant year (HBCA, D. 5/8: 159). By then some six hundred barrels were exported annually. See [Duflot De Mofras], *Duflot De Mofras' Travels on the Pacific Coast,*

trans. and ed. Marguerite Eyer Wilbur (Santa Ana, CA: Fine Arts Press, 1937), vol. 2, p. 105.
18. James Douglas, "Private Papers of Sir James Douglas," 1st series, Bancroft Library, Ms. P-C 12, p. 7; [James Douglas], "Private Papers of Sir James Douglas, First Series," PABC, B20 1858, p. 5.
19. Rich, *Letters of John McLoughlin*, vol. 1, p. 204.
20. George Foster Emmons, "Journal kept while attached to the Exploring Expedition...," Yale University Library, Western Americana Ms. 166, vol. 3, n.p.
21. HBCA, B.223/z/5: 276, D.4/59: 91.
22. Pierce and Winslow, *H.M.S. Sulphur*, p. 65.
23. HBCA, D.4/125: 78v.
24. Rich, *Letters of John McLoughlin*, vol. 1, p. 265.
25. British and American Joint Commission for the Final Settlement of the Claims of the Hudson's Bay and Puget's Sound Agricultural Companies [*Proceedings*] (Montreal: J. Lovell, 1868), vol. 2, p. 51.
26. Grassy openings in woodland were called prairies in the Old Northwest and plains in the New Northwest, particularly in the Willamette Valley, where they were oak openings. William Tolmie, a company physician, referred to small prairies as "prairions."
27. After Simpson's canoe trip down the Fraser in 1828 had demonstrated that river's unsuitability as a route to the interior, Fort Vancouver remained the departmental capital. In 1829 it was relocated one mile downstream and closer to the riverbank to facilitate the procurement of water and the handling of freight. See John Scouler, "Dr. John Scouler's Journal of a Voyage to N. W. America...," OHS, Ms. 949B, p. [126].
28. Cox was the name of the Hawaiian swineherd who tended the company's pigs there.
29. Sauvie's Island was named after J.B. Sauvé, a Canadian who managed the dairies there.

Wappatoo *(Sagittaria variablilis)* and camas *(Camassia quamash)* tubers and acorns also loomed large in the Indian diet. Paul Kane, a Canadian painter, observed in 1846 that:

> The camas is a bulbous root, much resembling the onion in outward appearance, but is more like the potato when cooked, and is very good eating. The wappatoo is somewhat similar, but larger, and not so dry or delicate in its flavour. They are found in immense quantities in the plains in the vicinity of Fort Vancouver, and in the spring of the year present a most curious and beautiful appearance, the whole surface presenting an uninterrupted sheet of bright ultramarine blue, from the innumerable blossoms of these plants. They are cooked by digging a hole in the ground, then putting down a layer of hot stones, covering them with dry grass, on which the roots are placed, they are then covered with a layer of grass, and on top of this they place earth with a small hole perforated through the earth and grass down to the vegetables. Into this water is poured, which, reaching the hot stones, forms sufficient steam to completely cook the roots in a short time, the hole being immediately stopped up on the introduction of the water. (Paul Kane, *Wanderings of an Artist* . . . . [Edmonton: M.G. Hurtig, 1968], pp. 127-28; see also Kate Ball Powers, Flora Ball Hopkins, and Lucy Ball, eds., *John Ball . . . Autobiography* [Glendale, CA: Arthur H. Clark, 1925], p. 96).

Camas, rated "queen" of the Columbia Plateau's roots by a Catholic missionary, was a staple of the interior Indians but was also exploited by those on the lower Northwest Coast. See Pierre-Jean De Smet, *Oregon Missions and Travels over the Rocky Mountains in 1845-46* (New York: Edward Dunigan, 1847), p. 117. Esquimalt on Vancouver Island meant "the place for gathering camas" (Kane, *Wanderings of an Artist*, p. 144).

The Indians also ate "Chinook olives," acorns fermented by human urine (ibid., p. 128).

The "swamp potato" (wappatoo) helped to sustain the Lewis and Clark expedition during its lean winter at Fort Clatsop in 1805-6. "Those roots are equal to the Irish potato, and is a tolerable substitute for bread," wrote the leaders (Thwaites, *Journals of the Lewis and Clark Expedition*, vol. 3, p. 244).

30. HBCA, A.6/23: 140d; Rich, *Letters of John McLoughlin*, vol. 1, p. 183.
31. Wilkes, "Oregon Territory," pp. 73-74.
32. Carl Landerholm, trans., *Notices & Voyages of the Famed Quebec Mission to the Pacific Northwest* (Portland: Oregon Historical Society, 1956), p. 177. Spring oats yielded well but not many were needed because the horses were not stabled during the mild winters.

33. Ibid. p. 27. In the late 1820's and early 1830's grain was trod by horses.
34. George B. Roberts, "Recollections of George B. Roberts," Bancroft Library, Ms. P-A 83, p. 24.
35. Rich, *Letters of John McLoughlin*, vol. 1, p. 283.
36. The mill on the Mill Plain could grind twenty thousand bushels of wheat yearly (HBCA, D.4/59:91). The mill at Willamette Falls was called the White Mill.
37. Silas Holmes, "Journal kept by Assist. Surgeon Silas Holmes...," Yale University Library, Western Americana Ms. 260, vol. 3, p. 5.
38. Rich, *Letters of John McLoughlin*, vol. 1, p. 283.
39. [Tolmie], *Journals of William Fraser Tolmie*, p. 173.
40. Emmons, "Journal," vol. 3, n.p.; Holmes, "Journal," vol. 3, p. 5.
41. In 1831 McLoughlin reckoned that the potato crop would be "enough for all the Kings Posts" (John McLeod, "John McLeod Papers 1811-1837," PAC, Ms. 19, series A23, vol. 1, p. 296).
42. Wilkes, "Oregon Territory," pp. 45-46, 74.
43. [Nathaniel J. Wyeth], *The Correspondence and Journals of Captain Nathaniel J. Wyeth 1831-6*, ed. F.G. Young (Eugene: University Press, 1899), p. 180.
44. Jason Lee, "Journal," p. [104], and "Letters on Oregon Mission, 1834-35," pp. 11-12, OHS, Ms. 1212.
45. Kane, *Wanderings of an Artist*, p. 117.
46. Edwards, *Sketch of the Oregon Territory*, p. 17; M.M. Quaife, ed., "Letters of John Ball, 1832-1833," *Mississippi Valley Historical Review* 5 (1919): 467.
47. In 1837, for example, McLoughlin bought eighty-six head from Ewing Young, who had brought them from California. John Warren Dease, "Diary," OHS, Ms. 560, p. 67.
48. HBCA, B.223/d/126: 63, B.223/d/136: 68v., B.223/d/144: 92, D.4/59: 90.
49. Wilkes, "Oregon Territory," p. 46.
50. McLoughlin, "Copy of a Document," 46; Rich, *Letters of John McLoughlin*, vol. 1, pp. 206-7.
51. Burt Brown Barker, ed., *Letters of Dr. John McLoughlin Written at Fort Vancouver 1829-1832* (Portland; Oregon Historical Society, 1948), p. 94; T.C. Elliott, ed., "Letters of Dr. John McLoughlin," *Oregon Historical Quarterly* 23 (1922): 370; McLoughlin to E. Ermatinger, 3 March 1837, "Ermatinger Papers," PABC, AB40 Er.62.3; C.O. Ermatinger, "The Columbia River Under Hudson's Bay Company Rule," *Washington Historical Quarterly* 5 (1914): 196; Merk, *Fur Trade and Empire*, p. 324; Rich, *Letters of John McLoughlin*, vol. 1, pp. 207, 228, 259.
52. HBCA, B.223/e/1: lv.; McLoughlin, "Copy of a Document," 46; Rich, *Letters of John McLoughlin*, vol. 1, pp. 61, 67. Elsewhere, however, McLoughlin stated in 1830 that "this is the first year since I am here...in which I have been able to supply our people adequately" (HBCA, D.4/125: 77).
53. Merk, *Fur Trade and Empire*, p. 301.
54. HBCA, A.6/22: 48d.
55. Barker, *Letters of Dr. John McLoughlin*, p. 145: Mrs. David Harvey, "Life of John McLoughlin...," 1878, Bancroft Library, Ms. P-B 12, p. 9; Rich, *Letters of John McLoughlin*, vol. 1, pp. 92-93, 128.
56. Merk, *Fur Trade and Empire*, p. 310; Madge Wolfenden, ed., "John Tod: 'Career of a Scotch Boy,'" *British Columbia Historical Quarterly* 28 (1954): 169.
57. Rich, *Letters of John McLoughlin*, vol. 2, p. 158; Wilkes, "Oregon Territory," p. 40.
58. HBCA, D. 4/100: 6-6v.
59. Rich, *Letters of John McLoughlin*, vol. 1, pp. 205, 284.
60. Fort Colvile was named in honour of Andrew Wedderburn Colvile, a prominent London broker who bought stock in the company in 1809 and joined its committee in 1810. One of the company's ablest directors, he authored the "Retrenching System" (1810), which emphasized economy and care and payment by results. He was responsible for the choice of Simpson as governor of the Northern Department.
61. "Information Concerning Fort Colville, Washington," Washington State Universtiy Library, Ms. 979: 725 H869i: p. [1].
62. Merk, *Fur Trade and Empire*, p. 139.
63. Elkanah Walker and Mary Walker, "Diaries," OHS, Ms. 1204, pp. 9-10. Fort Colvile was generally considered to be the most comfortable post in the Columbia. On Fort Colvile and Archibald McDonald see Jean Murray Cole, *Exile in the Wilderness: The Biography of Chief Factor Archibald McDonald 1790-1853* (Don Mills, Ont.: Burns & MacEachern, 1979), esp. p. 173.
64. HBCA, D.4/59: 49; E.E. Rich, ed., *Part of Dispatch from George Simpson Esqr Governor of Ruperts Land to the Governor & Committee of the Hudson's Bay Company London March*

1, 1829 (Toronto: Champlain Society, 1947), p. 227.
65. "Information Concerning Fort Colville," p. [2].
66. HBCA, B.45/e/3: 12-12v.
67. Samuel Black, "List of the Mineral Samples Sent to Geo. Simpson by Sam Black," British Museum of Natural History, p. [14]; W.D. Brackenridge, *The Brackenridge Journal for the Oregon Country*, ed. O.B. Sperlin (Seattle: University of Washington Press, 1931), p. 30; British and American Joint Commission [*Proceedings*], vol. 2, p. viii; Clifford Merrill Drury, ed., *Nine Years with the Spokane Indians: The Diary 1838-1848, of Elkanah Walker* (Glendale, CA: Arthur H. Clark, 1976), p. 75; T.C. Elliott, ed., "British Values in Oregon, 1847," *Oregon Historical Quarterly* 32 (1931): 38; McLeod, "Papers," vol. 1, p. 225; [John Work] "Work Correspondence," PABC, Ms. AB40, p. 13; McDonald to McLoughlin, 15 August 1842, [Fort Colvile], "Correspondence of Archibald McDonald relating to Fort Colvile," PABC, AB20 C72M.1; HBCA, A.11/17: 1v., B.223/d/49: 49v; "Papers relative to the Expedition of Lieuts. Warre and Vavasour to the Oregon Territory," PRO, Ms. F.O. 4/457, p. 86; United States, Congress, Senate, "Message from the President of the United States . . . ," *Senate Document No. 39*, 21st Congress, 2d Session, p. 9; Walker and Walker, "Diaries," p. 9; Charles Wilkes, *Narrative of the United States Exploring Expedition During the Years 1838, 1839, 1840, 1841, 1842* (Philadelphia: Lea & Blanchard, 1845), vol. 4, p. 445.
68. British and American Joint Commission [*Proceedings*], vol. 4, p. 170; HBCA, B.223/z/4: 196v.; "Papers relative to the Expedition of Lieuts. Warre and Vavasour," p. 67v.
69. Brackenridge, *Brackenridge Journal*, p. 31; T.C. Elliott, ed., "Journal of John Work [1825-1826] . . . ," *Washington Historical Quarterly* 5 (1914): 284; HBCA, D.4/120: 63; [Archibald McDonald], *Peace River. A Canoe Voyage from Hudson's Bay to Pacific . . .*, ed. Malcolm McLeod (Edmonton: M.G. Hurtig, 1971), pp. 94, 108.
70. "Information Concerning Fort Colville," p. [3].
71. Elliott, "Journal of John Work," 37, 98, 169-70.
72. HBCA, D.4/120: 63, 64v.-65.
73. Rich, *Part of Dispatch from George Simpson*, p. 49.
74. [Work], "Work Correspondence," pp. 2, 5.
75. HBCA, B.45/e/3: 10, 11.
76. Barker, *Letters of Dr. John McLoughlin*, p. 131.
77. HBCA, B.45/e/3: 14.
78. Ibid., D.4/100: 7v.
79. M. Whitman to Greene, 10 May 1839, "Papers of the American Board of Commissioners for Foreign Missions," Houghton Library, Ms. ABC: 18.5.3, vol. 1.
80. Brackenridge, *Brackenridge Journal*, p. 31.
81. McDonald to McLoughlin, 8 June 1842, [Fort Colvile], "Correspondence of Archibald McDonald."
82. McDonald to E. Ermatinger, 1 April 1836, [Work], "Work Correspondence."
83. McDonald to Douglas, 19 April 1842, McDonald to McLoughlin, 21 October 1841, 8 June 1842, [Fort Colvile], "Correspondence of Archibald McDonald."
84. HBCA, D.4/59: 82; D.4/90: 195v.
85. McMillan to McLeod, 21 January 1828, McLeod, "Papers," vol. 1.
86. McLoughlin to Simpson, 3 March 1835, Douglas to Yale, 21 November 1838, [Fort Langley], "Correspondence Relating to Fort Langley from the Hudson's Bay Company Archives, 1830-1859," PABC, AB20 L3A; HBCA, D.4/127: 70; Robie L. Reid, ed., "Fort Langley Correspondence," *British Columbia Historical Quarterly* 1 (1937): 188, 189.
87. Anonymous, "Documents," *Washington Historical Quarterly* 1 (1907): 259, 265; Hartwell Bowsfield, ed., *Fort Victoria Letters 1846-1851* (Winnipeg: Hudson's Bay Record Society, 1979), p. 6; [Fort Langley], "Correspondence," pp. [24, 29].
88. Glyndwr Williams, ed., *Hudson's Bay Company Miscellany 1670-1870* (Winnipeg: Hudson's Bay Record Society, 1975), pp. 183-84.
89. Merk, *Fur Trade and Empire*, p. 117.
90. Rich, *Part of Disptach from George Simpson*, p. 41.
91. [McDonald], *Peace River. A Canoe Voyage*, p. 39; [Fort Langley], "Fort Langley Journal June 27, 1827-July 30, 1830," PABC, AB20 L2A2M, p. 81.
92. [Fort Langley], "Journal," pp. 121-22.
93. HBCA, D.4/125: 62v., 63v.; [Work], "Work Correspondence," p. 22.
94. HBCA, B.223/d/49: 36; [John Work], *The Journal of John Work January to October, 1835*, ed. Henry Drummond Dee (Victoria: Charles F. Banfield, 1945), pp. 84-85.
95. [Fort Langley], "Correspondence," p. [22];

Landerholm, *Notices & Voyages*, pp. 103-4; "Papers relative to the Expedition of Lieuts. Warre and Vavasour," p. 86.
96. Glyndwr Williams, ed., *London Correspondence Inward from Sir George Simpson 1841-42* (London: Hudson's Bay Record Society, 1973), pp. 73-74.
97. HBCA, D.4/125: 63.
98. [Work], *Journal of John Work*, p. 84.
99. HBCA, B.223/b/26: 2.
100. [James Douglas], "Journal of Sir James Douglas 1840-1841," Bancroft Library, Ms. P-C 11, pp. 62-63; HBCA, D.4/59: 82; [Otto Klotz], *Certain Correspondence of the Foreign Office and of the Hudson's Bay Company . . . .* (Ottawa: Government Printing Bureau, 1899), pt. 2, p. 69; Williams, *London Correspondence*, p. 74. As early as 1828 Simpson considered the post "in regard to the means of living ... already independent" (Rich, *Part of Dispatch from George Simpson*, p. 43).
101. [Wyeth], *Correspondence and Journals*, p. 184.
102. Townsend, *Narrative of a Journey*, p. 281.
103. Brackenridge, *Brackenridge Journal*, p. 36; Elliott, "British Values in Oregon," 39; "Papers relative to the Expedition of Lieuts. Warre and Vavasour," p. 86.
104. Rich, *Part of Dispatch from George Simpson*, p. 51. In 1832 the Indians on the lower Columbia sold horses for $8.00 each (Quaife, "Letters of John Ball," p. 467).
105. Rich, *Part of Dispatch from George Simpson*, pp. 30-31.
106. [McDonald], *Peace River. A Canoe Voyage*, p. 114; McDonald to Simpson, 30 September 1841, [Fort Colvile], "Correspondence of Archibald McDonald." Indians killed two of the commandants of Thompson's River: Pierre Charette in 1815 and Samuel Black in 1841.
107. [McDonald], *Peace River. A Canoe Voyage*, p. 114; Wolfenden, "John Tod," 220.
108. [Fort Kamloops], "Journal, Aug. 3, 1841-Dec. 18, 1843, kept by John Tod," PABC, AB20 K12A 1841-43: 39; "Papers relative to the Expedition of Lieuts. Warre and Vavasour," p. 86.
109. [John McLean], *John McLean's Notes of a Twenty-Five Year's Service in the Hudson's Bay Territory*, ed. W.S. Wallace (Toronto: Champlain Society, 1932), p. 171.
110. Ibid., p. 170; G.P. de T. Glazebrook, ed., *The Hargrave Correspondence 1821-1843* (Toronto: Champlain Society, 1938), p. 372; "Papers relative to the Expedition of Lieuts. Warre and Vavasour," p. 86; John Tod, "History of New Caledonia and the Northwest Coast," PABC, pp. 10-11.
111. [McLean], *John McLean's Notes*, p. 161.
112. HBCA, B.5/a/2: 18v.; B.5/a/4: 7; B.5/a/5: 30v., 54; B.5/a/6: 12; B.5/a/7: 3, 19, 52; W.E. Ireland, "Early Flour Mills in British Columbia," pt. 1, *British Columbia Historical Quarterly* 5 (1941): 91.
113. Glazebrook, *Hargrave Correspondence*, p. 276.
114. Ibid.; [McLean], *John McLean's Notes*, p. 170; "Papers relative to the Expedition of Lieuts. Warre and Vavasour," p. 86.
115. HBCA, B.188/a/17: 21v.; [McLean] *John McLean's Notes*, p. 151; "Papers relative to the Expedition of Lieuts. Warre and Vavasour," p. 86.
116. Glazebrook, *Hargrave Correspondence*, p. 256.
117. Anderson, "History of the Northwest Coast," pp. 29-30.
118. W.N. Sage, ed., "Peter Skene Ogden's Notes on Western Caledonia," *British Columbia Historical Quarterly* 1 (1937): 52.
119. HBCA, D.4/59: 50.
120. Rich, *Part of Dispatch from George Simpson*, p. 50.
121. Brackenridge, *Brackenridge Journal*, p. 28; Elliott, "British Values in Oregon," p. 40; McDonald, *Fur Trade Letters*, p. 67; Merk, *Fur Trade and Empire*, p. 50; "Papers relative to the Expedition of Lieuts. Warre and Vavasour," p. 86.
122. Clifford Merrill Drury, ed., *First White Women Over the Rockies* (Glendale, CA: Arthur H. Clark, 1963), vol. 1, p. 79; "Papers relative to the Expedition of Lieuts. Warre and Vavasour," p. 86; Theodore Talbot, "Journal, 1843-44," OHS, Ms. 773, vol. 1, pp. 118-19; Wilkes, "Oregon Territory," p. 16.
123. Elliott, "British Values in Oregon," 30, 41; "Papers relative to the Expedition of Lieuts. Warre and Vavasour," p. 86. For more details on the Siskiyou Trail, which was used by trapping expeditions, livestock drives, and migrant parties, see Richard Dillon, *Siskiyou Trail: The Hudson's Bay Company Route to California* (New York: McGraw-Hill, 1975).
124. Elliott Coues, ed., *The Manuscript Journals of Alexander Henry . . . and of David Thompson . . .*, (Minneapolis: Ross and Haines, 1965), vol. 2, p. 756; Ross Cox, *The Columbia River . . .*, ed. Edgar I. Stewart and Jane R. Stewart (Norman: University of Oklahoma Press, 1957), p. 70; Gabriel Franchère, *Journal of a Voyage on the North West Coast . . .*, trans. Wessie Tipping Lamb (Toronto: Champlain Society, 1969), p. 96; Michael

Leon Olsen, "The Beginnings of Agriculture in Western Oregon and Western Washington" (Ph.D. diss., University of Washington, 1970), pp. 7-8.
125. Peter Corney, *Early Voyages in the North Pacific 1813-1818* (Fairfield, WA: Ye Galleon Press, 1965), p. 177.
126. John H. Aulick, "Journal", OHS, Ms. 101: n.p.; John Scouler, "Dr. John Scouler's Journal of a Voyage to N.W. America . . . ," OHS, Ms. 949B, p. [99]; Josiah Sturgis, "Extracts from his Journal, 1818," OHS, Ms. 153, pp. 3, 4.
127. Townsend, *Narrative of a Journey*, p. 311.
128. Holmes, "Journal," vol. 2, p. 294.
129. Elliott, "British Values in Oregon," 41; "Papers relative to the Expedition of Lieuts. Warre and Vavasour," p. 86.
130. HBCA, D.4/59: 83. One year later it was proposed to abandon both Fort Nisqually and Fort Langley and replace them with a single new post, to be called Fort Langley, on Whidbey Island, which was "well calculated for a shipping Depot and large agricultural Establishment" (ibid., D.4/100: 10-10v.). This project was abandoned in 1835 when it was realized that the island was inconveniently situated for trade.
131. Tess E. Jennings, trans., *Mission of the Columbia* . . . . (Seattle: Works Progress Administration, 1937), vol. 2, p. 9.
132. Clarence B. Bagley, ed., "Journal of Occurences at Nisqually House, 1833 [1833-35]," *Washington Historical Quarterly* 6 (1915): 185-87; [Tolmie], *Journals of William Fraser Tolmie*, pp. 195-96, 199, 203, 207.
133. "Papers relative to the Expedition of Lieuts. Warre and Vavasour," p. 86; Rich, *Letters of John McLoughlin*, vol. 1, p. 281.
134. HBCA, F.8/1: 24v.
135. Merk, *Fur Trade and Empire*, p. 322.
136. HBCA, A.6/23: 154-154v.
137. Reid, "Fort Langley Correspondence," 188-89.
138. Rich, *Letters of John McLoughlin*, vol. 1, pp. 214, 267.
139. HBCA, D.4/67: 87v.-88.
140. Ibid., D.5/8: 144, 187v.
141. Ibid., 187-187v.
142. Ibid., B.223/z/4: 206v. Vavasour's 1846 report also stated that Fort Victoria's site had been selected solely on agricultural grounds (see ibid., B.223/z/4: 230v.; "Papers relative to the Expedition of Lieuts. Warre and Vavasour," pp. 47, 81).
143. HBCA, B.223/z/4: 206v.; "Papers relative to the Expedition of Lieuts. Warre and Vavasour," pp. 80v., 86.
144. Bowsfield, *Fort Victoria Letters*, pp. 5-6; HBCA, A.11/72: 16v., B.223/d/155: 108, B.223/d/160: 86v., B.223/d/165: 56v.-57.
145. HBCA, D.4/59: 62; Williams, *London Correspondence*, pp. 61-62.
146. Joseph Schafer, ed., "Letters of Sir George Simpson, 1841-1843," *American Historical Review* 14 (1908): 75; Williams, *London Correspondence*, p. 62.
147. [Work], *Journal of John Work*, p. 67.
148. Rich, *Letters of John McLoughlin*, vol. 1, p. 324.
149. HBCA, B.201/a/3: 104v., 147v., B.223/c/1: 124v.
150. Work to E. Ermatinger, 11 October 1841, [Work], "Work Correspondence."
151. "Papers relative to the Expedition of Lieuts. Warre and Vavasour," p. 86.
152. HBCA, B.201/a/3: 161, B.223/c/1: 124v.; Rich, *Letters of John McLoughlin*, vol. 1, p. 325.
153. HBCA, D.4/59: 61; Schafer, "Letters of Sir George Simpson," 75; Williams, *London Correspondence*, p. 61.
154. George Simpson, *Narrative of a Journey Round the World* (London: Henry Colburn, 1847), vol. 1, pp. 209-10.
155. HBCA, D.4/59: 65; Schafer, "Letters of Sir George Simpson," 76; Williams, *London Correspondence*, p. 63.
156. HBCA, D.4/59: 66-67; Schafer, "Letters of Sir George Simpson," 76; Williams, *London Correspondence*, p. 64.

NOTES TO CHAPTER 3

1. John Work made this reference to the heavy rainfall. Vulcanism was probably not a factor. Both Mount Baker and Mount St. Helens erupted in the late autumn of 1842 but there is no evidence in the sources that ash from either eruption harmed agriculture at Fort Langley or Fort Vancouver. (Given the prevailing winds, the latter post would have been more at risk.)
2. E.E. Rich, ed., *The Letters of John McLoughlin from Fort Vancouver to the Governor and Committee* (London: Hudson's Bay Record Society, 1941-44), vol. 1, p. 205.
3. Ibid.
4. [John Work], *The Journal of John Work January to October, 1835*, ed. Henry Drummond Dee (Victoria: Charles F. Banfield, 1945), p. 85.
5. Clarence B. Bagley, ed., "Journal of Oc-

curences at Nisqually House, 1833 [1833-35]," *Washington Historical Quarterly* 6 (1915): 182.
6. G.P. de T. Glazebrook, ed., *The Hargrave Correspondence 1821-1843* (Toronto: Champlain Society, 1938), p. 137.
7. Rich, *Letters of John McLoughlin*, vol. 1, pp. 154, 165. In August 1833 Heron had visited the island and discovered a plain measuring five miles by two miles with "excellent" soil, and in December William Tolmie found from eighty to one hundred acres of plain with "black loamy soil mixed with sand" to a depth of two feet ([William Fraser Tolmie], *The Journals of William Fraser Tolmie: Physician and Fur Trader* [Vancouver: Mitchell Press, 1963], pp. 229, 256).
8. Robie L. Reid, ed., "Fort Langley Correspondence," *British Columbia Historical Quarterly* 1 (1937): 188; see also Rich, *Letters of John McLoughlin*, vol. 1, p. 155.
9. Rich, *Letters of John McLoughlin*, vol. 1, p. 325.
10. [Work], *Journal of John Work*, pp. 16, 18.
11. Rich, *Letters of John McLoughlin*, vol. 1, p. 232.
12. Frederick Merk, ed., *Fur Trade and Empire: George Simpson's Journal . . . 1824-25* (Cambridge, MA: Belknap Press, 1968), p. 106.
13. Also, the fort's location on the so-called "American" (southern) side of the river discouraged company investment in an operation with a doubtful political future ([Charles Wilkes], *Diary of Wilkes in the Northwest*, ed. Edmund S. Meany [Seattle: University of Washington Press, 1926], p. 31).
14. Douglas believed that the "most eligible station" for Fort Vancouver's cattle was the "Fallatey [Tualatin] Plains" twenty miles downstream on the left bank of the Columbia (Rich, *Letters of John McLoughlin*, vol. 1, pp. 258, 259).
15. HBCA, D.4/121: 15, 16.
16. Ibid., B.113/a/1: 6.
17. John Dunn, *History of the Oregon Territory . . . .* 2d ed. (London: Edwards and Hughes, 1846), p. 245.
18. [Work], *Journal of John Work*, pp. 16, 44.
19. Ibid., p. 48.
20. HBCA, A.11/69: 83v.-84.
21. Rich, *Letters of John McLoughlin*, vol. 1, pp. 205, 283; Charles Wilkes, "Oregon Territory," OHS, Ms. 56, p. 17.
22. [Wilkes], *Diary*, pp. 38, 48.
23. HBCA, A.11/69: 83v.; D.4/125: 78v.; Rich, *Letters of John McLoughlin*, vol. 1, p. 228. See also HBCA, D.4/125: 50.
24. Rich, *Letters of John McLoughlin*, vol. 1, p. 257.
25. Ibid., p. 258; HBCA, B.223/b/27: 22.
26. [Fort Langley], "Correspondence Relating to Fort Langley, from the Hudson's Bay Company Archives, 1830-1859," PABC, AB20 L3A, pp. [1-2]; HBCA, D.4/123: 72.
27. [Fort Langley], "Correspondence," p. [23]; Reid, "Fort Langley Correspondence," 192.
28. [Fort Langley], "Correspondence," p. [21].
29. Ibid., pp. [23, 25, 26].
30. Rich, *Letters of John McLoughlin*, vol. 3, p. 184.
31. E.E. Rich, ed., *Part of Dispatch from George Simpson Esqr Governor of Ruperts Land to the Governor & Committee of the Hudson's Bay Company London March 1, 1829* (Toronto: Champlain Society, 1947), p. 226.
32. McDonald to McLoughlin, 24 August 1840, 5 January 1841, [Fort Colvile], "Correspondence of Archibald McDonald relating to Fort Colvile," PABC, AB20 C72M.1.
33. HBCA, B.146/e/2: 3v.
34. Burt Brown Barker, ed., *Letters of Dr. John McLoughlin Written at Fort Vancouver 1829-1832* (Portland: Oregon Historical Society, 1948), p. 245; HBCA, D.4/121: 37v.; Rich, *Letters of John McLoughlin*, vol. 1, pp. 50, 234.
35. Rich, *Letters of John McLoughlin*, vol. 1, p. 143; [Work], *Journal of John Work*, p. 85.
36. HBCA, B.223/b/24: 30, 41v.
37. Clarence B. Bagley, ed., *Early Catholic Missions in Old Oregon* (Seattle: Lowman & Hanford, 1932), vol. 2, p. 94; Jean Baptiste Zacharie Bolduc, *Mission of the Columbia*, ed. and trans. Edward J. Kowrach (Fairfield, WA: Ye Galleon Press, 1979), p. 127; Carl Landerholm, trans., *Notices & Voyages of the Famed Catholic Mission to the Pacific Northwest* (Portland: Oregon Historical Society, 1956), p. 240; Rich, *Letters of John McLoughlin*, vol. 3, pp. 37, 43-45.
38. [John McLean], *John McLean's Notes of a Twenty-Five Year's Service in the Hudson's Bay Territory*, ed. W.S. Wallace (Toronto: Champlain Society, 1932), p. 170.
39. Glazebrook, *Hargrave Correspondence*, p. 276.
40. [McLean], *John McLean's Notes*, p. 172.
41. [Wilkes], *Diary*, p. 72.
42. HBCA, D.4/121: 23v.; [John Work], "Work Correspondence," PABC, AB40, p. 2.
43. Clifford Merrill Drury, ed., *Nine Years with the Spokane Indians: The Diary 1838-1848, of Elkanah Walker* (Glendale, CA: Arthur H. Clark, 1976), p. 379; Paul Kane, *Wanderings*

of an Artist . . . . (Edmonton: M.G. Hurtig, 1968), p. 118.
44. [Fort Langley], "Correspondence," p. [2]; HBCA, A.11/69: 83, D.5/8: 143, D.4/120: 54, D.4/125: 63; [McLean], *John McLean's Notes*, p. 170; Merk, *Fur Trade and Empire*, pp. 270, 291; Rich, *Letters of John McLoughlin*, vol. 1, p. 44; [Work], *Journal of John Work*, p. 85.
45. HBCA, D.4/121: 14v.-15; [Work], *Journal of John Work*, p. 27.
46. McDonald to McLoughlin, 5 January 1841, McDonald to Simpson, 1 August 1841, McDonald to McLoughlin, 8 August 1842, [Fort Colvile], "Correspondence of Archibald McDonald."
47. George B. Roberts, "Recollections of George B. Roberts," Bancroft Library, Ms. P-A 83, p. 23.

NOTES TO CHAPTER 4

1. Thomas E. Jessett, ed., *Reports and Letters of Herbert Beaver 1836-1838* (Portland: Champoeg Press, 1959), pp. 78-79.
2. On the logistics of the Russian-American Company see James R. Gibson, *Imperial Russia in Frontier America: The Changing Geography of Supply of Russian America, 1784-1867* (New York: Oxford University Press, 1976).
3. Frederick Merk, ed., *Fur Trade and Empire: George Simpson's Journal . . . 1824-25* (Cambridge, MA: Belknap Press, 1968), p. 86.
4. E.E. Rich, ed., *The Letters of John McLoughlin from Fort Vancouver to the Governor and Committee* (London: Hudson's Bay Record Society, 1941-44), vol. 2, p. 24.
5. Burt Brown Barker, ed., *Letters of Dr. John McLoughlin Written at Fort Vancouver 1829-1832* (Portland: Oregon Historical Society, 1948), p. 17; Merk, *Fur Trade and Empire*, pp. 311-12; [Russian-American Company], "Records of the Russian-American Company 1802-67: Correspondence of Governors General," NARS, Mc. 11, roll 7: 26-26v.
6. E.E. Rich, ed., *Part of Dispatch from George Simpson Esqr Governor of Ruperts Land to the Governor & Committee of the Hudson's Bay Company March 1, 1829* (Toronto: Champlain Society, 1947), p. 85.
7. [Russian-American Company], "Records," roll 7: 20-22, roll 34: 103.
8. [William Fraser Tolmie], *The Journals of William Fraser Tolmie: Physician and Fur Trader* (Vancouver: Mitchell Press, 1963), pp. 178-79.
9. HBCA, D.4/100: 30-30v., 31v.
10. Ibid., 22v., 23v.
11. Ibid., A.6/23: 100-100v.; A.C. Anderson, "The Origin of the Puget Sound Agricultural Company," 1865, OHS, Ms. 1502, p. 1.
12. HBCA, A.6/24: 59v.-60; D.4/21: 64.
13. Ye. Chernykh, "O zemledelii v verkhney Kalifornii' ["Concerning Agriculture in Upper California"] *Zhurnal selskavo khozyaistva i ovtsevodstva*, 1841, p. 261; L.A. Shur, ed., *K. beregam Novovo Sveta* . . . . [*To the Shores of the New World* . . . .] (Moscow: "Nauka," 1971), p. 196.
14. HBCA, B.223/b/12: 23.
15. [Russian-American Company], "Records," roll 12: 427.
16. Gibson, *Imperial Russia in Frontier America*, pp. 170-71.
17. [Russian-American Company], "Records," roll 8: 327v.-328, 330. This was the same differential that favoured the Americans over the Russians in the coastal trade in the mid-1810's. See James R. Gibson, "Russian Expansion in Siberia and America," *Geographical Review* 70 (1980): 131.
18. [Russian-American Company], "Records," roll 8: 327v., roll 42: 446.
19. G.P. de T. Glazebrook, ed., *The Hargrave Correspondence 1821-1843* (Toronto: Champlain Society, 1938), p. 184.
20. HBCA, F.29/2: 144-47.
21. [Russian-American Company], "Records," roll 12: 278-278v.
22. For the full text of the agreement, see HBCA, F.29/2: 162-170, 174-177v. The agreement was renewed for ten years in 1849, for five years in 1859, and for two years in 1863 without the provisionment clause. As a result of the accord, the Hudson's Bay Company replaced St. Dionysius Redoubt with Fort Stikine in 1840. Also, Sitka placed its last American supply order in 1839; the shipment arrived in 1841. The Hudson's Bay Company gained control of the Stikine River otters, which were worth nearly twice as much as other land otters.
23. At the beginning of 1837 the Governor and Committee had complained to McLoughlin that the farm at Fort Vancouver still could not produce enough to supply both returning London ships with beef and pork and Sitka with grain and flour. There simply was insufficient farmland at the departmental headquarters to do so. That was the price that the company paid for its political decision to locate Fort Vancouver on the northern side of the Columbia rather than on its southern side,

where the Willamette Valley offered ample farmland (see HBCA, A.6/24: 66v.).

24. The contract, plus the Hudson's Bay Company's holdings at Cowlitz and Nisqually, were transferred to the Puget's Sound Agricultural Company.

In 1839 the Governor and Committee explained to Simpson why they had postponed the formation of McLoughlin's scheme:

> For several years past our attention has been directed to the formation of an agricultural Settlement with a view to the production of wool, hides, tallow and other farm produce for the English and other markets, in the District of Country situated between the head waters of the Cowlitz Portage and Pugets Sound, which in regard to soil and climate has been represented to us as highly favorable for such object, but we defer'd carrying it into effect until we were secured in the possession of the country by a renewal of the Exclusive Licence [monopoly charter] from Government, which was obtained last year for a term of twenty-one years. The Government is favorable to the object for political reasons, and after giving the subject mature consideration and obtaining the best legal advice, we are of the opinion that it can be better done under the protection and auspices of the Hudson's Bay Company by a separate Association than if under taken by the Company in conjunction with the Fur Trade [HBCA, A.6/25: 28].

NOTES TO CHAPTER 5

1. HBCA, F.8/1: 13. In accordance with their attempts at diversification, the Governor and Committee also formed a company in 1839 to exploit the Norwegian timber trade, but it was dissolved four years later at a loss of £15,165 to the "unfortunate Norway affair" ([Archibald McDonald], "Correspondence, etc. re Puget Sound Agricultural Company," PABC, AB25 M14).
2. HBCA, F.8/1: 31v.
3. Ibid., A.6/25: 28v.
4. A.C. Anderson, "The Origin of the Puget Sound Agricultural Company," 1865, OHS, Ms. 1502, p. 1.
5. HBCA, A.6/25: 28v.
6. Ibid., 123-123v.
7. Simpson testified in 1857 before the Select Committee of the British House of Commons investigating "British Oregon," or British Columbia, that the Puget's Sound Agricultural Company was "an offshoot of the Hudson's Bay Company; an agricultural establishment formed by the Hudson's Bay Company, or parties connected with or interested in the Hudson's Bay Company, encouraged by the Government of the day" (Great Britain, Parliament, House of Commons, Select Committee on the Hudson's Bay Company, *Report from the Select Committee on the Hudson's Bay Company* . . . . [London, 1857], p. 64).
8. HBCA, A.6/25: 28v.
9. Ibid., F.11/1: 3.
10. James Cooper, "Maritime Matters on the Northwest Coast . . . ," 1878, Bancroft Library, Ms. P-C 6, p. 23; William Fraser Tolmie, "History of Puget Sound and the Northwest Coast," 1878, Bancroft Library, Ms. P-B 25, p. 17.
11. HBCA, B.223/b/21: 45; E.E. Rich, ed., *The Letters of John McLoughlin from Fort Vancouver to the Governor and Committee* (London: Hudson's Bay Record Society, 1941-44), vol. 1, p. 264.
12. HBCA, F.11/1: 4.
13. In 1841 it was reported that Cowlitz was "entirely" and Nisqually was "principally" engaged in Puget's Sound Agricultural Company business. At Nisqually the company was to collect furs for the Hudson's Bay Company in return for an annual payment of £200 to £300. Ibid., D.4/59: 83, F.11/1: 5, 54.
14. Ibid., 21.
15. Ibid., 5-6. Douglas would have preferred to establish the "Pasture Farm" on the Tualatin Plains. He wrote:

> We would have given preference to the Falletey Plains a section of country situate[d] west of the Willamette, abounding with excellent herbage, well watered, enjoying in winter, a temperate climate and in short surpassing the Nisqually District in every natural advantage; but were deterred from making choice of it, by its distance from our Establishment and its being South of the Columbia, within American Territory [Rich, *Letters of John McLoughlin*, vol. 1, p. 284].

16. HBCA, F.11/1: 6.
17. Ibid., 4-5.

18. Ibid., 6-7, A.6/25: 29-29v.
19. Ibid., F.11/1: 5, 7.
20. Ibid., 3. In 1839 the Hudson's Bay Company dispatched two dairymen and their wives, two shepherds, two ploughmen, and one millwright, plus six Merino and six Leicester sheep, on the schooner *Vancouver* for the Puget's Sound Agricultural Company. The principal shepherd, James Steel, was not transferred with the others; instead, he accompanied McLoughlin on his return from furlough to Fort Vancouver from London in 1839. Ibid., A.6/25: 29.
21. Ibid., F.11/1: 7.
22. The agents recommended that British supplies for the Puget's Sound Agricultural Company be indented directly from England rather than from Hudson's Bay Company stores because freight and insurance costs by sea were less than advances on Hudson's Bay Company goods. Ibid., 7-10, 15, 24.
23. Ibid., 55-56.
24. Ibid., 14.
25. Ibid., 13.
26. Sometimes McLoughlin disregarded the instructions of the agents, such as by keeping sheep at Fort Vancouver instead of moving them to Nisqually Farm and by keeping more men at Nisqually and Cowlitz than had been recommended. This may explain why the agent's directives to McLoughlin were often direct and blunt, even rude. He was relieved of the superintendence of the Company's operations when his efforts had proved successful, that is, after the sheep breeds had been improved, the Indian labourers had been trained, and the rotation system had been perfected.
27. Paul Kane, *Wanderings of an Artist* .... (Edmonton, M.G. Hurtig, 1968), p. 142.
28. HBCA, F.12/1: 442v.
29. Tolmie, "History of Puget Sound," p. 18.
30. HBCA, B.223/z/4: 229.
31. [William Fraser Tolmie], *The Journals of William Fraser Tolmie: Physician and Fur Trader* (Vancouver: Mitchell Press, 1963), p. 190. According to another observer, however, Cowlitz Prairie measured only two miles by one mile. Carl Landerholm, trans., *Notices & Voyages of the Famed Quebec Mission to the Pacific Northwest* (Portland: Oregon Historical Society, 1956), p. 148.
32. HBCA, D.4/106: 11v.-12.
33. Ibid., 12v.
34. British and American Joint Commission for the Final Settlement of the Claims of the Hudson's Bay and Puget's Sound Agricultural Companies, [*Proceedings*] (Montreal: J. Lovell, 1868), vol. 3, p. 11; John Dunn, *History of the Oregon Territory* .... 2d ed. (London: Edwards and Hughes, 1846), pp. 236-37: Harriet Duncan Munnick, ed., *Catholic Church Records of the Pacific Northwest: Vancouver Volumes I and II and Stellamaris Mission*, trans. Mikell De Lores Wormell Warner (St. Paul, OR: French Prairie Press, 1972), pp. I-19, 20.
35. P.L. Edwards, *Sketch of the Oregon Territory: or, Emigrants' Guide* (Liberty, MO: The "Herald" Office, 1842), p. 9; Munnick, *Catholic Church Records*, pp. I-37, 38. A Catholic missionary counted four families in the autumn of 1838. Clarence B. Bagley, ed., *Early Catholic Missions in Old Oregon* (Seattle: Lowman & Hanford, 1932), vol. 1, pp. 53, 61.
36. HBCA, A.11/69: 82-82v., B.223/b/24: 16, 37, F.8/1: 24v., F.12/1: 442.
37. [Charles Wilkes], *Diary of Wilkes in the Northwest*, ed. Edmond S. Meany (Seattle: University of Washington Press, 1926), p. 26.
38. Ibid.; T.C. Elliott, ed., "British Values in Oregon, 1847," *Oregon Historical Quarterly* 32 (1931): 31, 36; HBCA, D.4/59: 85, D.4/110: 23; William Peel, "Lieutenant Peel's Report to Captain J. Gordon, September 27, 1845," PRO, Ms. A.D.M. 1/5564, no. 5: [1]; Joseph Schafer, ed., "Letters of Sir George Simpson, 1841-1843," *American Historical Review* 14 (1908): 78; Charles Wilkes, "Oregon Territory," OHS, Ms. 56: 19.
39. British and American Joint Commission, [*Proceedings*], vol. 3, pp. 106-7.
40. Ibid., vol. 2, p. 197.
41. A.C. Anderson, "A.C. Anderson's Memo Relating to the Cowlitz Farm, etc., 1841," OHS, Ms. 1502, p. 1.
42. HBCA, D.4/59: 85-86, D.4/110: 23.
43. Ibid., B.223/b/24: 4, 16v., 30.
44. Peel, "Lieutenant Peel's Report," [1].
45. HBCA, D.4/59: 85, D.4/110: 23, F.12/2: 39v.-40; [Wilkes], *Diary*, pp. 26, 77; Wilkes, "Oregon Territory," 19.
46. HBCA, F.12/2: 76-76v.
47. Rich, *Letters of John McLoughlin*, vol. 2, pp. 26, 238.
48. HBCA, D.4/67: 85v. According to McLoughlin in late 1840, fourteen thousand bushels of wheat were required annually to meet local needs one year in advance and to fulfil the "Russian contract." By 1841

production at both Cowlitz Farm and Fort Vancouver was sufficient to provide only ninety-five hundred, or two-thirds, of these fourteen thousand bushels. See Rich, *Letters of John McLoughlin*, vol. 2, p. 26.
49. Rich, *Letters of John McLoughlin*, vol. 2, pp. 26, 125.
50. Ibid., vol. 3, p. 148.
51. [Wilkes], *Diary*, pp. 26-27.
52. HBCA, F.11/1: 46, 67-68, F.12/2: 9A, 21v.
53. Ibid., F.8/1: 24v.; [Tolmie], *Journals of William Fraser Tolmie*, pp. 198-99.
54. HBCA, F.12/1: 443.
55. Ibid., D.4/59: 84, D.4/110: 22v.
56. Ibid., F.12/1: 562, 611.
57. [William Briskoe], "Journal of William Briskoe, Armorer Aboard the *Relief* and *Vincennes* August 18, 1838-March 23, 1842," NARS, Mc. 75, pp. 6-7.
58. HBCA, F.12/2: 146v. In 1865, however, before the British and American Joint Commission, the Puget's Sound Agricultural Company testified that the Nisqually tract stretched from the Nisqually River to the Puyallup River and from the coast to the mountains and totalled 261 square miles (British and American Joint Commission, [*Proceedings*], vol. 3, p. 26).
59. British and American Joint Commission, [*Proceedings*], vol. 3, pp. 107-8.
60. HBCA, D.4/59: 83, F.11/1: 53, 55; [Wilkes], *Diary*, p. 80.
61. HBCA, F.12/2: 48-49, 53v.
62. Ibid., F.12/1: 467v.; Rich, *Letters of John McLoughlin*, vol. 1, p. 251, vol. 2, p. 237.
63. HBCA, D.4/10: 22-22v., D.4/58: 139v., D.4/59: 83-84, F.12/1: 508v., 559; Rich, *Letters of John McLoughlin*, vol. 2, p. 256.
64. Clifford Merrill Drury, ed., *The Diaries and Letters of Henry H. Spalding and Asa Bowen Smith relating to the Nez Perce Mission 1838-1842* (Glendale, CA: Arthur H. Clark, 1958), p. 322; HBCA, A.11/69: 80, F.12/1: 561.
65. Ibid., F.15/29: 12-13.
66. Anderson, "Origin of the Puget Sound Agricultural Company," p. 4.
67. HBCA, F.12/2: 137.
68. Anderson, "Origin of the Puget Sound Agricultural Company," p. 4.
69. Ibid., p. 5.; [Wilkes], *Diary*, pp. 80, 88; Wilkes, "Oregon Territory," 18.
70. HBCA, F.8/1: 39.
71. [Wilkes], *Diary*, p. 80.
72. HBCA, B.223/b/26: 5; Rich, *Letters of John McLoughlin*, vol. 2, pp. 73, 230, 236.
73. Tolmie, "History of Puget Sound," p. 19. A member of the U.S. Exploring Expedition asserted that forty tons of butter were exported annually to Russian America, but he obviously erred grossly (Silas Holmes, "Journal kept by Assist. Surgeon Silas Holmes . . . ," Yale University Library, Western Americana Ms. 260, vol. 3, p. 5).

American observers usually overestimated the Puget's Sound Agricultural Company's performance. Similarly, American officials generally foresaw rosier prospects, economic and otherwise, for the Puget Sound Lowland than did their British counterparts. This optimism undoubtedly reinforced U.S. resolve to gain all of Puget Sound under the boundary settlement of 1846.
74. Elliott, "British Values in Oregon," 31, 37; HBCA, B.223/b/24: 25, D.4/59: 84, D.4/110: 22v., F.8/1: 40, F.12/2: 9Av.
75. [Wilkes], *Diary*, p. 80; Wilkes, "Oregon Territory," 18.
76. HBCA, A.11/69: 79v., F.12/2: 34v.; Rich, *Letters of John McLoughlin*, vol. 2, p. 147.
77. HBCA, B.223/b/26: 4v., D.4/58: 143.
78. [R.C. Clark, ed.], "Puget Sound Agricultural Company," *Washington Historical Quarterly* 18 (1927): 58; HBCA, F.12/2: 14v.
79. HBCA, F.12/2: 108.
80. [Clark], "Puget Sound Agricultural Company," 58; HBCA, F.8/1: 51, 54v., F.15/9: 53. Tolmie's accounts showed a profit of £1,577 in 1844, but much of it was uncollected, including £586/10/- in debts owed by freemen, a "great part" of which, Simpson feared, would be lost. HBCA, D.4/68: 105v.
81. [Clark], "Puget Sound Agricultural Company," 59; HBCA, D.4/68: 106, F.12/2: 149v., F.26/1: 22.
82. British and American Joint Commission, [*Proceedings*], vol. 3, p. 28.
83. HBCA, F.26/1: 22.
84. Ibid., A.12/2: 292v., D.4/66: 53.
85. Ibid., A.12/2: 292v. McLoughlin, however, stated in 1843 that the Willamette wheat cost the Hudson's Bay Company 4 shillings, 6 pence per bushel of 69 pounds. He believed that the company sold provisions too cheaply to the Russians and that they would bring higher prices in the Sandwich Islands. HBCA, D.5/8: 153; Rich, *Letters of John McLoughlin*, vol. 2, pp. 124-25.
86. HBCA, A.12/2: 292v.
87. Ibid., A.12/2: 292, D.4/66: 52v., D.5/8: 153.
88. Roderick Finlayson, "The History of Van-

couver Island and the Northwest Coast," 1878, Bancroft Library, Ms. P-C 15, p. 19.
89. HBCA, A.12/2: 293v., D.4/66: 54.
90. In 1841 Simpson and McLoughlin disagreed over the conduct of the coast trade, the former favouring a couple of vessels and the latter a string of posts. The rift was widened in 1842 by the murder of McLoughlin's son at Fort Stikine, with Simpson placing most of the blame on the victim and McLoughlin accusing Simpson of not conducting a thorough and impartial inquest. The animosity between the two men was aggravated by McLoughlin's claim to Willamette Falls and Simpson's less than wholehearted support of the claim. McLoughlin was further embittered by the belief on the part of some company officials that he had given too much aid and comfort to American settlers. Disillusioned and old, he retired to the Willamette in 1849 and became an American citizen but found little solace, being deprived of his land claim.
91. HBCA, B.223/z/5: 276, D.4/58: 140v., D.5/8: 161v.
92. See John S. Galbraith, "The Early History of the Puget Sound Agricultural Company," *Oregon Historical Quarterly* 55 (1954): 234-59, and John S. Galbraith, *The Hudson's Bay Company as an Imperial Factor, 1821-1869* (Berkeley: University of California Press, 1957), Ch. 10. Galbraith states unequivocally that the company was both an economic and a political failure; he also ignores the wider purposes of the company, particularly vis-à-vis the Russian-American Company and the coast trade. To prove his economic point Galbraith concentrates on the firm's early setbacks and neglects its subsequent successes. Galbraith's outlook is imbued with the belief (common in American historiography of westward expansion) that British monopoly capital was doomed to failure by American free enterprise in the form of freewheeling mountain men and self-reliant homesteaders.
93. G.P. de T. Glazebrook, ed., *The Hargrave Correspondence 1821-1843* (Toronto: Champlain Society, 1938), p. 382.
94. Ibid.
95. See Frederick Merk, "The Oregon Pioneers and the Boundary," *American Historical Review* 29 (1924): 681-99, and Frederick Merk, *The Oregon Question: Essays in Anglo-American Diplomacy and Politics* (Cambridge: MA: Belknap Press, 1967), Essay 8.
96. [Russian-American Company], "Records of the Russian-American Company 1802-67: Correspondence of Governors General," NARS, Mc. 11, roll 12: 287v.-290v.
97. Ibid., roll 52: 6v.-7.
98. Ibid., roll 51: 227v., roll 52: 7-7v.
99. Ibid., roll 46: 359. The company's director, Baron von Wrangell, lowered this estimate to 234, 735 pounds (ibid., roll 48: 566, 638).
100. Ibid., roll 46: 136, 253.
101. P.A. Tikhmenev, *A History of the Russian-American Company,* trans. and ed. Richard A. Pierce and Alton S. Donnelly (Seattle: University of Washington Press, 1978), p. 236.
102. Finlayson, "History of Vancouver Island," p. 32.
103. [Russian-American Company], "Records," roll 55: 146.
104. Ibid., 150v.
105. Kenneth L. Holmes, ed. and comp., *Covered Wagon Women: Diaries & Letters from the Western Trails 1840-1890* (Glendale, CA: Arthur H. Clark, 1983), vol. 1, p. 38.
106. [Russian-American Company], "Records," roll 18: 102.
107. Ibid., roll 19: 579-582, roll 57: 580v.-581.
108. For the later years of the Puget's Sound Agricultural Company, see Barry M. Gough, "Corporate Farming on Vancouver Island: The Puget's Sound Agricultural Company, 1846-1857," *Canadian Papers in Rural History* 4 (1984): 72-82, and Michael L. Olsen, "Corporate Farming in the Northwest: The Puget's Sound Agricultural Company," *Idaho Yesterdays* 14 (1970): 18-23.

NOTES TO CHAPTER 6

1. HBCA, F.11/1: 53, F.12/2: 18v., 21.
2. Ibid., F.8/1: 25v., 31v.
3. Ibid., F.11/1: 1-2.
4. [Charles Wilkes], *Diary of Wilkes in the Northwest,* ed. Edmond S. Meany (Seattle: University of Washington Press, 1926), pp. 25-26, 77.
5. HBCA, D.4/25: 57v.-58v.; see also A.6/25: 68v., A.6/26: 14. According to Simpson, the Red River settlement contained three thousand Indians and *Métis* and two thousand whites in 1837 (ibid., A.8/2: 50-50v.).
6. Ibid., D.4/25: 58-58v.
7. Ibid., A.6/25: 51-51v., 91, D.4/25: 58.
8. Ibid., B.235/b/2: 7v.-8, D.4/25: 103v.
9. Ibid., F.11/1: 55-56.

10. Ibid., D.4/25: 103v.
11. Ibid., 103v.-104v.
12. Ibid., 105-105v.
13. Ibid., A.6/25: 123v.-124.
14. Ibid., D.5/6: 137.
15. Ibid., 137v.
16. Ibid., 137-137v., D.4/58: 142; John Flett, "A Sketch of the Emigration from Selkirk's Settlement to Puget Sound in 1841," *Tacoma Daily Ledger*, 18 February 1885, n.p.; Henry Harmon Spalding, "Papers," OHS, Ms. 1201, p. 76. For the names of the twenty-one heads of families who agreed on 31 May 1841 to resettle on halves, as well as the terms of their agreement, see HBCA, A.12/7: 391-392v. Another list of nineteen names is found in ibid, F.26/1: 35v. On James Sinclair, see D. Geneva Lent, *West of the Mountains: James Sinclair and the Hudson's Bay Company* (Seattle: University of Washington Press, 1963), especially Ch. 6 for his role in escorting the Red River migrants to the lower Columbia.
17. HBCA, F.8/1: 28v., F.12/1: 508; Flett, "Sketch of the Emigration," n.p.
18. HBCA, D.4/58: 142, F.12/1: 507v.-508.
19. Ibid., B.223/b/28: 59.
20. Ibid., D.4/110: 23v.-24.
21. Ibid., D.4/59: 86-87.
22. Ibid., 86.
23. Ibid., 142.
24. Ibid., F.12/1: 613.
25. Ibid., B.223/b/41: 78, D.4/58: 87, D.4/59: 87, D.4/110: 24; Flett, "Sketch of the Emigration," n.p.; Joseph Schafer, ed., "Letters of Sir George Simpson, 1841-1843," *American Historical Review* 14 (1908): 79.
26. HBCA, B.223/b/41: 78, D.4/58: 87-88, D.4/59: 87-88, D.4/110: 24-24v.; Schafer, "Letters of Sir George Simpson," 79.
27. Flett, "Sketch of the Emigration," n.p.
28. HBCA, F.8/1: 31-31v., F.12/1: 563; McNeil to Douglas, 14 December 1841, [Fort Nisqually], "Correspondence outward, May 23, 1841-Sept. 17, 1842," PABC, AB20 Ni2.
29. G.P. de T. Glazebrook, ed., *The Hargrave Correspondence 1821-1843* (Toronto: Champlain Society, 1938), p. 385.
30. HBCA, B.223/b/29: 81-81v., B.223/b/41: 78, F.8/1: 40, F.12/1: 612-613, F.12/2: 14-14v., 19; E.E. Rich, ed., *The Letters of John McLoughlin from Fort Vancouver to the Governor and Committee* (London: Hudson's Bay Record Society, 1841-44), vol. 2, pp. 77-79, 120. The settlers' contracts were not cancelled, and some of them subsequently sued the Puget's Sound Agricultural Company but to no avail (HBCA, B.223/b/41: 78v.).
31. Ibid., B.223/b/29: 81v., F.12/1: 163; F.12/2: 19-19v.
32. Ibid., D.5/9: 67v.
33. A.J. Allen, comp., *Ten Years in Oregon: Travels and Adventures of Doctor E. White and Lady West of the Rocky Mountains* (Ithaca: Andrus, Gauntlett, & Co., 1850), p. 197; HBCA, B.223/z/4: 208, 229; Tess E. Jennings, trans., *Mission of the Columbia* . . . . (Seattle: Works Progress Administration, 1937), vol. 1, p. 110; Carl Landerholm, trans., *Notices & Voyages of the Famed Quebec Mission to the Pacific Northwest* (Portland: Oregon Historical Society, 1956), pp. 94, 148, 171, 173, 185n; White to the Commissioner of Indian Affairs, 1 April 1843, [Office of Indian Affairs], "Letters Received by the Office of Indian Affairs 1824-81," NARS, Mc. 234, roll 607; Joel Palmer, *Journal of Travels over the Rocky Mountains* . . . . (Cincinnati: J.A. and U.P. James, 1847), p. 115; "Papers relative to the Expedition of Lieuts. Warre and Vavasour to the Oregon Territory," PRO, Ms. F.O. 5/457: 46, 82v.; Charles Wilkes, "Oregon Territory," OHS, Ms. 56, 19.
34. HBCA, A.6/25: 51v., 69v., 100v., F.8/1: 40-40v.
35. Ibid., D.4/58: 5-6v., D.4/67: 72, D.5/8: 143v., F.12/2: 20-20v., 35-37. See also Joseph Heath, *Memoirs of Nisqually* (Fairfield, WA: Ye Galleon Press, 1979).
36. HBCA, A.6/25: 14v., F.8/1: 25.
37. Ibid., F.12/2: 48-49, 53, 53v.
38. Ibid., D.4/106: 27v.
39. Rich, *Letters of John McLoughlin*, vol. 1, p. 240.
40. HBCA, A.6/25: 52; see also 100v.
41. Ibid., 52-52v., D.4/25: 107v.
42. Ibid., D.4/66: 84v.-85, F.11/1: 73, 76.
43. Ibid., A.11/69: 79v., F.12/1: 562, 611.
44. [Wilkes], *Diary*, p. 77.
45. Rich, *Letters of John McLoughlin*, vol. 2, pp. 15-17.
46. HBCA, B.223/z/4: 208, F.8/1: 57-57v., F.11/1: 80-81.
47. British and American Joint Commission for the Final Settlement of the Claims of the Hudson's Bay and Puget's Sound Agricultural Companies, [*Proceedings*] (Montreal: J. Lovell, 1868), vol. 3, pp. 32, 109, 111.
48. James Cooper, "Maritime Matters on the Northwest Coast . . . ," 1878, Bancroft Library, Ms. P-C 6, pp. 2-3.
49. HBCA, F.11/1: 87, F.26/1: 22.
50. Ibid., D.4/67: 73v., F.12/1: 562v., F.12/2: 87v.,

138; Heath, *Memoirs of Nisqually*, p. 21; McNeil to Douglas, 12 January 1842, 8 February 1842, [Fort Nisqually], "Correspondence outward, May 23, 1841-Sept. 17, 1842."
51. [William Briskoe], "Journal of William Briskoe, Armorer Aboard the *Relief* and The *Vincennes* August 18, 1838-March 23, 1842," NARS, Mc. 75, p. 7.
52. HBCA, F.11/1: 51, F.12/2: 78v., 80v., 84v.
53. Heath, *Memoirs of Nisqually*, pp. 14-15, 30-31.
54. HBCA, F.11/1: 13; [Briskoe], "Journal of William Briskoe," p. 7.
55. HBCA, D.5/8: 143-143v., F.8/1: 39, F.12/2: 13v., 87, 136v.
56. Ibid., D.5/8: 143, F.11/1: 62-63, 85, F.12/2: 13v.
57. Ibid., F.8/1: 39, F.11/1: 85-86, F.12/1: 561.
58. Ibid., F.12/1: 612, F.12/2: 13v., 37, F.15/6: 68.
59. Ibid., D.5/9: 67v.
60. Ibid., D.4/67: 72v.-73, F.12/1: 560v.
61. Ibid., D.4/66: 85-85v., D.4/67: 73, F.11/1: 74-75.
62. Ibid., D.4/68: 105.
63. Ibid., F.11/1: 57, F.12/2: 30, 34v., 50.
64. Ibid., D.4/66: 83v., D.5/9: 67-67v., F.11/1: 72, F.12/2: 30, 33v., 34v.-35, 39v., 50-51v.
65. Ibid., D.4/66: 83v.-84, F.11/1: 73, 75-76, 85.
66. Ibid., D.4/68: 105.
67. Ibid., D.4/66: 85v., F.11/1: 75, F.12/2: 52v.
68. Ibid., D.4/66: 84, F.11/1: 14. If the wool were damp when it was shipped, it could heat and ignite at sea and burn the ship, (ibid., F.11/1: 21).
69. Ibid., F.11/1: 13-14.
70. Ibid., F.8/1: 28, F.12/1: 561v.
71. Ibid., F.8/1: 32, F.11/1: 38, 57. In the autumn of 1842 fourteen thousand of the company's sheep were kept at Fort Vancouver and the rest at Nisqually Farm (ibid. F.12/1: 613v.-614v.). The Fort Vancouver sheep were the purebred English stock, which McLoughlin wanted to oversee personally (ibid., F.12/1: 443v., 563-563v., F.12/2: 9B).
72. Ibid., D.5/9: 67v., F.8./1: 39v., F.11/1: 64, F.12/2: 11.
73. Ibid., F.11/1: 73, F.12/2: 52v., 109v.
74. Ibid., D.4/66: 84-84v., F.11/1: 73-74.
75. Ibid., F.8/1: 46, 51; [Clark], "Puget Sound Agricultural Company," p. 58.
76. HBCA, D.4/8: 105v., F.8/1: 57, F.11/1: 82. It seems that few skins or hides were exported to England, apparently owing to the erratic drying weather following the summer killing season (ibid., F.11/1: 58).
77. Ibid., A.11/69: 82v., B.223/b/24: 34, 37 D.223/b/26: lv., F.8/1: 39v., F.11/1: 63, F.12/1: 442, 611v., F12/2: 8Bv.; [Wilkes], *Diary*, p. 26.
78. Ibid., F.8/1: 28, F.12/1: 468, F.12/2: 18v., 39; [William Fraser Tolmie], *The Journals of William Fraser Tolmie: Physician and Fur Trader* (Vancouver: Mitchell Press, 1963), pp. 212, 237.
79. Landerholm, *Notices & Voyages*, pp. 48-49.
80. HBCA, D.4/67: 55.

NOTES TO CHAPTER 7

1. John Dunn, *History of the Oregon Territory* . . . . 2d ed. (London: Edwards and Hughes, 1846), p. 169. The Willamette Valley was particularly favoured by those men with "country [Indian or halfbreed]" wives, who faced less discrimination on the frontier than in the Canadas or the British Isles.
2. Meredith Gairdner, "Observations during a Voyage from England to Fort Vancouver, on the North-West Coast of America," *Edinburgh New Philosophical Journal* 16 (1834): 290n.
3. P.L. Edwards, *Sketch of the Oregon Territory: or, Emigrants' Guide* (Liberty, MO: The "Herald" Office, 1842), p. 12; Kenneth L. Holmes, ed. and comp., *Covered Wagon Women: Diaries & Letters from the Western Trails 1840-1890* (Glendale, CA: Arthur H. Clark, 1983), vol. 1, p. 37.
4. Edwards, *Sketch of the Oregon Territory*, pp. 11-12.
5. M.P. Deady, "History & Progress of Oregon after 1845," 1878, Bancroft Library, Ms. P-A 24, pp. 70-71; Carl L. Johannessen, William A. Davenport, Artimus Millett, and Steven McWilliams, "The Vegetation of the Willamette Valley," *Annals of the Association of American Geographers* 61 (1971): 291.
6. Samuel Hancock, "Thirteen Years' Residence on the north-west coast . . . ," Bancroft Library, Ms. P-B 29, p. 87.
7. William Peel, "Lieutenant Peel's Report to Captain J. Gordon, September 27, 1845," PRO, Ms. A.D.M. 1/5564, no. 5, p. [6].
8. [Nathaniel J. Wyeth], *The Correspondence and Journals of Captain Nathaniel J. Wyeth 1831-6*, ed. F.G. Young (Eugene: University Press, 1899), pp. 52, 178-79.
9. Geo. H. Himes, ed., "Mrs. Whitman's Letters," *Transactions of the Oregon Pioneer Association 1893*, p. 201.
10. Ibid., p. 203.
11. John McLeod, "John McLeod Papers, 1811-

1837," PAC, Ms. 19, series A23, vol. 1, p. 338.
12. HBCA, D.4/120: 50. These hides were much in demand as armour by the Indians of the coast, where they were known as "clamons" or "clemmels."
13. On "country" wives see Jennifer S.H. Brown, *Strangers in Blood: Fur Trade Company Families in Indian Territory* (Vancouver: University of British Columbia Press, 1980) and Sylvia Van Kirk, *"Many Tender Ties": Women in Fur-Trade Society in Western Canada, 1670-1870* (Winnipeg: Watson and Dwyer, 1980).
14. Clarence B. Bagley, ed., *Early Catholic Missions in Old Oregon* (Seattle: Lowman & Hanford, 1932), vol. 1, p. 62; HBCA, B.223/z/4: 4, 201; "Papers relative to the Expedition of Lieuts. Warre and Vavasour to the Oregon Territory," PRO, Ms. F.O. 5/457:73v.; M.M. Quaife, ed., "Letters of John Ball, 1832-1833," *Mississippi Valley Historical Review* 5 (1919): 466: [Wyeth], *Correspondence and Journals*, pp. 176, 179.
15. Lois Halliday McDonald, ed., *Fur Trade Letters of Francis Ermatinger . . . .* (Glendale, CA: Arthur H. Clark, 1980), p. 163.
16. E. E. Rich, ed., *The Letters of John McLoughlin from Fort Vancouver to the Governor and Committee* (London: Hudson's Bay Record Society, 1941-44), vol. 1, pp. 172-73.
17. Ibid., pp. 173-74; John McLoughlin, "Copy of a Document Found among the Private Papers of the late Dr. John McLoughlin," *Transactions of the Oregon Pioneer Association 1880*, p. 50.
18. Rich, *Letters of John McLoughlin*, vol. 1, pp. 173-74; McLoughlin, "Copy of a Document," pp. 48-49, 51. It should be noted that all of the settlers repaid their loans within three years. The company, incidentally, was forbidden by law to release engagés in the Indian Country (HBCA, B.223/b/24: 45v.).
19. Daniel Lee, "Articles on the Oregon Mission," OHS, Ms. 1211, pt. 1: 12v.; D. Lee and J. H. Frost, *Ten Years in Oregon* (New York: The Authors, 1844), p. 125; Willamette Settlers, "Letters to the Bishop of Juliopolis," Bancroft Library, Ms. P-A 305, n.d. In 1834 the Methodist missionary Jason Lee counted twenty families, "mostly French" (Jason Lee, "Letters on Oregon Mission, 1834-35," OHS, Ms. 1212, pt. 4, p.3). By the fall of 1838 the Canadian settlement consisted of twenty-six families (Bagley, *Early Catholic Missions*, vol. 1, p. 53).
20. [Wyeth], *Correspondence and Journals*, pp. 144, 148-49, 233, 255-56.
21. On the decline of this fur trade, see David J. Wishart, *The Fur Trade of the American West 1807-1840: A Geographical Synthesis* (London: Croom Helm, 1979). In 1836 there were three hundred to four hundred American mountain men west of the Missouri, down from as many as one thousand in 1832 (Spalding to Greene, 16 February 1837, "Papers of the American Board of Commissioners for Foreign Missions," Houghton Library, Ms. ABC: 18.5.3, vol. 1; Wishart, *Fur Trade of the American West*, p. 142).
22. George Foster Emmons, "Journal kept while attached to the Exploring Expedition...," Yale University Library, Western Americana Ms. 166, vol. 3, n.p.
23. On migration to the Oregon territory, see William A. Bowen, *The Willamette Valley: Migration and Settlement on the Oregon Frontier* (Seattle: University of Washington Press, 1978), Chs. 2-3. Francis Parkman's popular book, *The Oregon Trail* (originally published as *The California and Oregon Trail: Being Sketches of Prairie and Rocky Mountain Life . . . .* 1st ed. [New York and London: G. P. Putnam, 1849]), really says very little about the overland route. For an exhaustive study of American westward overland migration, see John D. Unruh, Jr., *The Plains Across: The Overland Emigrants and the Trans-Mississippi West, 1840-60* (Urbana: University of Illinois Press, 1979).
24. Bowen, *Willamette Valley*, pp. 17-21.
25. Geo. H. Himes, "Letters Written by Mrs. Whitman from Oregon to her Relations in New York," *Transactions of the Oregon Pioneer Association 1891*, p. 179.
26. M. Whitman to S. Prentiss, 16 May 1844, Marcus Whitman and Narcissa Whitman, "Correspondence," OHS, Ms. 1203.
27. Peel, "Lieutenant Peel's Report," p. [4].
28. Joel P. Walker, "Narrative of Adventures...," 1878, Bancroft Library, Ms. C-D 170, pp. 9-10, 14-15.
29. N. Whitman to J. Prentiss, 1 October 1841, N. Whitman to S. Prentiss, 6 October 1841, Whitman and Whitman, "Correspondence"; Spalding to Allen, 29 July 1843, Henry Harmon Spalding, "Papers," OHS, Ms. 1201; Waller to Cooke, 2 August 1843, Amos S. Cooke Papers, OHS, Ms. 1223.
30. Peter H. Burnett, *Recollections and Opinions of an Old Pioneer* (New York: D. Appleton,

1880) p. 141; Spalding to the Presbyterian Church of Kinsman, Ohio, 15 September 1845, Spalding, "Papers."
31. Burnett, *Recollections and Opinions*, p. 141; G. P. de T. Glazebrook, ed., *The Hargrave Correspondence 1821-1843* (Toronto: Champlain Society, 1938), p. 426; HBCA, B.223/z/4: 202.
32. Thomas Baillie, "Enclosure No. 1 to Letter No. 18, 4 March 1845," PRO, Ms. A.D.M. 1/5550, p. [6]; Rich, *Letters of John McLoughlin*, vol. 2, pp. 140-41.
33. Bowen, *Willamette Valley*, pp. 59-64, 69-71, 73-78.
34. Thomas J. Farnham, *Travels in the Great Western Prairies, the Anahuac and Rocky Mountains, and in the Oregon Territory* (London: Richard Bentley, 1843), vol. 2, p. 175; [Charles Wilkes], *Diary of Wilkes in the Northwest*, ed. Edmond S. Meany (Seattle: University of Washington Press, 1926), pp. 59, 66.
35. W. D. Brackenridge, *The Brackenridge Journal for the Oregon Country*, ed. O. B. Sperlin (Seattle: University of Washington Press, 1931), p. 56.
36. HBCA, D.4/59: 95, 109; Peel, "Lieutenant Peel's Report," pp. [6-7].
37. HBCA, B.223/z/4: 204v., 231; "Papers relative to the Expedition of Lieuts. Warre and Vavasour," 47v., 78v.-79.
38. [Wilkes], *Diary*, pp. 28, 56; [Wyeth], *Correspondence and Journals*, p. 178.
39. Mrs. Anne Abernathy, "The Mission Family and Governor Abernethy the Mission Steward," 1878, Bancroft Library, Ms. P-A 1, p. 5; Foster to E. and H. Grimes, 1 December 1843, Foster "Papers," OHS, Ms. 996; "Papers relative to the Expedition of Lieuts. Warre and Vavasour," 47,78; F. W. Pettygrove, "Oregon in 1843," 1878, Bancroft Library, Ms. P-A 60, pp. 16-17.

NOTES TO CHAPTER 8

1. William A. Bowen, *The Willamette Valley: Migration and Settlement on the Oregon Frontier* (Seattle: University of Washington Press, 1978), pp. 65-69.
2. William Fraser Tolmie, "History of Puget Sound and the Northwest Coast," 1878, Bancroft Library, Ms. P-B 25, p. 21.
3. [John Charles Frémont], *The Expeditions of John Charles Frémont*, ed. Donald Jackson and Mary Lee Spence (Urbana: University of Illinois Press, 1970), vol. 1, p. 567.
4. HBCA, A.11/70: 185; Tolmie, "History of Puget Sound," p. 20.
5. Waller to Cooke, 2 August 1843, Amos S. Cooke, "Papers," OHS, Ms. 1223.
6. Lois Halliday McDonald, ed., *Fur Trade Letters of Francis Ermatinger* . . . . (Glendale, CA: Arthur H. Clark, 1980), p. 267; Joseph Schafer, ed.,"Letters of Sir George Simpson, 1841-1843," *American Historical Review* 14 (1908): 80-81.
7. E.E. Rich, ed., *The Letters of John McLoughlin from Fort Vancouver to the Governor and Committee* (London: Hudson's Bay Record Society, 1941-44), vol. 1, p. 241.
8. Willamette Settlers to the Bishop of Juliopolis, 22 March 1836, Willamette Settlers, "Letters to the Bishop of Juliopolis," Bancroft Library, Ms. P-A 305.
9. Henry Eld, Jr., "Journal Statistics etc. in Oregon & California," Yale University Library, Western Americana Ms. 161, n.p.
10. National Intelligencer, "From Oregon," *New York Herald Tribune*, 18 January 1842, p. 1.
11. Charles Wilkes, "Oregon Territory," OHS, Ms. 56, p. 20.
12. Peter H. Burnett, *Recollections and Opinions of an Old Pioneer* (New York: D. Appleton and Company, 1880), pp. 139-40.
13. Carl Landerholm, trans., *Notices & Voyages of the Famed Quebec Mission to the Pacific Northwest* (Portland: Oregon Historical Society, 1956), pp. 149, 177.
14. Bowen, *Willamette Valley*, p. 68.
15. John McBride, "Oregon in 1846," OHS, Ms. 458, pp. 13-14.
16. Bowen, *Willamette Valley*, p. 88.
17. W.D. Brackenridge, *The Brackenridge Journal for the Oregon Country*, ed. O.B. Sperlin (Seattle: University of Washington Press, 1931), p. 56; P.L. Edwards, *Sketch of the Oregon Territory: or, Emigrants' Guide* (Liberty, MO: The "Herald" Office, 1842), p. 11; Thomas J. Farnham, *Travels in the Great Western Prairies, the Anahuac and Rocky Mountains, and in the Oregon Territory* (London: Richard Bentley, 1843), vol. 2, p. 186; LeRoy R. Hafen and Ann W. Hafen, eds., *To the Rockies and Oregon 1839-1842* (Glendale, CA: Arthur H. Clark, 1955), p. 298; Neil M. Howison, "Report of Lieutenant Neil M. Howison on Oregon, 1846," *Oregon Historical Quarterly* 14 (1913): 29; Joel Palmer, *Journal of Travels over the Rocky Mountains* . . . . (Cincinnati: J.A. and U.P. James, 1847), p. 100; Richard A. Pierce and John H. Winslow, eds., *H.M.S. Sulphur on the Northwest and California Coasts, 1837 and*

1839 (Kingston, Ont.: Limestone Press, 1979), p. 67; [Charles Wilkes], *Diary of Wilkes in the Northwest*, ed. Edmond S. Meany (Seattle: University of Washington Press, 1926), p. 59; Charles Wilkes, "Oregon Territory," OHS, Ms. 56, p. 20.

18. Kenneth L. Holmes, ed. and comp., *Covered Wagon Women: Diaries & Letters from the Western Trails 1840-1890* (Glendale, CA: Arthur H. Clark, 1983), vol. 1, p. 37; Bowen, *Willamette Valley*, pp. 88-89; Wilkes, "Oregon Territory," p. 20.

19. John Ball, "Oregon Expedition," *Zion's Herald*, 15 January 1834, n.p.; Tess E. Jennings, trans., *Mission of the Columbia* . . . . (Seattle: Works Progress Administration, 1937), vol. 1, p. 106; Landerholm, *Notices & Voyages*, p. 145; Jason Lee, "Journal" and "Letters on Oregon Mission, 1834-35," OHS, Ms. 1212, pt. 4, p. 13.

20. National Intelligencer, "From Oregon," p. 1; [Office of Indian Affairs], "Letters Received by the Office of Indian Affairs 1824-81," NARS, Mc. 234, roll 607, pp. 112-19.

21. T.C. Elliott, ed., "Letters of John McLoughlin," *Oregon Historical Quarterly* 23 (1922): 367, 370; McLoughlin to E. Ermatinger, 1 February 1835, 3 March 1837, "Ermatinger Papers," PABC, AB40 Er. 62.3; Edward Ermatinger, "Edward Ermatinger Papers 1820-1874," PAC, Ms. 19, series A2(2), vol. 1, pp. 178, 187; G.P. de T. Glazebrook, ed., *The Hargrave Correspondence 1821-1843* (Toronto: Champlain Society, 1938), p. 385; John McLeod, "John McLeod Papers 1811-1837," PAC, Ms. 19, series A23, vol. 1, p. 361; *Oregon Spectator*, 7 January 1847.

22. HBCA, D.4/59: 110; Landerholm, *Notices & Voyages*, p. 77; [Office of Indian Affairs], "Letters Received," pp. 112-19; Schafer, "Letters of Sir George Simpson," 82.

23. HBCA, A.11/70: 229v., D.4/59: 94, D.4/110: 28; *Oregon Spectator*, 6 August 1846; Rich, *Letters of John McLoughlin*, vol. 1, p. 241; Schafer, "Letters of Sir George Simpson," 80.

By 1845 the company did not want to buy any more wheat from the settlers than was absolutely necessary because "such trade would necessarily lead to Credit dealings with the settlers, to which we are decidedly averse" and because "considering the state of our present relations with the United States & the troublesome population by whom we are surrounded, we are indisposed to embark more largely in that branch of trade" (HBCA, D.4/67: 66v.-67). The company was already having difficulty collecting debts from some settlers, partly because collection could entail the cooperation of the local authorities and thereby the recognition of the provisional government.

24. John McLoughlin, "Copy of a Document Found among the Private Papers of the late Dr. John McLoughlin," *Transactions of the Oregon Pioneer Association 1880*, p. 50; Charles Wilkes, *Narrative of the United States Exploring Expedition During the Years 1838, 1839, 1840, 1841, 1842* (Philadelphia: Lea & Blanchard, 1845), vol. 4, p. 365.

This scrip was known as "Ermatinger money" after Frank Ermatinger, the clerk in charge of the company's store at Willamette Falls in 1844-46. Its solid value, backed by the high quality and wide variety of the company's goods, contrasted with the softness of notes drawn on the much smaller and poorer stock of goods of the competing American independent merchant at Willamette Falls. This latter scrip was called "Abernethy money" and often could only be traded at a discount (Bowen, *Willamette Valley*, p. 69). On Ermatinger, see McDonald, *Fur Trade Letters*.

25. George B. Roberts, "Recollections of George B. Roberts," Bancroft Library, Ms. P-A 83, pp. 22-23, 28.

26. Bowen, *Willamette Valley*, pp. 91, 93.

27. Joseph Williams, *Narrative of a Tour from the State of Indiana to the Oregon Territory in the Years 1841-2* (New York: Edward Eberstadt, 1921), p. 57. By 1841 there were already feral cattle in the Willamette Valley deriving from Hudson's Bay Company drives from California over the Siskiyou Trial.

28. Bowen, *Willamette Valley*, p. 83.

29. Theressa Gay, *Life and Letters of Mrs. Jason Lee* (Portland: Metropolitan Press, 1936), p. 163.

30. Ibid.; Charles Henry Carey, ed., "The Mission Record Book of the Methodist Episcopal Church, Willamette Station, Oregon Terrtitory, North America, Commenced 1834," *Oregon Historical Quarterly* 23 (1922); 258; Edwards, *Sketch of the Oregon Territory*, pp. 17, 20-22; Thomas E. Jessett, ed., *Reports and Letters of Herbert Beaver 1836-1838* (Portland: Champoeg Press, 1959), p. 78; Daniel Lee, "Articles on the Oregon Mission," OHS, Ms. 1211, pt. 2, p. 11; D. Lee and J.H. Frost, *Ten Years in Oregon* (New York: The Authors, 1844), p. 146; Rich, *Letters of John*

McLoughlin, vol. 1, p. 216; [John Work], "Work Correspondence," PABC, AB40, p. 122. Two Hudson's Bay Company sources state that in the 1837 drive only 250 or 251 cattle arrived (James Douglas, "Private Papers of Sir James Douglas," 3d series, Bancroft Library, Ms. P-C 14, pt. 2, p. 7; HBCA, B.223/e/4: 2v.).
31. Waller to Cooke, 2 August 1843, Cooke, "Papers"; Joel P. Walker, "Narrative of Adventures ...," 1878, Bancroft Library, Ms. C-D 170, p. 12.
32. Bowen, *Willamette Valley*, pp. 80, 86-87.
33. HBCA, D.4/59: 110; [Office of Indian Affairs], "Letters Received," pp. 112-19; Rich, *Letters of John McLoughlin*, vol. 1, p. 241; Schafer, "Letters of Sir George Simpson," 82.
34. Rich, *Letters of John McLoughlin*, vol. 1, p. 240.
35. Ibid., p. 242. In fact, once capital had been obtained and production commenced another conundrum arose — the lack of a sufficient market. Even the company's "Russian contract" could not absorb all of the burgeoning output of the Willamette settlers. As Frank Ermatinger noted in 1844, "the principal thing however is still wanting, that is a good market for lumber and farm produce" (McDonald, *Fur Trade Letters*, p. 260).
36. Joseph William McKay, "Recollections of a Chief Trader in the Hudson's Bay Company," 1878, Bancroft Library, Ms. P-C 24, p. 2.
37. HBCA, D.4/25: 107v.-108.
38. Silas Holmes, "Journal kept by Assist. Surgeon Silas Holmes ...," Yale University Library, Western Americana Ms. 260, vol. 3, p. 2.
39. White to Crawford, 1 April 1843, Elijah White, "Papers," OHS, Ms. 1217.
40. Burnett, *Recollections and Opinions*, p. 142.
41. "Papers relative to the Expedition of Lieuts. Warre and Vavasour to the Oregon Territory," PRO, Ms. F.O. 5/457: 25-25v.

NOTES TO CHAPTER 9

1. Charles Henry Carey, ed., "The Mission Record Book of the Methodist Episcopal Church, Willamette Station, Oregon Territory, North America, Commenced 1834," *Oregon Historical Quarterly* 23 (1922): 234.
2. Charles Henry Carey, ed., "Methodist Annual Reports Relating to the Willamette Mission (1834-1848)," *Oregon Historical Quarterly* 23 (1922): 306.
3. Jason Lee, "Journal" and "Letters on the Oregon Mission 1834-35," OHS, Ms. 1212, p. [112]; John McLoughlin, "Copy of a Document Found among the Private Papers of the late Dr. John McLoughlin," *Transactions of the Oregon Pioneer Association 1880*, p. 51.
4. Spalding to Greene, 15 March 1838, "Papers of the American Board of Commissioners for Foreign Missions," Houghton Library, Ms. ABC: 18.5.3, vol. 1; E. E. Rich, ed, *The Letters of John McLoughlin from Fort Vancouver to the Governor and Committee* (London: Hudson's Bay Record Society, 1941-44), vol. 1, pp. 240-41.
5. Thomas J. Farnham, *Travels in the Great Western Prairies, the Anahuac and Rocky Mountains, and in the Oregon Territory* (London: Richard Bentley, 1843), vol. 2, pp. 165-67.
6. J. L. Parrish, "Anecdotes of Intercourse with the Indians," 1878, Bancroft Library, Ms. P-A 59, pp. 13-15.
7. Carey, "Methodist Annual Reports," 350; Robert J. Loewenberg, *Equality on the Oregon Frontier: Jason Lee and the Methodist Mission 1834-43* (Seattle: University of Washington Press, 1976), p. 64.
8. J. Lee to White, 23 March 1843, [Office of Indian Affairs], "Letters Received by the Office of Indian Affairs 1824-81," NARS, Mc. 234, roll 607; Carey, "Methodist Annual Reports," 337; Charles Henry Carey, ed., "Diary of Reverend George H. Gray," *Oregon Historical Quarterly* 24 (1923): 154.
9. Loewenberg, *Equality on the Oregon Frontier*, p. 109; Carey, "Mission Record Book," 265-66.
10. Carey, "Methodist Annual Reports," 330-31.
11. Ibid, 331, 335-36, 348, 361.
12. Ibid, 334.
13. Clifton Jackson Phillips, *Protestant America and the Pagan World: The First Half Century of the American Board of Commissioners for Foreign Missions, 1810-1860* (Cambridge, MA: Harvard University East Asian Research Center, 1969), pp. 1-31.
14. Clifford Merrill Drury, ed., *First White Women Over the Rockies* (Glendale, CA: Arthur H. Clark, 1963), vol. 3, p. 105; N. Whitman to her parents, 14 March 1838, Marcus Whitman and Narcissa Whitman, "Correspondence," OHS, Ms. 1203.
15. N. Whitman to her brother, 5 December 1836, Whitman and Whitman, "Correspondence."
16. Charles Wilkes, *Narrative of the United States Exploring Expedition During the Years 1838, 1839, 1840, 1841, 1842*, (Philadelphia: Lea

& Blanchard, 1845), vol. 4, p. 394.
17. Clifford Merrill Drury, ed., *The Diaries and Letters of Henry H. Spalding and Asa Bowen Smith relating to the Nez Perce Mission 1838-1842* (Glendale, CA: Arthur H. Clark, 1958), p. 87; Drury, *First White Women*, vol. 3, p. 158.

The thick sod of the virgin grassland was difficult to till, however. The same missionary reported that "the land here is generally very hard to break, especially at this station. It is covered with a large kind of grass, 5 or 6 feet high with very strong roots so that it requires a strong plough & a strong team to break the soil" (Drury, *First White Women*, vol. 3, p. 163).

18. Geo. H. Himes, ed., "Mrs. Whitman's Letters," *Transactions of the Oregon Pioneer Association 1893*, pp. 64-65.
19. Ibid., pp. 200-201. This statement was made in early November 1846, after the settlement of the boundary dispute in Washington, DC in June, but before the outcome was known in the Oregon Country in December.
20. Drury, *First White Women*, vol. 1, p. 108.
21. M. Walker to J. Richardson, 25 March 1839, Elkanah Walker and Mary Walker, "Diaries," OHS, Ms. 1204.
22. Drury, *Diaries and Letters*, p. 146.
23. Carl Landerholm, trans., *Notices & Voyages of the Famed Quebec Mission to the Pacific Northwest* (Portland: Oregon Historical Society, 1956), p. 10.
24. Gregory Mengarini, *Recollections of the Flathead Mission*, ed. and trans. Gloria Ricci Lothrop (Glendale, CA: Arthur H. Clark, 1977), pp. 63, 67.
25. "Papers of the Americn Board of Commissioners," vol. 1, folders 1-2, p. 5.
26. Drury, *First White Women*, vol. 3, p. 164.
27. Carey, "Methodist Annual Reports," 312-13.
28. Joseph Williams, *Narrative of a Tour from the State of Indiana to the Oregon Territory in the Years 1841-2* (New York: Edward Eberstadt, 1921), p. 57.
29. [Nicolas Point], *Wilderness Kingdom: Indian Life in the Rocky Mountains: 1840-1847: The Journals & Paintings of Nicolas Point, S.J.*, trans. Joseph P. Donnelly (New York: Holt, Rinehart and Winston, 1967), p. 43.
30. Spalding to Greene, 11 September 1838, "Papers of the American Board of Commissioners," vol. 1.
31. Ibid., E. Walker to Greene, 15 October 1838.
32. Drury, *Diaries and Letters*, p. 110.

NOTES TO CHAPTER 10

1. Theressa Gay, *Life and Letters of Mrs. Jason Lee* (Portland: Metropolitan Press, 1936), p. 156; Daniel Lee, "Articles on the Oregon Mission," OHS, Ms. 1211, pt. 2, p. 2.
2. Charles Henry Carey, ed., "The Mission Record Book of the Methodist Episcopal Church, Willamette Station, Oregon Territory, Commenced 1834," *Oregon Historical Quarterly* 23 (1922): 248; LeRoy R. Hafen and Ann W. Hafen, eds., *To the Rockies and Oregon 1838-1842* (Glendale, CA: Arthur H. Clark, 1955). p. 255.
3. [Office of Indian Affairs], "Letters Received by the Office of Indian Affairs 1824-81," NARS, Mc. 234, roll 607, p. 114; Jason Lee, "Journal" and "Letters on Oregon Mission, 1834-35," OHS, Ms. 1212, p. 7.
4. Hafen and Hafen, *To the Rockies and Oregon*, pp. 62, 255; Lee, "Articles on the Oregon Mission," pt. 2, p. 11; Lee "Journal" and "Letters," p. 5; David Leslie, "Oregon Mission," *Christian Advocate and Journal*, 30 September 1840, n.p.
5. Thomas J. Farnham, *Travels in the Great Western Prairies, the Anahuac and Rocky Mountains, and in the Oregon Territory* (London: Richard Bentley, 1843), vol. 2, p. 169.
6. Henry Bridgman Brewer, "Log of the Lausanne — V," *Oregon Historical Quarterly* 30 (1929): 112, 115; Hafen and Hafen, *To the Rockies and Oregon*, p. 255; Lee, "Articles on the Oregon Mission," pt. 6, 2v.; Brewer to Waller, 1 November 1841, Alvan F. Waller, "Papers," OHS, Ms. 1210; "Diary Extracts 1845," Waller, "Papers," p. 12; Joseph Williams, *Narrative of a Tour from the State of Indiana to the Oregon Territory in the Years 1841-2* (New York: Edward Eberstadt, 1921), p. 62.
7. Nellie B. Pipes, ed., "Journal of John H. Frost, 1840-43," *Oregon Historical Quarterly* 35 (1934): 353-55.
8. Clifford Merrill Drury, ed., *The Diaries and Letters of Henry H. Spalding and Asa Bowen Smith relating to the Nez Perce Mission 1838-1842* (Glendale, CA: Arthur H. Clark, 1958), p. 86; Clifford Merrill Drury, ed., *First White Women Over the Rockies* (Glendale, CA: Arthur H. Clark, 1963), vol. 3, pp. 158-59; Paul Kane, *Wanderings of an Artist....* (Edmonton: M. G. Hurtig, 1968), p. 195.
9. M. Whitman to Greene, 6 July 1840, N. Whitman to Greene, 11 November 1841, "Papers of the American Board of Commis-

sioners for Foreign Missions," Houghton Library, Ms. ABC: 18.5.3, vol. 1; N. Whitman to S. Prentiss, 11 April 1838, N. Whitman to C. Hinny, 20 May 1844, M. and N. Whitman to S. Prentiss, 8 April 1845, Marcus Whitman and Narcissa Whitman, "Correspondence," OHS, Ms. 1203.
10. Drury, *Diaries and Letters,* pp. 85. 88.
11. Asahel Munger and Eliza Munger, "Diary of Asahel Munger and Wife," *Oregon Historical Quarterly* 8 (1907): 404; M. Whitman to Greene, 15 October 1840, "Papers of the American Board of Commissioners," vol. 1.
12. [Medorem Crawford], *Journal of Medorem Crawford,* ed. F. G. Young (Eugene: Star Job Office, 1897), p. 20.
13. Spalding to Hinsdale, 17 August 1842, Henry Harmon Spalding, "Papers," OHS, Ms. 1201.
14. Drury, *Diaries and Letters,* pp. 359, 366; Spalding to White, n.d., [Office of Indian Affairs], "Letters Received"; Spalding to Allen, 9 October 1845, Spalding to Greene, 17 October 1845, Spalding, "Papers."
15. Spalding to Allen, 29 July 1843, Spalding, "Papers." The wheat was threshed on the bare ground with the help of fifteen or twenty trampling horses.
16. Drury, *Diaries and Letters,* pp. 359-61, 366.
17. Spalding to Hinsdale, 17 August 1842, Spalding, "Papers."
18. Spalding to White, n.d., Walker and Eells to White, 9 January 1843, [Office of Indian Affairs], "Letters Received."
19. Drury, *Diaries and Letters,* p. 186; Clifford Merrill Drury, ed., *Nine Years with the Spokane Indians: The Diary 1838-1848, of Elkanah Walker* (Glendale, CA: Arthur H. Clark, 1976) p. 381n; A. Smith to Greene, 28 September 1840, "Papers of the American Board of Commissioners," vol. 1.
20. [Office of Indian Affairs], "Letters Received," p. 115.
21. Drury, *Diaries and Letters,* p. 149.
22. Ibid., pp. 184, 186; Drury, *First White Women,* vol. 3, p. 193; A. Smith to Greene, 28 September 1840, "Papers of the American Board of Commissioners," vol. 1.
23. [Office of Indian Affairs], "Letters Received," p. 112. One source states that occasionally 554 bushels of wheat were harvested in the early 1840's (Carl Landerholm, trans., *Notices & Voyages of the Famed Quebec Mission to the Pacific Northwest* [Portland: Oregon Historical Society], p. 146).
24. Tess. E. Jennings, trans. *Mission of the Columbia ....* (Seattle: Works Progress Administration, 1937), vol. 1, p. 109; Landerholm, *Notices & Voyages,* p. 148.
25. [Nicolas Point], *Wilderness Kingdom: Indian Life in the Rocky Mountains: 1840-1847: The Journals and Paintings of Nicolas Point, S. J.,* trans. Joseph P. Donnelly (New York: Holt, Rinehart and Winston, 1967), p. 45: Pierre-Jean De Smet, *Oregon Missions and Travels over the Rocky Mountains, in 1845-46* (New York: Edward Dunigan, 1847), pp. 259, 289.

NOTES TO CHAPTER 11

1. N. Whitman to Mrs. Perkins, 23 March 1839, N. Whitman to S. Prentiss, 6 October 1841, Marcus Whitman and Narcissa Whitman, "Correspondence," OHS, Ms. 1203.
2. Spalding to Bridges, 5 May 1840, Henry Harmon Spalding, "Papers," OHS, Ms. 1201.
3. Ibid., Spalding to Greene, 3 February 1847.
4. McDonald to E. Ermatinger, 22 March 1844, [John Work], "Work Correspondence," PABC, AB40.
5. Clifford Merrill Drury, ed., *Nine Years with the Spokane Indians: The Diary 1838-1848, of Elkanah Walker* (Glendale, CA: Arthur H. Clark, 1976), p. 336.
6. M. Whitman to Greene, 13 July 1841, "Papers of the American Board of Commissioners for Foreign Missions," Houghton Library, Ms. ABC: 18.5.3, vol. 1.
7. Ibid., M. Whitman to Greene, 29 October 1840.
8. Ibid., M. Whitman to Greene, 28 March 1841.
9. Drury, *Nine Years with the Spokane Indians,* p. 125.
10. Clifford Merrill Drury, ed., *First White Women Over the Rockies* (Glendale, CA: Arthur H. Clark, 1963), vol. 3, p. 200.
11. Jason Lee, "Diary," *Oregon Historical Quarterly* 17 (1916): 262.
12. N. Whitman to S. Prentiss, 11 April 1838, Whitman and Whitman, "Correspondence."
13. Geo. H. Himes, ed., "Mrs. Whitman's Letters," *Transactions of the Oregon Pioneer Association 1893,* pp. 64-65.
14. McLoughlin to Greene, 24 October 1838, "Papers of the American Board of Commissioners," vol. 1.
15. N. Whitman to C. Prentiss, 2 May 1840, Whitman and Whitman, "Correspondence."
16. Ibid., N. Whitman to S. Prentiss, 9 October 1844.
17. Himes, "Mrs. Whitman's Letters," p. 69. To

make matters worse, the flour mill at Waiilatpu and two hundred to five hundred bushels of wheat and corn were burned in late 1842 or early 1843 (see Spalding to Allen, 29 July 1843, Spalding, "Papers," and N. Whitman to S. Prentiss, 7 February 1843, Whitman and Whitman, "Correspondence").

18. Clifford Merrill Drury, ed., *The Diaries and Letters of Henry H. Spalding and Asa Bowen Smith relating to the Nez Perce Mission 1838-1842* (Glendale, CA: Arthur H. Clark, 1958), p. 150.
19. Walker to Greene, October 1841, "Papers of the American Board of Commissioners," vol. 1.
20. For 5 June and 31 August 1845, see Drury, *First White Women*, vol. 2, pp. 214 and 285; for 30 August 1841, 21 June 1846, 22 July 1846, and 9 September 1846, see Drury, *Nine Years with the Spokane Indians*, pp. 165, 356, 359-60, and 365; for 1841 see Eells to Greene, 2 September 1841, "Papers of the American Board of Commissioners," vol. 1; and for 4 June 1841, see M. Walker to E. Walker, 9 June 1841, Elkanah Walker and Mary Walker, "Diaries," OHS, Ms. 1204.
21. Drury, *Diaries and Letters*, p. 185.
22. Drury, *Nine Years with the Spokane Indians*, p. 379.
23. Drury, *Diaries and Letters*, p. 119; Spalding to Allen, 29 July 1843, Spalding, "Papers."
24. Spalding to the Presbyterian Church of Kinsman, Ohio, 15 September 1845, Spalding, "Papers." In the Flathead Country at the Catholic Mission of St. Mary's, frost, drought, and rocky soil hindered farming (Gregory Mengarini, *Recollections of the Flathead Mission*, ed. and trans. Gloria Ricci Lothrop [Glendale, CA: Arthur H. Clark, 1977], p. 187).
25. Drury, *Diaries and Letters*, p. 149.
26. Walker to Greene, 12 September 1839, "Papers of the American Board of Commissioners," vol. 1.
27. Drury, *Diaries and Letters*, p. 268n.
28. Spalding to Greene, 16 March 1840, "Papers of the American Board of Commissioners," vol. 1; Spalding to Allen, 18 February 1842, Spalding, "Papers."
29. Drury, *Diaries and Letters*, p. 174.

NOTES TO CHAPTER 12

1. Clifford Merrill Drury, ed., *The Diaries and Letters of Henry H. Spalding and Asa Bowen Smith relating to the Nez Perce Mission 1838-1842* (Glendale, CA: Arthur H. Clark, 1958), pp. 274-75.
2. Clifford Merrill Drury, ed., *First White Women Over the Rockies* (Glendale, CA: Arthur H. Clark, 1963), vol. 3, p. 112.
3. Drury, *Diaries and Letters*, pp. 149-50.
4. Ibid., pp. 133-40.
5. Drury, *First White Women*, vol. 3, p. 199.
6. Spalding to White, n.d., [Office of Indian Affairs], "Letters Received by the Office of Indian Affairs 1834-81," NARS, Mc. 234, roll 607.
7. Pierre-Jean De Smet, *Origin, Progress, and Prospects of the Catholic Mission to the Rocky Mountains* (Fairfield: WA: Ye Galleon Press, 1972), p. 11.
8. Carl Landerholm, trans. *Notices & Voyages of the Famed Quebec Mission to the Pacific Northwest* (Portland: Oregon Historical Society, 1956), p. 168.
9. Spalding to Allen, 18 February 1842, Henry Harmon Spalding, "Papers," OHS, Ms. 1201.
10. Whitman and Spalding to Greene, 21 April 1838, "Papers of the American Board of Commissioners for Foreign Missions," Houghton Library, Ms. ABC: 18.5.3, vol. 1.
11. Ibid, M. Whitman to Greene, 7 April 1843.
12. N. Whitman to S. Prentiss, 11 April 1838, Marcus Whitman and Narcissa Whitman, "Correspondence," OHS, Ms. 1203.
13. N. Whitman to Greene, 7 April 1843, "Papers of the American Board of Commissioners," vol. 1.
14. Spalding to White, n.d., [Office of Indian Affairs], "Letters Received"; M. Whitman to Greene, 7 April 1843, "Papers of the American Board of Commissioners," vol. 1; Spalding to Hinsdale, 17 August 1842, Spalding, "Papers."
15. Pierre-Jean De Smet, *Oregon Missions and Travels over the Rocky Mountains, in 1845-46* (New York: Edward Dunigan, 1847), pp. 104-5; Drury, *First White Women*, vol. 3, p. 158.
16. N. Whitman to Mrs. J. Parker, 25 July 1842, M. Whitman to Mr. and Mrs. Allen, 23 August 1842, Whitman and Whitman, "Correspondence."
17. Ibid., M. Whitman to Greene, 8 April 1845, M. and N. Whitman to S. Prentiss, 8 April 1845.
18. Whitman to Greene, 7 April 1843, Spalding to Greene, 22 September 1838, 2 October 1839, "Papers of the American Board

of Commissioners," vol. 1; Drury, *Diaries and Letters*, p. 252n; Spalding to Allen, 22 September 1838, Spalding, "Papers."
19. Drury, *Diaries and Letters*, p. 336; Drury, *First White Women*, vol. 1, p. 216; Spalding to White, n.d., [Office of Indian Affairs], "Letters Received"; Spalding to Greene, 26 February 1843, "Papers of the American Board of Commissioners," vol. 1; Spalding to Hinsdale, 17 August 1842, Spalding to Allen, 29 April, 29 July 1843, Spalding, "Papers."
20. Spalding to Hinsdale, 17 August 1842, Spalding, "Papers."
21. Ibid., Spalding to Greene, 26 August 1843, 3 February 1847; Drury, *Diaries and Letters*, p. 338; Drury, *First White Women*, vol. 1, p. 216.
22. Drury, *Diaries and Letters*, p. 185.
23. Walker to Greene, October 1841, "Papers of the American Board of Commissioners," vol. 1; Spalding to Hinsdale, 17 August 1842, Spalding, "Papers"; Walker to Greene, 3 October 1842, Elkanah Walker and Mary Walker, "Diaries," OHS, Ms. 1204.
24. J. Lee to White, 23 March 1843, [Office of Indian Affairs], "Letters Received."
25. Thomas E. Jessett, ed., *Reports and Letters of Herbert Beaver 1836-1838* (Portland: Champoeg Press, 1959), pp. 58-59.
26. Landerholm, *Notices & Voyages*, pp. 44, 65; Charles Wilkes, "Oregon Territory," OHS, Ms. 56, p. 18; [John Work], *The Journal of John Work January to October 1835*, ed. Henry Drummond Dee (Victoria: Charles F. Banfield, 1945), p. 85.
27. Peter H. Burnett, *Recollections and Opinions of an Old Pioneer* (New York: D. Appleton and Company, 1880), pp. 146-50.
28. Ibid., p. 150.
29. N. Whitman to C. Hinny, 20 May 1844, Whitman and Whitman, "Correspondence."
30. Walker to Greene, 23 January 1843, Walker and Walker, "Diaries."
31. Walker to Greene, October 1841, "Papers of the American Board of Commissioners," vol. 1.
32. Alvin M. Josephy, Jr., *The Nez Perce Indians and the Opening of the Northwest* (New Haven: Yale University Press, 1971), pp. xvi, 426.

NOTES TO THE EPILOGUE

1. "Joint occupation" is misleading in that the overwhelming majority of the Euroamerican occupants — hunters, traders, farmers, guides, explorers, scientists, officials, officers, — were British and Canadian, not American, at least until 1843, just three years before the boundary settlement. Only 150 of the seven to eight hundred whites in the Oregon Country in 1841 were Americans. Even after the large influx of migrants from "Jonathan Land" in 1843, nearly all of the Americans were concentrated in the Willamette Valley, where they had been preceded by retired Canadian servants and where they became established with the help of the Hudson's Bay Company, the principal buyer of thier output until at least the mid-1840's (see Charles Wilkes, "Oregon Territory," OHS, Ms. 56, p. 29).
2. In 1857 Simpson testified to the Select Committee on the Hudson's Bay Company that the firm possessed five thousand acres of ploughland below the forty-ninth parallel at the time of the boundary settlement (Great Britain, Parliament, House of Commons, Select Committee on the Hudson's Bay Company, *Report from the Select Committee on the Hudson's Bay Company* . . . . [London, 1857], p. 64).
3. Anonymous, "Documents," *Washington Historical Quarterly* 2 (1908): 164; E. E. Rich, ed., *The Letters of John McLoughlin from Fort Vancouver to the Governor and Committee* (London; Hudson's Bay Record Society, 1941-44), vol. 2, p. 41.
4. HBCA, F.12/1: 443v; Rich, *Letters of John McLoughlin*, vol. 2, p. 41.
5. HBCA, A.11/70: 185v.; [Office of Indian Affairs], "Letters Received by the Office of Indian Affairs 1824-81," NARS, Mc. 234, roll 607, p. 115.
6. On the symbiotic relationship of fur trading and farming in British North America, see D. W. Moodie, "Agriculture and the Fur Trade," in *Old Trails and New Directions: Papers of the Third North American Fur Trade Conference*, ed. Carol M. Judd and Arthur J. Ray (Toronto: University of Toronto Press, 1980), pp. 272-90.
7. Rich, *Letters of John McLoughlin*, vol. 1, p. 242.
8. Ibid.
9. Wilkes, "Oregon Territory," p. 72.
10. Ibid., p. 52.
11. Kate Ball Powers, Flora Ball Hopkins, and Lucy Ball, eds., *John Ball... Autobiography* (Glendale, CA: Arthur H. Clark, 1925), p. 93.
12. McLoughlin to Tolmie, 15 October 1843,

McLoughlin to Simpson, 20 March 1845, John McLoughlin, "Correspondence," University of Washington Library, Ms. 271 C and E.
13. See Frederick Merk, *The Oregon Question: Essays in Anglo-American Diplomacy and Politics* (Cambridge, MA: Belknap Press, 1967).
14. American willingness to wage war was tempered by more serious difficulties with Mexico over Texas and California. In 1846, just before the signing of the Oregon Treaty, the United States declared war on Mexico.
    Captain Gray did not really discover the Columbia River in 1792; Bruno de Hezeta had done that in 1755, but Gray was the first Euroamerican to sail into the river. However, the Spanish surrendered to the United States any territorial claims stemming from their discoveries when they signed the Transcontinental Treaty of 1819.
15. For a cogent summary of the British position, see Henry Commanger, "England and [the] Oregon Treaty of 1846," *Oregon Historical Quarterly* 28 (1927): 18-38. See also Walter N. Sage, "The Oregon Treaty of 1846," *Canadian Historical Review* 27 (1946): 349-67.
16. William Sturgis, *The Oregon Question: Substance of a Lecture Before the Mercantile Library Assocation, Delivered January 22, 1845*, (Boston: Jordan, Swift & Wiley, 1845), p. 28.
17. Wilkes, "Oregon Territory," p. 10.
18. Ibid, p. 11.
19. Charles Wilkes, *Narrative of the United States Exploring Expedition During the Years 1838, 1839, 1840, 1841, 1842* (Philadelphia: Lea & Blanchard, 1845), vol. 4, p. 305; Wilkes, "Oregon Territory," p. 5. In spite of his negative comments, however, Wilkes recommended that his country acquire the entire territory (see ibid., pp. 31-35).
20. HBCA, D.4/67: 36.
21. Ibid., B.223/z/4: passim.
22. John Gordon, "Captain J. Gordon's Report to the Secretary of the Admiralty, October 19, 1845," PRO, Ms. A.D.M. 1/5564, no. 8; George B. Roberts, "Recollections of George B. Roberts," Bancroft Library, Ms. P-A 83, pp. 62, 74.
23. Roderick Finlayson, "The History of Vancouver Island and the Northwest Coast," 1878, Bancroft Library, Ms. P-C 15, pp. 41-42; Gordon, "Captain J. Gordon's Report."
24. Frederick Merk, ed., *Fur Trade and Empire: George Simpson's Journal...1824-25* (Cambridge, MA: Belknap Press, 1968), pp. 338-39.
25. One exception was Archibald McDonald, who praised the Columbia Department as a "blessed country." Moreover, he believed that the Oregon Country as a frontier region with new opportunities attracted the ambitious: "being appointed to this side of the mountains is reckoned by the Sanguine as a sure step to their promotion, in as much as it is thought to be the only field where a young man can exert himself" (Lois Halliday McDonald, ed., *Fur Trade Letters of Francis Ermatinger*....[Glendale, CA: Arthur H. Clark, 1980], p. 63).
26. Work to E. Ermatinger, 10 September 1838, John Work, "Journals," and "Letters," OHS, Ms. 319; Work to E. Ermatinger, 28 March 1829, 5 August 1832, [Work], "Work Correspondence," PABC, AB40.
27. Madge Wolfenden, ed., "John Tod: 'Career of a Scotch Boy,'" *British Columbia Historical Quarterly* 28 (1954): 157. One of the few who actually liked New Caledonia was Peter Ogden, who headed the district from 1835 to 1844 (Anonymous, "Documents," *Washington Historical Quarterly* 2 [1908]: 259-60).
28. E. E. Rich, ed., *Part of Dispatch from George Simpson Esqr. Governor of Ruperts Land to the Governor & Committee of the Hudson's Bay Company London March 1, 1829* (Toronto: Champlain Society, 1947), p. 26; Work to E. Ermatinger, 19 March 1830, [Work] "Work Correspondence."
29. HBCA, D.4/120: 44v.
30. Ibid., 45, 64. In the late 1820's New Caledonia required at least forty-four men (ibid., D.4/121: 28v., 32).
31. Ibid., D.4/88: 79. At the end of 1828, the veteran trader John Harriott rated New Caledonia "the worst country for game he ever saw" (John Edward Harriott, "Memoirs of Life and Adventure in the Hudson's Bay Company's Territories, 1819-1825," Yale University Library, Western Americana Ms. 245, n.p.).
32. Rich, *Part of Dispatch from George Simpson*, p. 25; see also HBCA, D.4/88:187, 213.
33. John Tod, "History of New Caledonia and the Northwest Coast," PABC, p. 52; [John McLean], *John McLean's Notes of a Twenty-five Year's Service in the Hudson's Bay Territory*, ed. W. S. Wallace (Toronto: Champlain Society, 1932), p. 170. As mentioned earlier, miscreant employees were banished to New Caledonia as punishment. In 1829 Connolly

beseeched the Governor and Council of Rupert's Land as follows: "I would beg it as a favour that no more Convicts be transported hither, we have outdoor rogues enough to guard against without having any among ourselves, and I fear that we have already too many of the latter description" (HBCA, D.4/121: 34).
34. [John Work], *The Journal of John Work January to October, 1835*, ed. Henry Drummond Dee (Victoria: Charles F. Banfield, 1945), pp. 16, 73, 76, 78.
35. HBCA, D.4/125: 78v.-79; Rich, *Letters of John McLoughlin*, vol. 1, p. 228.
36. McDonald to McLeod, 15 January 1832, [Archibald McDonald], "McDonald Correspondence," PABC, AB20 M142A; [Work], "Work Correspondence," p. 82.
37. Merk, *Fur Trade and Empire*, p. 338; Rich, *Letters of John McLoughlin*, vol. 2, p. 118. Other notable violent deaths were those of Fort Nez Percés's Chief Trader Pierre Pambrun, who was mortally injured in a horseriding accident in 1841; John McLoughlin Jr., who was killed in a fight with a fellow Canadian in 1842 at Fort McLoughlin; and Chief Trader William Rae, who committed suicide in 1845 in San Francisco, where he was the company's agent.

Father Pierre De Smet asserted in 1842 that "of a hundred men who inhabit this [Oregon] country, there are not ten who do not die by some or other fatal accident" (Pierre-Jean De Smet, *Letters and Sketches: With a Narrative of a Year's Residence Among the Indian Tribes of the Rocky Mountains* [Philadelphia: M. Fithian, 1843], p. 378).
38. Elia Spalding and Henry Spalding,"Letters of Reverend H. Spalding and Mrs. Spalding . . . ," *Oregon Historical Quarterly* 13 (1912): 375. Father De Smet heartily concurred. "No river probably on the globe, frequented as much, could tell of more disastrous accidents," he observed in 1846 (Pierre-Jean De Smet, *Oregon Missions and Travels over the Rocky Mountains, in 1845-46* [New York: Edward Dunigan, 1847], p. 235).
39. Clifford Merrill Drury, ed., *First White Women Over the Rockies* (Glendale, CA: Arthur H. Clark, 1963), vol. 1, p. 121; G.P. de T. Glazebrook, ed., *The Hargrave Correspondence 1821-1843* (Toronto: Champlain Society, 1938), p. 71; Rich, *Letters of John McLoughlin*, vol. 1, pp. 85, 97.
40. M. Walker to Richardson, 15 September 1838, Elkanah Walker and Mary Walker, "Diaries," OHS, Ms. 1204; N. Whitman to her parents, 14 March 1838, Marcus Whitman and Narcissa Whitman, "Correspondence," OHS, Ms. 1203.
41. Isaac Cowie, *The Minutes of the Council of the Northern Department of Rupert's Land 1830 to 1843* . . . . (State Historical Society of North Dakota, n.d.), p. 787; HBCA, D.4/106: 29; Daniel Lee, "Articles on the Oregon Mission," OHS, Ms. 1211, pt. 5: 5v.; Rich, *Letters of John McLoughlin*, vol. 1, p. 293.
42. John Kirk Townsend, *Narrative of a Journey Across the Rocky Mountains to the Columbia River* (Fairfield, WA: Ye Galleon Press, 1970), p. 307. Another virulent disease was "bloody flux," or dysentery, which decimated the Indians below the Dalles in September and October 1844. See Jean Baptiste Zacharie Bolduc, *Mission of the Columbia*, ed. and trans. Edward J. Kowrach (Fairfield, WA: Ye Galleon Press, 1979), p. 126, and De Smet, *Oregon Missions and Travels*, pp. 82-83. See also Robert T. Boyd, "Another Look at the 'Fever and Ague' of Western Oregon," *Ethnohistory* 22 (1975): 135-54; Sherburne Cook, "The Epidemic of 1830-33 in California and Oregon," *University of California Publications in American Archaeology and Ethnology* 43 (1955): 303-26; Edward Hodgson, "The Epidemic on the Lower Columbia," *The Pacific Northwesterner* 1 (1957): 1-8; and Herbert Taylor and Lester Hoaglin, "The Intermittent Fever Epidemic of the 1830's on the Lower Columbia River," *Ethnohistory* 9 (1962): 160-78.
43. Burt Brown Baker, ed., *Letters of Dr. John McLoughlin Written at Fort Vancouver 1829-1832* (Portland: Oregon Historical Society, 1943), p. 132; John Dunn, *History of the Oregon Territory* . . . . (London: Edwards and Hughes, 1846), p. 115; HBCA, D.4/126: 48v.-49. On Ogden, see Archie Binns, *Peter Skene Ogden: Fur Trader* (Portland: Binfords & Mort, 1967), and Gloria Griffen Cline, *Peter Skene Ogden and the Hudson's Bay Company* (Norman: University of Oklahoma Press, 1974).
44. Anonymous, "Documents," *Washington Historical Quarterly* 1 (1907): 40; John McLeod, "John McLeod Papers, 1811-1837," PAC, Ms. 19, series A23, vol. 1, p. 297; Rich, *Letters of John McLoughlin*, vol. 1, p. 233.
45. Anonymous, "Documents," *Washington Historical Quarterly* 2 (1908): 164; Rich, *Letters of John McLoughlin*, vol. 1, p. 104; [Work], "Work Correspondence," p. 36.

46. Cook, "The Epidemic of 1830-1833," 322; Carl Landerholm, trans., *Notices & Voyages of the Famed Quebec Mission to the Pacific Northwest* (Portland: Oregon Historical Society, 1956), p. 18. De Smet, however, reported that up to two-thirds of the Indians died (De Smet, *Oregon Missions and Travels*, pp. 22-23).
47. Tess E. Jennings, trans., *Mission of the Columbia* .... (Seattle: Works Progress Administration, 1937), vol. 2, p. 29; Rich, *Letters of John McLoughlin*, vol. 1, p. 88. See also Dunn, *History of the Oregon Territory*, pp. 115-16.
48. HBCA, E.12/2: 24v.; McLeod, "Papers," vol. 1, p. 353.
49. Silas Holmes, "Journal kept by Assist. Surgeon Silas Holmes ...," Yale University Library, Western Americana Ms. 260, vol. 3, p. 6; [William A. Slacum], "Mr. Slacum's Report," *Reports Committees*, vol. 1 (1838-39), 25th Congress, 3d Session, p. 201.
50. Thomas E. Jessett, ed., *Reports and Letters of Herbert Beaver 1836-1838* (Portland: Champoeg Press, 1959), p. 88; Rich, *Letters of John McLoughlin*, vol. 1, p. 271; [Russian-American Company], "Records of the Russian-American Company 1802-67: Correspondence of Governors General," NARS, Mc. 11, roll 43: 333-333v.; [Work], *Journal of John Work*, p. 27. See also James R. Gibson, "Smallpox on the Northwest Coast, 1835-1838," *BC Studies* (Winter 1982-83): 61-81. Earlier smallpox may have ravaged the Chinooks in 1776 or 1778 and again in 1801 or 1802 (Rueben Gold Thwaites, ed., *Original Journals of the Lewis and Clark Expedition 1804-1806* [New York: Arno Press, 1969], vol. 4, pp. 50-51, 241).
51. Harriott to McLeod, 25 February 1831, McLeod, "Papers", vol. 1.
52. Ibid.
53. HBCA, D.4/67: 52-52v.
54. Ibid., D.4/67: 52v., D.5/8: 151.
55. Rich, *Part of Dispatch from George Simpson*, p. 1.
56. Ibid., pp. li-lii.
57. T.C. Elliott, ed., "Letters of Dr. John McLoughlin," *Oregon Historical Quarterly* 23 (1922): 368, 370; see also Anonymous, "Documents," (1908): 166.
58. Merk, *Fur Trade and Empire*, p. 339.
59. Rich, *Letters of John McLoughlin*, vol. 2, p. 271.
60. William Fraser Tolmie, "History of Puget Sound and the Northwest Coast," 1878, Bancroft Library, Ms. P-B 25, p. 21.
61. *Oregon Spectator*, 19 February 1846; Edmund Sylvester, "Founding of Olympia," 1878, Bancroft Library, Ms. P-B 22, p. 10.
62. Merk, *Fur Trade and Empire*, p. 321.
63. Anonymous, "Documents," (1908), 161-62, 254, 260.
64. Holmes, "Journal," vol. 3, p. 7.
65. Douglas to Ross, 12 March 1844, [James Douglas], "Correspondence outward, 1844-1857," PABC, B40, pt. 1.
66. [William Fraser Tolmie], *The Journals of William Fraser Tolmie: Physician and Fur Trader* (Vancouver: Mitchell Press, 1963), p. 334.
67. Tod, "History of New Caledonia," p. 21. See also J.F. Crean, "Hats and the Fur Trade," *Canadian Journal of Economics and Political Science* 28 (1962): 373-86 and M. Lawson, "The Beaver Hat and the Fur Trade," in *People and Pelts*, ed. M. Bolus (Winnipeg: Peguis Publishers, 1972), pp. 27-38.
68. Lois Halliday McDonald, ed., *Fur Trade Letters of Francis Ermatinger* .... (Glendale, CA: Arthur H. Clark, 1980), p. 261; Work to E. Ermatinger, 23 November 1847, Work, "Journals" and "Letters." By 1847 the "depreciation of Beaver in the English market" was such that the company was forced to pay the Columbia Department's Indians thirty per cent less for beaver pelts that year (Hartwell Bowsfield, ed., *Fort Victoria Letters 1846-1851* [Winnipeg: Hudson's Bay Record Society, 1979], pp. 13-14).
69. HBCA, D.4/120: 64v.
70. Ibid., D.4/125: 78v.
71. Anonymous, "Documents," (1907), 262-64.
72. [Tolmie], *Journals of William Fraser Tolmie*, p. 179.
73. Rich, *Letters of John McLoughlin*, vol. 1, p. 150.
74. [Fort Vancouver], "Fort Vancouver — Fur Trade Returns — Columbia District & New Caledonia, 1825-1857," PABC, AB20 V3, passim.
75. HBCA, D.4/58: 138v.; Wilkes, "Oregon Territory," p. 45.
76. HBCA, D.4/59: 51-52.
77. Rich, *Letters of John McLoughlin*, vol. 2, p. 306. The fact that beaver prices continued to fall, rather than rise, as supply declined indicates, of course, that changing taste had weakened demand.
78. Tod to E. Ermatinger, March 1844, "Ermatinger Papers," PABC, AB40 Er62.3.
79. Merk, *Fur Trade and Empire*, p. 341.
80. [Tolmie], *Journals of William Fraser Tolmie*,

p. 334.
81. McLeod, "Papers," vol. 1, p. 365; McDonald, *Fur Trade Letters*, p. 234.
82. Work to E. Ermatinger, 6 February 1844, [Work], "Work Correspondence."
83. Anonymous, "Documents," (1908), 161; HBCA, D.4/67: 64; Rich, *Letters of John McLoughlin*, vol. 3, p. lvii.
84. HBCA, B.201/a/3: 111, 130.
85. Rich, *Letters of John McLoughlin*, vol. 2, pp. xv-xx.
86. HBCA, D.4/60: 24.
87. Ibid., D.4/67: 62v. Merk has argued that the Hudson's Bay Company's decision in 1845 to shift its base from Fort Vancouver to Fort Victoria signalled to the British government its devaluation of the lower reaches of the Columbia River. This "surrender of the Columbia," he asserts, "was the key to the peaceful settlement of the Oregon boundary." In fact, however, this was not a surrender; the Fort Vancouver headquarters were moved but the post itself was not abandoned, nor were the company's other posts on the river. It seems more likely that the situation was reversed: the company decided to relocate its headquarters after Peel's administration indicated as early as 1842 that it would be willing to cede the river's right bank (see Merk, *The Oregon Question*, p. 251).
88. Ogden and Douglas to Tolmie, 4 November 1846, William Fraser Tolmie, "Correspondence," University of Washington Library, Ms. AIA 8/3.
89. HBCA, A.11/72: 13-13v.
90. Ball to Cass, 15 September 1835, John Ball, "Papers," OHS, Ms. 195.
91. Peter H. Burnett, *Recollections and Opinions of an Old Pioneer* (New York: D. Appleton and Company, 1880), p. 239.
92. Ibid., p. 240.
93. McLoughlin to Simpson, 20 March 1845, McLoughlin, "Correspondence," Ms. 271 E.
94. Even a boundary at 48° 20' would still have given the United States more than one-half of the area of the Oregon Country because of its greater longitudinal extent towards the south. It would also have received the more favourable part for agriculture. Although Great Britain was left with the poorer northern part, one recent writer glibly stated that in 1846 "Oregon had been lost to the United States (Cline, *Peter Skene Ogden*, p. 204).
95. F.W. Howay, W.N. Sage, and H.F. Angus, *British Columbia and the United States: The North Pacific Slope From Fur Trade to Aviation*, ed. H.F. Angus (New York: Russell & Russell, 1970), p. 129. Merk has confidently stated that "the line of 49° thus conformed with Utrecht, with contiguity, with exploration, and in general with settlement" (Merk, *The Oregon Question*, p. 399). In fact, however, the Treaty of Utrecht (1713) said nothing about a boundary line west of the Rocky Mountains: contiguity would have been realized at any parallel between 54° and 42°; there was virtually no American land exploration to the north of 49° but considerable British exploration to the south; and there was no American settlement at all north of 47° and almost none north of the Columbia-Snake. This represents what Merk has called the "very persuasive case" of the United States (ibid., p. 398). And as for the validity of Spanish claims inherited by the United States in 1819, they were not very substantial, being based upon some brief and unpublicized landfalls made on the coast. Only sixteen Spanish vessels landed between 1774 and 1796; there were only two short-lived settlements at Santa Cruz de Nootka on Nootka Sound, from 1789 to 1795, and Núñez Gaona (Neah Bay) in 1792, both on the coast and both below 50° N.
96. Before the boundary negotiations were concluded, Peel's government had fallen over the repeal of the Corn Laws, and the new Foreign Secretary, crusty Lord Palmerston, admitted that the treaty "gave the Americans everything that they had ever really wanted" (E.E. Rich, *The Fur Trade and the Northwest to 1857* [Toronto: McClelland and Stewart, 1967], p. 283).

Index

Aberdeen, Lord, 191
Abernethy, George, 153
Acorns, as mast, 37, 129
Adams-Onís Treaty. *See* Transcontinental Treaty
Agriculture, 4, 5
    and antagonism of fur traders, 17-18
    encouragement, 18-19
    and missionaries, 163, 164, 166
    obstacles, 5, 17, 67, 73, 175, 176
    purposes, 4, 18, 19, 22, 23, 25, 26, 27
    successes, 4-5, 67, 73-74, 77, 184
Alaska, 26, 61, 105
"Albany beef," 25
Albatross, 14
Aleutian Islands, 1
Aleuts, as sea otter hunters, 2, 81
Alexander Archipelago, 105, 184
Alexandria, 56. *See also* Fort Alexandria
Allen, Dudley, 163
Alta California, 2, 80
American Board of Commissioners for Foreign Missions, 154, 158, 160, 164, 167, 169, 172, 173, 179, 185
American Fur Company, 134
Anderson, Alexander, 57-58, 100
Andover Theological Seminary, 154
Annance, Françoise, 49
Assiniboia. *See* Red River Settlement
Astor, John Jacob, 3, 130
Astoria, 3, 14, 15, 17, 22, 59, 130
Athabasca Country. *See* Red River Country
Australia, 194

Babine Lake, 56
    and Indians, 25
Bailey, William 145, 196
Baker's Bay, 22
Baranov, Alexander. *See* Von Baranoff, Alexander
Bardeau, 24
Bathurst, Earl of, 10
Batteaus, 22, 58
Bear Lake, 26, 56

Beaver, Herbert, 77, 187
Beaver House, 63, 79, 80, 85-86, 110, 117, 144, 201
Beaver, 9, 17, 199, 200
Benton, Thomas, 133
Berens, Joseph, 10
Bering, Vitus, 1
Bishop, Charles, 14
Bitterroot Mountains, 159
Black, Samuel, 195
Black River, 114
Blackfeet Indians, 159, 161
Blanchet, François, 90, 117, 124, 158, 159, 181
Board of Managers of the Missionary Society of the Methodist Episcopal Church, 152, 153
Bolduc, Jean, 60, 196
Boston, 3, 17, 158
"Boston men," 2, 81
Boundary settlement, 55, 106, 107, 119, 137, 145, 147, 187-205 passim. *See also* Oregon Treaty, Washington, treaty of
Brewer, Henry, 153, 174
British and American Joint Commission, 92, 97, 106
British Columbia, 3
British North America, 4, 114, *See also* Canada
Bruce, William, 37
Buchanan, James, 119
Buffalo. *See* Bison
Burnett, Peter, 133, 141, 146, 185, 203, 204

California, 15, 17, 59, 79, 80, 81, 118, 130, 134, 135, 141, 143, 145
    and agriculture and livestock, 33, 34, 37, 79, 80, 88, 92, 100-101, 120, 133, 143
Camas, 96, 137
Camas Plain. *See* Vancouver Plain
Campment du Sable, 131, 135, 136, 143. *See also* Champoeg
Canada, 5, 12, 37, 111, 118, 130, 131
Canton, 2
Cape Disappointment, 21

260   Index

Cape Flattery, 191
Cape Horn, 9, 17, 22
Cape Mendocino, 1, 191
Cape of Good Hope, 123
Cape of St. Elias, 1
Capendal, Mr. and Mrs. William, 37
Cascade Mountains, 32, 89, 127, 129, 135, 177
Cattlepoodle River, 121
Cayuse Indians, 53, 154, 155, 181-82
Champoeg, 130, 131
Chatham Sound, 82
Chaudiere Falls. *See* Kettle Falls
Chehalis River, 114
Chemeketa, 143, 153
Chilcotin's Lake, 56
Chile, 82, 108
Chinook Indians, 2, 34
Clatsop Station, 152-53, 164
Clearwater River, 157
Clearwater Mission. *See* Lapwai
Coast Range, 89, 127
Coastal trade. *See* Fur trade
Cognoir, Michel, 92
Columbia, 22, 101
Columbia brigade, 4, 18, 22-23, 43, 44, 53, 55, 56, 72
Columbia Department, 4, 18, 20, 22, 23, 27, 29, 41, 52, 60, 67, 73, 77, 78-79, 100-101, 105, 106, 108, 113, 115, 127, 187, 193, 195, 198, 199, 200. *See also* Oregon Territory
Columbia District, 15, 16, 17, 18, 71
Columbia River, 4, 5, 9, 14, 18, 20, 23, 24, 29, 32, 68, 69, 70, 72, 79, 80, 89, 110, 117, 123, 127, 133, 135, 151, 152, 154, 157, 184, 187, 195
 and bar, 3, 21-22, 63
 as international boundary, 27, 61, 80, 87, 106, 109, 111, 114, 116, 118, 144, 169, 191, 192
 lower, 79-80, 110, 119, 130, 140
 middle, 71
Colvile, Andrew, 11, 15, 85, 86
Colvile District, 44
Colvile House. *See* Fort Colvile
Connolly, William, 194
Connolly's Lake, 26, 44, 56, 57
Continental Divide, 24
Convention of 1818, 4
Convention of 1824, 2
Convention of 1825, 82
Cook, James, 2, 14, 21
"Country produce," 14, 23, 24, 25, 44, 47, 64, 65
Coureurs de bois, 9

Coutenais Portage. *See* Kootenay Portage
Cowlitz, 87, 88, 89, 92, 124, 158, 198
Cowlitz Farm, 90, 92-94, 95-96, 100, 102, 104, 106-7, 109, 110, 112, 114, 115, 116, 119, 120, 122, 123, 124, 189
Cowlitz Indians, 137
Cowlitz Portage, 79, 86-90, 101, 113-14, 117, 118, 119, 124, 145, 189
Cowlitz Prairie, 90-92, 119
Cowlitz River, 87, 90, 92, 93, 117, 121, 124, 159
Cowlitz, 22, 101, 124
Cushing, Caleb, 133

Dalles, the, 22-23, 24, 152, 195
Dears, Thomas, 25-26
Dease, Peter, 16-17, 44, 46, 47, 55
Demers, Modeste, 158, 159
De Smet, Father Pierre-Jean, 24, 159, 180
Douglas, David, 203
Douglas, James, 115, 116, 145, 147, 152, 189, 190, 199, 203
 and agriculture and livestock, 35, 43, 48, 70, 87, 96, 100, 102, 103, 118-19
 on Cowlitz Prairie, 88-93
 and Fort Vancouver, 68, 69-70
 and Fort Victoria, 61-63
 and fur trade, 144-45
 and Puget's Sound Agricultural Company, 106, 116
 and settlers, 115, 117, 140-41
Dryad, 62, 82-83
Duke of York's Island, 66

Eagle, 50
East India Company, 2
Edmonton, 112
Edwards, Philip, 127-28, 143
Eells, Cushing and Myra, 156, 167, 172, 175, 176
El Dorado, 113
England, 2, 3, 4, 5, 9, 17, 20, 22, 32, 37, 41, 77, 82, 86, 87, 96, 101, 105, 106, 107, 118, 120-22, 151. *See also* London
Ermatinger, Edward, 33, 169, 179, 198, 199
Ermatinger, Frank, 24, 199
Esquimalt Harbour, 119

Faignant, François (a.k.a. François Failland), 92
Farnham, Thomas, 145, 152
Finlayson, Duncan, 14, 110, 112-13, 196
Finlayson, Roderick, 29, 80
First North Plain, 36
Flathead Country, 4, 44, 151, 159
Flathead Indians, 151, 159, 161
Flathead Mission, 159

Index

Forrest, Charles, 93, 95, 100
Fort Alexandria, 22, 56, 71, 72
Fort Boise, 58, 71, 133
Fort Bridger, 133
Fort Chilcotin, 56
Fort Colvile, 5, 18, 25, 27, 43, 44, 45, 46-48, 58, 71-73, 101, 156, 159, 172
Fort Durham, 66
Fort Fraser, 56
Fort Garry, 112
Fort George (Columbia), 4, 17, 18, 21, 22, 23, 26, 29, 32, 33, 40, 59, 60, 68, 130
Fort George (New Caledonia), 56, 58, 71-72
Fort Hall, 44, 58, 59, 71, 133, 159
Fort Kamloops, 44, 53, 195
Fort Langley, 5, 18-19, 27, 43, 48-51, 67-73 passim, 83, 102, 107, 109, 184
Fort Laramie, 133
Fort Lewis, 86
Fort McLoughlin, 65-69 passim, 81, 106, 184, 196, 203
Fort Nez Percés, 16, 24, 47, 52-53, 59, 71, 73, 133, 135, 154, 189
Fort Nisqually, 60-61, 184, 193, 198
Fort Okanagan, 22, 24, 54, 56, 58, 101
Fort Plain, 36
Fort Prince of Wales, 23
Fort Ross, 1, 81
Fort St. James, 14, 56
Fort Simpson, 64, 65, 68-69, 73, 81, 184
Fort Stikine, 66, 106
Fort Taku, 66, 106, 203
Fort Umpqua, 59, 145
Fort Vancouver, 20, 22, 23, 47, 48, 53, 59, 63, 79, 86, 87, 101, 107, 113, 114, 115, 116, 121, 123, 127, 133, 137, 140, 141, 142, 143, 144, 145, 158, 159, 164, 184, 189, 191, 196, 198, 202
 and agriculture and livestock, 5, 18, 27, 32-43 passim, 67-73 passim, 77, 83, 120
 settlement of, 32, 34, 36, 61, 69, 203
Fort Victoria, 61-63, 107, 193, 203
Fort Walla Walla, 33, 52
Fort William (Columbia River), 131-33
Fort William (Lake Superior), 11
Franchère, Gabriel, 17
Fraser, Paul, 145
Fraser, Simon, 3
Fraser Canyon, 3, 24
Fraser River, 3, 4, 18, 22, 24, 48, 49, 50, 56, 64, 68, 69, 119, 192, 202
Fraser Valley, 48, 49, 51
Fraser's Lake, 56, 57, 71
Frémont, John, 140
French Prairie, 130, 131, 159
Friendly Cove, 14

Frost, John, 152-53
Fur, as currency, 43
Fur trade
 continental, 3, 9-10, 130-31
 maritime, 1, 2, 19, 49, 64, 77-78, 81-82

Gairdner, Meredith, 127
Galbraith, John, 106
Gary, George, 153-54
Georgia Strait, 191
Ghent, Treaty of, 29
Gordon, John, 193
Grand Prairie, 96-97
Grande Prairie, 60
Gray, Robert, and discovery of Columbia River, 3, 191
Gray, William, 156-58
Great Britain. *See* England
Great Lakes, 9
Great Plains, 4, 133
Greene, David, 172, 175, 179, 181, 185
Gulf of Alaska, 1, 81

Haida Indians, 81, 184, 196
Hargrave, James, 106
Harmon, Daniel, 14
Hawaii, 35, 48, 61, 83, 130, 134, 174. *See also* Sandwich Islands
Hawaiians. *See* Kanakas
Heath, Joseph, 116
Heron, Francis, 47, 67
Hoecken, Adrian, 159
Holmes, Silas, 145-46
Hopkins, Samuel, 154
Howison, Neil, 192
Hudson, William, 192
Hudson Bay, 9, 10, 18, 21
Hudson Straits, 10
Hudson's Bay Company, 9, 14, 16, 17, 22, 35, 60, 63, 67, 83, 92, 100, 101, 111, 129, 133, 146, 158, 160, 161, 184, 185, 187-93 passim, 199, 202
 and agriculture and livestock, 4-5, 73-74, 77, 79, 94-95, 104-5, 142-45, 177, 189
 and boundary dispute, 197, 199, 204
 and charter, 10, 11
 employees, 10, 15, 24, 26, 81, 82
 and fur trade, 105
 governor and committee, 11, 13, 15, 19, 20, 47, 61, 79, 94, 96, 106, 113-15
 and Puget's Sound Agricultural Company, 80, 82, 85-88 passim, 93-94, 104
 and Russian America, 107-8, 119
 and servants, 5, 12, 19-20, 24, 25, 73, 85, 86, 117-18, 127, 130-31, 139
 and settlers, 110, 112, 137, 139, 144, 198

Index

Indian corn, 4, 9, 16, 19, 32, 33, 44, 46, 53
Indian Country, 19, 92, 131
Indians, 3, 4, 17, 19, 20, 24, 29, 34, 47, 56, 80, 81, 82, 86, 93, 97, 110, 116, 179, 180, 184-85. *See also* individual tribes
   and agriculture and livestock, 5, 23, 37, 64-66, 164, 179-86 passim
   and missionaries, 173, 174-75, 180
Influenza, 114, 195
Iroquois Indians, 50
Irving, Washington, 3
Isabella, 21

Jackson, John, 198
Jesuits, 158
Jolie Prairie, 29, 36
Juan de Fuca, Strait of, 61, 87, 114, 192
Juliopolis, Bishop of, 141

Kamchatka, 94, 107
Kamiah, 157, 158, 160, 167, 174-80 passim
Kanakas, 35, 37, 50, 60, 117
Kane, Paul, 40, 89
Keith, James, 17
Kennedy, Alexander, 23, 32
Kettle Falls, 18, 43
"King George's men," 2, 34
Kittson, William, 17
Klikitat Indians, 184
Kootenay Country, 44
Kootenay Portage, 112
Kupreyanov, Ivan, 81

Lake Superior, 11
Lake Winnipeg, 11, 13
Langley Prairie, 45, 49-50, 67
Lausanne, 152
Lapwai, 156-58, 161, 164-67, 173, 176, 177, 180, 183, 195
Lee, Daniel, 151-54
Lee, Jason, 39, 151-54, 163, 174, 184
Leslie, David, 152
Lewis and Clark Expedition, 3, 191
Lewis River (Washington), 183. *See also* Cattlepoodle River
Lisière, 82, 105, 107
London, 2, 11, 16, 17, 21, 25, 32, 41, 63, 79, 81, 85, 86, 102, 120, 122, 123
"London ship," 20, 21, 41, 61, 77, 88, 101
Louisiana Territory, 2
Lucier, Etienne, 130, 131

McDonald, Archibald, 4, 24, 44, 45, 48-50, 54, 67, 70, 71, 73, 156, 172, 194, 202
McDonell, Miles, 12
McIntosh, John, 195

McKay's Farm, 40
McKay's Fort, 59
Mackenzie, Alexander, 3, 56
Mackenzie District, 78
McKinlay, Archibald, 56, 57, 72, 135
McLean, John, 15, 24, 56, 71-72
McLeod, John, 59
McLeod's Lake, 25, 56, 195
McLoughlin, John, 29, 32, 44, 61, 73, 83, 110, 117, 118, 120, 122, 154, 174, 179, 184-91 passim, 196-98, 200, 204
   and agriculture and livestock, 20, 37, 38-39, 40-41, 43, 47, 48, 79, 87, 94-96, 102, 103, 116, 142, 152
   and fur trade, 78
   and Puget's Sound Agricultural Company, 79, 80, 85, 86, 88, 90, 94, 104, 119
   retirement, 147
   and settlers, 140-146 passim
   and trade, 21, 35, 67-68, 105
   and Willamette Valley, 113-14, 116, 119, 130-31, 147, 151
McMillan, James, 48, 49, 68, 73
McNeill, William, 63, 115, 202
MacTavish, Dugald, 106, 115
Malaria, 61, 114, 195
Manson, Donald, 69
Meek, Joe, 133
Mengarini, Gregory, 159
Merk, Frederick, 5, 106, 191
Methodism, 151, 172
Métis, 11, 12, 13, 110, 116
Milbanks Sound, 65, 81
Mill Creek, 45
Mill Plain, 36, 37
Mission Board. *See* Board of Managers of the Missionary Society of the Methodist Episcopal Church
Mission Bottom, 152, 153
Missionaries and missions, 4, 5, 110, 158, 159-61, 172. *See also* individual missions
   and agriculture and livestock, 163, 164
   Congregational Presbyterians, 154, 157-58, 168
   and farming, 163-78 passim
   and Indians, 184
   Methodist, 151-53, 163, 169, 172, 174, 177
   Roman Catholic, 4, 158-59, 168, 169
Missionary Society of New York City, 151
Mississippi River, 133, 154
Missouri River, 3, 4, 133, 134, 145
Monterey, 2, 80
Mountain Portage, 89
Mountain men, 15, 16, 58, 105, 131-33
Mud Mountain, 89
Multnomah Island, 131. *See also* Wappatoo

Island and Sauvie's Island
Multnomah River. *See* Willamette River

Nass River, 64, 81
Neah Bay, 14
Nechalco River, 56
Nereide, 100
New Archangel, 3, 22, 26, 28. *See also* Sitka
New Caledonia, 3, 4, 15, 18, 58, 67, 71, 78, 200
  and agriculture and livestock, 24-25, 26, 41, 47, 48, 54, 55-57, 72, 73, 194
New England, 110, 141, 154
New France, 158
New South Wales, 123
New Spain, 2, 158
New York City, 3, 151, 152
New Zealand, 129
Nez Percés Indians, 72, 156, 157, 159, 183
Nez Percés Mission. *See* Lapwai
Nisqually, 60, 86-89, 97, 119, 120-24. *See also* Fort Nisqually
Nisqually Farm, 109, 114, 116, 117, 120, 122
  and agriculture and livestock, 93-96, 97, 100-103, 115, 121, 123
  and settlers, 106-7, 119
Nisqually Plain, 96, 116, 118-19, 120, 123. *See also* Grand Prairie
Nisqually River, 89, 97
Nisqually Station, 152-53
Nootka Convention, 2
Nootka Sound, 2, 14
Norfolk Sound, 78
North West Company, 3-4, 9, 14, 15, 130, 202
  and agriculture and livestock, 10-11, 15
  and fur trade, 3
  and Hudson's Bay Company, 4, 11, 14, 15
Northern Departments, 4, 13, 15
Northwest Coast, 1-2, 5, 21, 26, 63, 67, 77, 78, 81, 104, 106, 110, 114, 129
Northwest Coast Indians, 81
Nor'Westers, 3, 10-11, 14, 18, 22, 130
Norway House, 12-13
Nuñez, Gaona, 14
Nuttall, Thomas, 33

Ogden, Peter, 58, 68, 70, 72, 116, 196, 200, 203
Okanagan River, 43, 58
Okhotsk, 79, 81, 107
Olympic Peninsula, 14
Oregon, 9, 14, 108, 133, 134, 135, 143, 151, 152, 155, 158, 159. *See also* Oregon Country
Oregon City, 130, 136-37
Oregon Country, 20, 26, 29, 32, 35, 133, 134, 151, 158
  and agriculture and livestock, 3, 5, 14, 72, 73-74, 107-8, 189
  and boundary dispute, 187-205 passim
  and economic development, 3, 4, 36, 15-19, 23, 72, 77, 189, 193
Oregon Mission, 152-58 passim
Oregon River. *See* Columbia River
Oregon Territory, 92, 108, 119, 120, 147, 154, 198-99
Oregon Trail, 5, 59, 124, 133-35, 139, 140, 165, 174, 190, 195
Oregon Beef and Tallow Company, 79
"Oregon fever," 133, 175
Oregon Treaty, 187, 203, 204
Overland migrants. *See* Settlers and settlement
Overland Express, 45. *See also* York Express

Pacific Fur Company, 3, 14-15, 22, 59, 130
Pacific Northwest. *See* Oregon Country
Pacific Ocean, 3, 4, 24, 26, 32, 56, 63, 114, 127, 133
Pacific Slope, 3, 15
Palatine Plains, 115
Palmerston, Lord, 191
Pambrun, Pierre, 52
Peace River, 3
Peel, Robert, 191
Peel, William, 129, 134, 193
Pelly, John, 82, 85, 86
Pemmican, 3-4, 9, 10-11, 24
"Pemmican War," 11
Pend Oreilles, 159
Pilcher, Joshua, 13
Pittman, Anna, 163
Plamondeau, Simon (a.k.a. Plamondou), 92
Plamondeau's Plain, 121
Point, Nicholas, 159, 161
Point Adams, 21
Polk, James, 191, 205
Port Camosun, 63
Prairie de Bute, 89, 98
Prairie de Thé, 37
Priest's Rapids, 26
Pudding River, 130
Puget Sound, 5, 60, 61, 79, 80, 90, 96, 103, 110, 113, 114, 122, 184, 191, 192
Puget Sound Portage, 90, 110. *See also* Cowlitz Portage
Puget's Sound Agricultural Company, 4, 5, 74, 83, 90, 92, 97, 100, 104-9 passim, 122, 127, 187, 190
  and agriculture and livestock, 61, 86-88, 93-96, 101-2, 105, 120-24
  and employees, 86-87, 88, 109, 117, 118, 188
  establishment and management of, 85, 86, 104, 106, 107, 119, 123

Index

and settlers, 88, 109-10, 112-18, 124
Puyallup River, 101

Quebec, 158
Queen Charlotte Islands, 184

Red River, 11, 13, 110, 115, 116, 130, 158
Red River Colony, 11-14
Red River Country, 10-11
Red River Settlement, 11-14, 109, 112
Richmond, John, 198
Roberts, George, 143
Rocky Mountains, 4, 9, 14, 27, 44, 79, 83, 86, 111, 112, 131, 133, 146, 159
Rosenberg, Nicholas, 108
Ross, Charles, 87
Ruby, 14
Rupert's Land, 3, 4, 10, 48, 189, 194
Russia, 1-2, 82, 107
Russian Alaska, 5, 51
Russian America, 15, 26, 35, 77-88 passim, 94, 102, 107, 108, 188
Russian-American Company, 3, 26, 64, 66, 77-87 passim, 94, 102, 104, 105, 107, 184
"Russian contract," 82-87, passim, 94, 104, 107-8, 139-40, 143, 145

Sacramento River, 88, 131
Sacred Heart Mission, 159
St. Dionysius Redoubt, 66, 82
St. Francis Xavier Mission, 90, 159, 159, 168
St. Ignatius Mission, 159
St. Joseph Mission, 159
St. Lawrence River, 9
St. Louis Mission, 159
Ste. Marie Mission, 159
St. Mary's Mission, 159
St. Michael Mission, 159
St. Paul Mission, 158-59, 168
St. Paul's Mission, 159
St. Petersburg, 2, 78, 79, 81, 82, 107
St. Pierre Mission, 159
Salem, Oregon, 130
Salmon, 3, 23, 24, 25, 41, 48-49, 57, 58, 60, 64, 71
Salmon River, 49, 50
San Francisco, 80, 100, 200
San Francisco Bay, 1, 143
Sandwich Islands, 35, 37, 48, 51, 83, 86, 88, 94, 97, 118, 133, 145, 146, 158, 188
Saskatchewan River, 4, 18, 121
Sauvie's Island, 37, 131
Scotland, 116
Scurvy, 26
Sea otters, 2, 26, 81, 107

Seattle, Chief, 186
Segwalitchew Creek, 97
Selkirk, Earl of, 10, 11
Selkirk Settlement. *See* Red River Colony, Red River Settlement
Settlers and settlement, 4, 59, 110, 118, 124, 130-35, 139-40, 145
Seven Oaks, 11
Shimnap, 157, 158
Shingles, 140
Siberia, 26, 79
Simmons, Michael, 198
Simpson, Aemelius, 78-79
Simpson, Alexander, 100
Simpson, George, 4, 15, 16, 33, 35, 37, 70, 100, 130, 187, 194, 197, 200, 202, 204
  and agriculture and livestock, 12, 17, 18, 19, 20, 22, 23, 25, 26, 27, 41-48, 79-80, 94, 114, 118-23
  and boundary dispute, 114, 198
  and economics and company development, 23, 25, 26, 29-32, 43-44, 49-58 passim, 64-66, 77-79, 114, 116, 118
  and Puget's Sound Agricultural Company, 85, 86, 123
  and settlers, 11-12, 63, 110-16, 135-36, 140
  and trade, 19, 64, 77-78, 81, 82, 85, 105-6, 145
Sinclair, James, 113, 114
Sioux Indians, 159
Siskiyou Trail, 59
Sitka, 3, 78, 81, 83, 94, 102, 105, 107, 108, 109, 140, 184
Slacum, William, 131, 196
Smith, Asa, 156, 157, 158, 160, 161, 165, 173-74, 179, 180, 183
Smith, Sarah, 154-60 passim
Snake Country, 4, 15, 58, 67
Snake River, 3, 4, 15, 18, 58, 183
Snake Expedition, 15, 52
South Pass, 133
South Sea Company, 2
Southern Department, 15
Southern Expedition, 15, 52, 59
Spain, 1, 2
Spalding, Eliza, 156, 157
Spalding, Henry, 156, 157, 161, 165, 167, 169, 172, 173, 176, 179, 180, 181, 183, 195
Spokane Indians, 156
Spokane House, 15, 18, 43, 45, 47
Spokane River, 157
Steel, James, 116
Steilicoom River, 97
Stephen's Passage, 66
Stikine River, 66, 82
Strange, James, 14

Stuart's Lake, 14, 24, 56-57, 71, 72
Sturgis, William, 191

Tebenkov, Michael, 107-8
Thompson, David, 3
Thompson's River, 3, 22, 24, 25, 48, 52-58, 71
Tlingit Indians, 81, 184, 196
Tod, John, 92, 193, 202
Tolmie, William, 38, 79, 100, 116-8, 121-23, 199, 200
Tonquin, 14
Townsend, John, 21-22, 33-34, 52-53, 60, 195
Trade, 49, 64, 82, 86, 105-6. See also Fur trade
Transcontinental Treaty, 2, 205
Tshimakain, 156-58, 161, 167, 172, 173, 175, 176, 183-84, 195
Tshimsian Indians, 196
Tsimean Peninsula, 64
Tualatin Plains, 133, 136. See also Palatine Plains

Umpqua River, 59
Unalaska Island, 1
United States, 2, 3, 4, 11, 29, 34, 81, 104, 111, 114, 118, 119, 128, 141, 142, 145, 146, 151, 152, 154, 156
  and agriculture and livestock, 37
  and boundary dispute, 106
U.S. Exploring Expedition, 23, 47-48, 58, 96, 120, 141, 189, 192
Upshur, Abel, 192

Vancouver, George, 21
Vancouver Island, 14, 48, 50, 63, 107, 114, 119, 158, 193, 194, 198, 204
Vancouver Plain, 36
Vancouver, 22, 101
Vavasour, Mervyn, 34-35, 45, 63, 146-47, 192
Venereal disease, 73, 195
Von Baranoff, Alexander, 78
Von Wrangell, Ferdinand, 79, 80, 82, 85
Vos, Peter, 159
Voyageurs, 9, 60

Waiilatpu, 154-56, 164-67, 173-80 passim
Walker, Elkanah, 156, 161, 167, 173, 175, 184, 185
Walker, Mary, 156, 157, 195
Walla Walla River, 53, 154
Wallace House, 130
Wallamette Valley. See Willamette Valley
Waller, Alvan, 153
Wappatoo Island, 35, 40
Wappatoos, as mast, 37, 129
Warre, Henry, 34-35, 45, 63, 146-47, 192
Wascopan, 152-53, 157, 164, 174

Washington Territory, 92
Washington, Treaty of, 187. *See also* Oregon Treaty
West, John, 12
Western, Lord, 122
Wheat, 43, 105, 107, 140-42
Whidbey Island, 67-68, 80, 114, 184
White, Elijah, 146, 152, 153, 184
Whitehorse Plain, 113
White Meadow Farm, 45
White Mud Farm, 45
White Pass, 101
Whitman, Marcus, 129-30, 133, 154-58, 165, 167, 172-82 passim
Whitman, Marcissa, 34, 154-56, 158, 163, 169, 174-75, 181-85 passim, 195
Whitman Massacre, 158, 174, 185
Wilkes, Charles, 23, 39, 40, 41, 72, 95, 101-3, 110, 119, 135, 141, 142, 184, 190, 192, 196, 200
Willamette Falls, 37, 129, 130, 136-37, 142, 143, 146, 159
Willamette Falls Mission, 152-53
Willamette River, 18, 29, 40, 80, 105, 114, 115, 130, 131, 152, 159, 160
Willamette Settlement, 104, 109, 118, 141, 145, 146, 147, 152, 183, 199
  and agriculture and livestock, 77, 94, 135-37, 144
  and settlers, 130-31, 135-36, 144, 189, 198
Willamette Station, 136, 152, 153, 163, 164
Willamette Valley, 59, 63, 110, 158, 177, 190, 193, 194, 196
  and agriculture and livestock, 5, 124, 127-29, 133, 134, 141, 151, 163
  and fur trade, 130
  and settlers, 94, 104, 106, 112-19, passim, 124, 129, 130-37 passim, 139, 140-45 passim, 151, 198, 203, 204
William and Ann, 21
Williams, Joseph, 160
Wilson, William, 153
Work, John, 25, 26, 33, 44-51 passim, 64-70 passim, 169, 193, 199, 200, 202
Wrangel Island, 66
Wyeth, Nathaniel, 39, 52, 58, 128, 129, 131-33, 151

Yakima River, 157
Yakima Valley, 101
Yale, James, 48
Yamhill, 135
York Factory, 13, 14, 23, 33
York Boats, 9, 11, 15

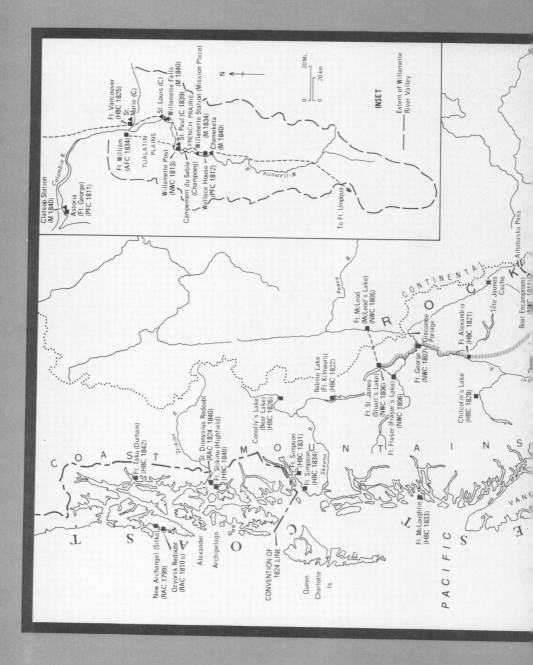